THE COLLECTED WORKS OF JOHN W. TUKEY

Volume VIII

MULTIPLE COMPARISONS (1948 - 1983)

- The goal of statistical analysis – not to be highly busy to control the probability of error p. 3

- mathematician vs. statistician p. 56

- difference between "tests of sig" – (sig or non sig) v. confidence procedures (how much?) p. 61

- a graphical method of depicting contrasts p. 82-83

- an approach to -wise contrasts p. 100

- mathematically v. applicationally indistinguishable assumptions p. 111

- Statistician's insight ↱ aha – I see it now p. 162
  ↳ New math – it'd be so nice to teach

  ↳ leading to significance v. confidence pp. 161-181
  p. 166, p. 175 p. 166, p. 175
  sanctification v. indication
  p. 172

- "tightness" as a confidence interval equivalent of power to significance testing.

- The natural basis for error rates p. 350 (i.e. what the experimenter would be willing to make a separate experiment for)

★ Cf chapter 6: confidence procedures are better for Tukey's main argument in non-math terms. (pp 347-3

THE COLLECTED WORKS OF JOHN W. TUKEY

Volume VIII

MULTIPLE COMPARISONS: 1948 - 1983

Edited by
Henry I. Braun
Educational Testing Service
Princeton, New Jersey

with the assistance of
Bruce Kaplan
Kathleen M. Sheehan
Min-Hwei Wang
Educational Testing Service
Princeton, New Jersey

CHAPMAN & HALL
New York • London

Published in 1994 by
Chapman & Hall
One Penn Plaza
New York, NY 10119

Published in Great Britain by
Chapman & Hall
2-6 Boundary Row
London SE1 8HN

Printed in the United States of America

Library of Congress Cataloging-in-Publication Data

Tukey, John Wilder, 1915-
The collected works of John W. Tukey.

(v. 1-2: The Wadsworth statistics/probability series)
(v. 3- : The Wadsworth & Brooks/Cole statistics/probability series)
Vol. 7 edited by D.R. Cox.
Vols. 2-4 published in Monterey Calif,; vols. 5-7 in Pacific Grove, Calif.
Vol. 3- published by Wadsworth & Brooks/Cole Advanced Books & Software; vol. 8 published by Chapman & Hall.
Includes bibliographies and indexes.
Contents: v. 1. Time series: 1949-1964 — v. 2. Time series: 1965-1984 — [etc.] — v. 8. Multiple comparisons, 1948-1983 / edited by Henry I. Braun.
1. Mathematical statistics. 2. Time-series analysis. 3. Tukey, John Wilder, 1915- Bibliography. I. Brillinger, David R. II. Cox, D. R. (David Roxbee). III. Series. IV. Series: Wadsworth & Brooks/Cole statistics/probability series. V. Title.
QA276.A12T85 1984 519.5 84-15277
ISBN 0-412-05121-4

British Library Cataloguing in Publication Data also available.

EDITORIAL BOARD
for
THE COLLECTED WORKS
OF JOHN W. TUKEY

Contents of Volume VIII
Multiple Comparisons: 1948 - 1983

Preface to The Collected Works

William S. Cleveland

To John Tukey we can pay the highest tribute possible to a statistician: he has profoundly affected the way we analyze data. *The Collected Works* demonstrate this.

The breadth and impact of his writings are stunning. Whole areas of statistics have their foundations laid and others are inspired; examples are spectrum estimation, multiple comparisons, quick and dirty methods, exploratory data analysis, graphics, robust methods, and Monte Carlo. The list of applications to science and technology goes on and on; seismology, education, band spectroscopy, halothane, federal statistics, indexing, astrophysics, the Kinsey report, pollution, molecular physics, and halocarbons are a small part of the list. Raw brilliance is everywhere — the fast Fourier transform, stem-and-leaf diagrams, coining the word "bit," one degree of freedom for nonadditivity, components of variance, the jackknife, Tukey's Lemma, binomial probability paper, 3RSRH, and so on.

The Collected Works are meant to be more than a collection for the record or for future historians to sort out scientific movements. They can be used right now by researchers, by data analysts, and by students. To help readers, *The Collected Works* have been divided into subjects with an editor for each. The editors have selected just enough papers to convey the philosophies and methodologies of Tukey and have provided comments to guide readers through the writings. Many of John Tukey's writings, including some of his most influential work, have never appeared before in published form. These volumes give the scientific and technological community easy access to this important material.

A testimony to the high regard that so many have for John Tukey was the willingness and enthusiasm with which people accepted the invitation to help assemble *The Collected Works*. The creation of these volumes has been the effort of a team: David Andrews, Melanie Belkin, Mary Bittrich, Henry Braun, David Brillinger, Nancy Clark, William Cleveland, David Cox, James Dolby, David Hoaglin, Ruth Hoover, Anna Jacobsen, Lyle Jones, John Kimmel, Colin Mallows, Patricia Martin, Marylyn McGill, Lincoln Moses, Frederick Mosteller, Carmela Patuto, Charles Smith, and Paul Tukey. The System Development Foundation also joined the team, providing a grant for project management and indexing.

It has taken ten subject editors to cover all of John Tukey's work; there is no one in the world with the energy and the breadth of knowledge to edit these works alone except, of course, John Tukey.

November 1, 1992

Biography of John W. Tukey

Frederick Mosteller

John Wilder Tukey, born 16 June 1915 in New Bedford, Massachusetts, was the only child of two educators, Ralph H. and Adah M. (Tasker) Tukey. Recognizing John as a prodigy, they largely educated him at home until he matriculated at Brown University. In 1937 he completed a master's degree in chemistry at Brown and then moved to Princeton University, where he completed a doctorate in mathematics in two years. His thesis, *Convergence and Uniformity in Topology*, was so outstanding that it was published as the second book in the *Annals of Mathematics Studies* of the Princeton University Press, and John was appointed an Instructor in the Department of Mathematics in 1939. When I met him that fall, his work had been in abstract mathematics, and he had done no work in statistics.

World War II rerouted his career. He joined the Fire Control Research Office, directed by Merrill Flood. Many statistical problems confronted this weapons study group. John worked with George W. Brown, who had just completed a doctorate in statistics in the Department of Mathematics at Princeton, and with Charles P. Winsor, a well-known biostatistician from (then) Iowa State College, and with Wilfrid Dixon of Princeton and Paul Dwyer of the University of Michigan. The community of statisticians in the town of Princeton soon grew because, in addition, S. S. Wilks assembled a war-time Statistical Research Group with W. G. Cochran, A. M. Mood, and Philip McCarthy, as well as a branch led by John Williams in New York, where I worked. Tukey enjoyed statistics and began spewing out a torrent of inventions and mixing with all these statisticians.

The origins of some of his exploratory data analysis appeared in memorandum form at this time, especially his recognition of the sensitivity of the sample mean and standard deviation to contamination in the tails of distributions. These were not published for decades, but many in the invisible college were aware of that work.

Wilks recognized Tukey's great talent and brought him into the statistical wing of the Department of Mathematics at Princeton University, up to then a one-man show, on his return to teaching in 1945. When Wilks edited the *Annals of Mathematical Statistics*, he once told me that he had watched Tukey thoroughly referee four papers in mathematical statistics on the railway trip from Princeton Junction to Philadelphia; "He ate them up like peanuts!" John often helped me with my thesis, tossing off more ideas than I could use.

In 1945, Tukey also joined the staff of AT&T Bell Laboratories at Murray Hill and has led at least a dual life ever since between there and Princeton.

His mother, a no-nonsense lady, finally got her wish that he marry "a real nice girl" when Elizabeth L. Rapp and he wed in 1950.

John loves to work with others, and many have had the pleasure of participating in his genius. Variety and breadth mark his accomplishments. He works successfully on both large- and small-scale problems and on both practical and theoretical problems. This brief biography cannot begin to discuss all his contributions or the role and contributions of his many co-workers. In statistics, Tukey revolutionized time series analysis, exploratory data analysis, multiple comparisons, and through his work with others on indexing has nearly brought the literature of probability and statistics into order. Beyond this he has published hundreds of scientific articles on many topics, with a vast number of co-authors.

In the field of public service, John has been a member of the President's Science Advisory Committee, served as a delegate in international working groups, and chaired major committees dealing with the environment and with the output of the nation's school system. He was a member of Technical Working Group 2 of the Conference on the Discontinuance of Nuclear Weapon Tests in Geneva, Switzerland and a member of the U.S. delegation to the U.N. Conference on the Human Environment in Stockholm, Sweden. He chaired the PSAC Panel on Chemicals and Health, which studied the impact of fluorocarbons on the stratosphere in increasing ultraviolet radiation at the surface of the earth and its effect on life there.

He is always eager to respond to new questions, and he gives generously of his time and ideas. At one award ceremony they said of John, "His work is his fun." The statisticians in the National Halothane Study met with him wherever he happened to be -- Phoenix, Denver, Palo Alto -- as he and Elizabeth drove across the country. He and I blocked out our first paper together driving from Princeton to North Carolina and back.

Although most people think of him primarily as an inventor and not as an organizer, when he wants something done, he does temporarily become a successful organizer, as I can testify, having often been so organized. Two examples must suffice: When he decided it was time for a Department of Statistics at Princeton University, he arranged it briskly as Ivy League speeds go. When he launched the indexing project, he had a nationwide army of statisticians and probabilists filling out forms for him. His unfailing good humor helps in organizing projects, as does being blessed with the long-time support of Mary Bittrich at AT&T Bell Laboratories.

Happily, his contributions have been recognized in profusion by medals, fellowships, honorary doctorates, and memberships in honorary societies. To name a few awards from outside statistics, he was elected to the National Academy of Sciences in 1961, and he received the National Medal of Science from President Nixon in 1973 and the Medal of Honor from the IEEE in 1982. *The Collected Works* should give the reader some notion of the appropriateness of these awards.

June 4, 1984

BIBLIOGRAPHY OF JOHN W. TUKEY

Published Works

Entries for works that appear in previous volumes of *The Collected Works of John W. Tukey*, abbreviated *CWJWT*, include information on the volume and page numbers in *CWJWT*.

1938a R. P. Boas, Jr. and J. W. Tukey. A note on linear functionals, *Bull. Amer. Math. Soc.* **44**, 523-528. [Also in *CWJWT*, VI (1990a), 1-8.] [See also 1940b]

1938b On the distribution of the fractional part of a statistical variable, *Mat. Sb.* **4**, 561-562. [Also in *CWJWT*, VI (1990a), 237-238.]

1939a The intrinsic metric of a polytope, *Proc. Nat. Acad. Sci. U.S.A.* **25**, 51.

1939b Denumerability in topology. Ph.D. Thesis, Princeton, University, Princeton, NJ (Pam 4572).

1940a Convergence and uniformity in topology, *Annals of Mathematics Studies*, Number 2, Princeton University Press, Princeton, NJ.

1940b R. P. Boas, Jr. and J. W. Tukey. A correction to "A note on linear functionals," *Bull. Amer. Math. Soc.* **46**, 566. [Also in *CWJWT*, VI (1990a), 9-10.]

1941a Compactness in general spaces. *Lectures in Topology*. (R. L. Wilder and W. L. Ayres, eds.) 307-309. University of Michigan Press, Ann Arbor, MI. [Also in *CWJWT*, VI (1990a), 11-13.]

1942a Some notes on the separation of convex sets, *Portugaliae Mathematica* **3**, 95-102. [Also in *CWJWT*, VI (1990a), 15-23.]

1942b A. H. Stone and J. W. Tukey. Generalized "sandwich" theorems, *Duke Math. J.* **9**, 356-359. [Also in *CWJWT*, VI (1990a), 25-28.]

1944a H. Scheffé and J. W. Tukey. A formula for sample sizes for population tolerance limits, *Ann. Math. Statist.* **15**, 217.

1945a H. Scheffé and J. W. Tukey. Non-parametric estimation, I. Validation of order statistics, *Ann. Math. Statist.* **16**, 187-192. [Also in *CWJWT*, VI (1990a), 239-247.]

1946a G. W. Brown and J. W. Tukey. Some distributions of sample means, *Ann. Math. Statist.* **17**, 1-12.

1946b An inequality for deviations from medians, *Ann. Math. Statist.* **17**, 75-78.

1946c J. W. Tukey and S. S. Wilks. Approximation of the distribution of the product of beta variables by a single beta variable, *Ann. Math. Statist.* **17**, 318-324. [Also in *CWJWT*, VI (1990a), 277-286.]

1946d J. W. Tukey and C. P. Winsor. Industrial statistics, *J. Amer. Statist. Assoc.* **41**, 406-411.

1946e Review of *Industrial Experimentation* by K. A. Brownlee, *J. Amer. Statist. Assoc.* **41**, 125-127.

1947a Linearization of solutions in supersonic flow, *Quart. Appl. Math.* **5**, 361-365. [Also in *CWJWT*, VI (1990a), 29-34.]

1947b C. Hastings, F. Mosteller, J. W. Tukey and C. P. Winsor. Low moments for small samples: a comparative study of order statistics, *Ann. Math. Statist.* **18**, 413-426.

1947c P. S. Olmstead and J. W. Tukey. A corner test for association, *Ann. Math. Statist.* **18**, 495-513.

1947d Non-parametric estimation, II. Statistically equivalent blocks and tolerance regions — the continuous case, *Ann. Math. Statist.* **18**, 529-539. [Also in *CWJWT*, VI (1990a), 249-262.]

1947e Discussion of symposium papers (Symposium on Statistical Methods in Experimental and Industrial Chemistry), *Analytical Chemistry* **19**, 956-957.

1947f Review of *Multiple Factor Analysis* by L. L. Thurstone, *Amer. Math. Monthly* **54**, 613-615.

1947g Review of *Statistical Analysis in Biology*, 2nd edition, by K. Mather, *Quart. Appl. Math.* **5**, 367-368.

1947h Review of *A Teoria da Inducão Estadistica* by Jorge Kingston, *J. Amer. Statist. Assoc.* **42**, 190-191.

1947i Review of *The Factorial Analysis of Human Ability*, Second Edition, by Godfrey H. Thomson, *J. Amer. Statist. Assoc.* **42**, 199-200.

1947j Review of *Sequential Analysis of Statistical Data: Applications* by Statistical Research Group, Columbia University for the Applied Mathematics Panel, *Ann. Math. Statist.* **18**, 142-144.

1948a Nonparametric estimation, III. Statistically equivalent blocks and multivariate tolerance regions — the discontinuous case, *Ann. Math. Statist.* **19**, 30-39. [Also in *CWJWT*, VI (1990a), 263-275.]

1948b Approximate weights, *Ann. Math. Statist.* **19**, 91-92.

1948c Some elementary problems of importance to small sample practice, *Human Biology* **20**, 205-214.

1948d (with many others). *Biometrical Clinic on Entomological Problems.* (C. I. Bliss, ed.) Biometric Society, New Haven, CT.

1949a Dyadic ANOVA, an analysis of variance for vectors, *Human Biology* **21**, 65-110.

1949b Memorandum on statistics in the Federal Government, Parts I and II, *American Statistician* **3**, No. 1, 6-17 and No. 2, 12-16.

1949c L. Spitzer, Jr. and J. W. Tukey. Interstellar polarization, galactic magnetic fields, and ferromagnetism, *Science* **109**, 461-462.

1949d H. W. Bode, F. Mosteller, J. W. Tukey and C. P. Winsor. The education of a scientific generalist, *Science* **109**, 553-558. [Also in *CWJWT*, III (1986c), 1-13.]

1949e Sufficiency, truncation and selection, *Ann. Math. Statist.* **20**, 309-311.

1949f Comparing individual means in the analysis of variance, *Biometrics* **5**, 99-114. [Also in *CWJWT*, VIII (1992d), 315-321.]

1949g F. Mosteller and J. W. Tukey. The uses and usefulness of binomial probability paper, *J. Amer. Statist. Assoc.* **44**, 174-212.

1949h One degree of freedom for non-additivity, *Biometrics* **5**, 232-242. [Also in *CWJWT*, VII (1992a), 1-13.]

1949i F. Mosteller and J. W. Tukey. Practical applications of new theory. A Review. Part I — location and scale: Tables, *Industrial Quality Control* **6**, September, 5-8.

1949j F. Mosteller and J. W. Tukey. Practical applications of new theory. A Review. Part II — counted data — graphical methods, *Industrial Quality Control* **6**, November, 5-7.

1949k Moments of random group size distributions, *Ann. Math. Statist.* **20**, 523-539. [Also in *CWJWT*, VI (1990a), 287-311.]

1949l Review of *Theory of Experimental Inference* by C. West Churchman, *J. Amer. Statist. Assoc.* **44**, 136-139.

1949m Review of *Probability Theory for Statistical Methods* by F. N. David, *J. Amer. Statist. Assoc.* **44**, 567-568.

1949n Answer to question 21, *American Statistician* **3**, No. 4 (October 1949), 12. [Also in *CWJWT*, VI (1990a), 313-316.]

1950a Discussion [Symposium: Statistics for the Clinician], *J. of Clinical Psych.* **6**, 61-74.

1950b F. Mosteller and J. W. Tukey. Practical applications of new theory. A Review. Part III — analytical techniques, *Industrial Quality Control* **6**, January, 5-7.

1950c F. Mosteller and J. W. Tukey. Practical applications of new theory. A Review. Part IV — gathering information, *Industrial Quality Control* **6**, March, 5.

1950d F. Mosteller and J. W. Tukey. Significance levels for a k-sample slippage test, *Ann. Math. Statist.* **21**, 120-123.

1950e J. W. Tukey, E. H. Dufour and F. Seibert. Lack of sensitization following repeated skin tests with standard tuberculin (PPD-S), *Amer. Rev. of Tuberculosis* **62**, 77-86.

1950f Some sampling simplified, *J. Amer. Statist. Assoc.* **45**, 501-519. [Also in *CWJWT*, VII (1992a), 107-129.]

1950g T. E. Harris, P. Meier and J. W. Tukey. Timing of the distribution of events between observations, *Human Biology* **22**, 249-270.

1950h M. F. Freeman and J. W. Tukey. Transformations related to the angular and the square root, *Ann. Math. Statist.* **21**, 607-611. [Also in *CWJWT*, VI (1990a), 149-155.]

1950i Index in *Contributions to Mathematical Statistics* by R. A. Fisher. Wiley, New York.

1950j Chairman's closure. *Acceptance Sampling. A symposium.* 150-155. American Statistical Association, Washington, DC.

1950k The sampling theory of power spectrum estimates. *Symposium on Applications of Autocorrelation Analysis to Physical Problems.* 47-67. (NAVEXOS P-735) Office of Naval Research, Washington, DC. [Also in *CWJWT*, I (1984b), 129-160.]

1950l M. P. Peisakoff and J. W. Tukey. Review of *Sequential Analysis* by Abraham Wald, *Bull. Amer. Math. Soc.* **56**, 200-202.

1951a Components in regression, *Biometrics* **7**, 33-69.

1951b Standard methods of analyzing data, *Proceedings, Computation Seminar December 1949*, 95-112. International Business Machines Corporation, Armonk, NY. [Also in *CWJWT*, III (1986c), 15-63.]

1951c L. Spitzer, Jr. and J. W. Tukey. A theory of interstellar polarization, *Astrophysical Journal* **114**, 187-205.

1951d Quick and dirty methods in statistics. Part II — Simple analyses for standard designs. *Quality Control Conference Papers 1951*. 189-197. American Society for Quality Control, New York.

1952a Review of *The Extrapolation, Interpolation and Smoothing of Stationary Time Series with Engineering Applications* by Norbert Wiener, *J. Amer. Statist. Assoc.* **47**, 319-321. [Also in *CWJWT*, I (1984b), 161-164.]

1952b Review of *Statistical Methods for Research Workers*, 11th edition, by R. A. Fisher, *Econometrica* **20**, 511-512.

1952c Review of *Statistische Methoden für Naturwissenschaftler, Mediziner, und Ingenieure* by A. Linder, *J. Amer. Statist. Assoc.* **47**, 554-556.

1952d Estimation in the alternative family of distributions. *Proceedings of the International Congress of Mathematicians 1950* **1**, 586.

1953a Some selected quick and easy methods of statistical analysis, *Trans. N. Y. Academy of Sciences* **16**, 88-97.

1953b W. G. Cochran, F. Mosteller and J. W. Tukey. Statistical problems of the Kinsey report, *J. Amer. Statist. Assoc.* **48**, 673-716.

1953c The growth of experimental design in a research laboratory. *Research Operations in Industry*. 303-313. King's Crown Press, New York. [Also in *CWJWT*, III (1986c), 65-75.]

1953d See **1954e**.

1954a W. G. Cochran, F. Mosteller and J. W. Tukey. Principles of sampling, *J. Amer. Statist. Assoc.* **49**, 13-35.

1954b Causation, regression and path analysis. *Statistics and Mathematics in Biology*. (O. Kempthorne, T. A. Bancroft, J. W. Gowen and J. L. Lush, eds.) 35-66. Iowa State College Press, Ames, IA.

1954c W. G. Cochran, F. Mosteller and J. W. Tukey. *Statistical Problems of the Kinsey Report on Sexual Behavior in the Human Male*. American Statistical Assoc., Washington, DC.

1954d Unsolved problems of experimental statistics, *J. Amer. Statist. Assoc.* **49**, 706-731. [Also in *CWJWT*, III (1986c), 77-105.]

1954e Peter G. Moore and J. W. Tukey. Answer to query 112, *Biometrics* **10**, 562-568 and **11**, 113 [errata].

1955a Systematization of journal practices, *Science* **122**, 246.

1955b Mathematical consultants, computational mathematics and mathematical engineering, *Amer. Math. Monthly* **62**, 565-571.

1955c Interpolations and approximations related to the normal range, *Biometrika* **42**, 480-485.

1955d W. G. Cochran, F. Mosteller and J. W. Tukey. Response to review article [on 1954c above], *J. Amer. Statist. Assoc.* **50**, 829.

1955e Review of *A Million Random Digits with 100,000 Normal Deviates* by the Rand Corporation, *J. Oper. Res. Soc. Amer.* **3**, 568-571.

1955f Answer to query 113, *Biometrics* **11**, 111-113. [Also in *CWJWT*, VII (1992a), 15-18.]

1956a Keeping moment-like sampling computations simple, *Ann. Math. Statist.* **27**, 37-54. [Also in *CWJWT*, VII (1992a), 131-156.]

1956b H. F. Trotter and J. W. Tukey. Conditional Monte Carlo for normal samples. *Symposium on Monte Carlo Methods*. (H. A. Meyer, ed.) 64-79. Wiley, New York; Chapman and Hall, London.

1956c H. J. Arnold, B. D. Bucher, H. F. Trotter and J. W. Tukey. Monte Carlo techniques in a complex problem about normal samples. *Symposium on Monte Carlo Methods*. (H. A. Meyer, ed.) 80-88. Wiley, New York; Chapman and Hall, London.

1956d Variances of variance components: I. Balanced designs, *Ann. Math. Statist.* **27**, 722-736. [Also in *CWJWT*, VII (1992a), 157-178.]

1956e W. C. Krumbein and J. W. Tukey. Multivariate analysis of mineralogic, lithologic and chemical composition of rock bodies, *J. Sedimentary Petrology* **26**, 322-337.

1956f J. Cornfield and J. W. Tukey. Average values of mean squares in factorials, *Ann. Math. Statist.* **27**, 907-949. [Also in *CWJWT*, VII (1992a), 179-239.]

1956g C. I. Bliss, W. G. Cochran and J. W. Tukey. A rejection criterion based upon the range, *Biometrika* **43**, 418-422.

1956h W. G. Cochran, F. Mosteller and J. W. Tukey. Fundamentos de muestreo, *J. Inter. Amer. Statist. Inst.*, June, 235-258. (Translation of 1954a)

1956i H. Press and J. W. Tukey. Power spectral methods of analysis and their application to problems in airplane dynamics. *Flight Test Manual, NATO Advisory Group for Aeronautical Research and Development*. Chapter IV-c, 1-41. [Also in *CWJWT*, I (1984b), 185-255.]

1957a Variances of variance components: II. The unbalanced single classification, *Ann. Math. Statist.* **28**, 43-56. [Also in *CWJWT*, VII (1992a), 241-260.]

1957b Variances of variance components: III. Third moments in a balanced single classification, *Ann. Math. Statist.* **28**, 378-384.

1957c Some examples with fiducial relevance, *Ann. Math. Statist.* **28**, 687-695. [Also in *CWJWT*, VI (1990a), 41-54.]

1957d On the comparative anatomy of transformations, *Ann. Math. Statist.* **28**, 602-632. [Also in *CWJWT*, VI (1990a), 167-209.]

1957e Antithesis or regression? *Proc. Camb. Phil. Soc.* **53**, 923-924.

1957f Sums of random partitions of ranks, *Ann. Math. Statist.* **28**, 987-992. [Also in *CWJWT*, VI (1990a), 373-381.]

1957g Approximations to the upper 5% points of Fisher's B distribution and non-central χ^2, *Biometrika* **44**, 528-530.

1957h The sampling theory of power spectrum estimates, *J. Cycle Research* **6**, 31-52. [1950k with comments] [Also in *CWJWT*, I (1984b), 129-160.]

1958a H. O. Gulliksen and J. W. Tukey. Reliability for the law of comparative judgment, *Psychometrika* **23**, 95-110.

1958b A smooth invertibility theorem, *Ann. Math. Statist.* **29**, 581-584.

1958c A problem of Berkson, and minimum variance orderly estimators, *Ann. Math. Statist.* **29**, 588-592. [Also in *CWJWT*, VI (1990a), 383-389.]

1958d R. B. Blackman and J. W. Tukey. The measurement of power spectra from the point of view of communications engineering, Part I, *Bell System Tech. J.* **37**, 185-282.

1958e R. B. Blackman and J. W. Tukey. The measurement of power spectra from the point of view of communications engineering, Part II, *Bell System Tech. J.* **37**, 485-569.

1958f Teaching of concrete mathematics, *Amer. Math. Monthly* **65**, 1-9.

1958g Bias and confidence in not-quite large samples [abstract], *Ann. Math. Statist.* **29**, 614. [Also in *CWJWT*, VI (1990a), 391.]

1959a A quick, compact, two-sample test to Duckworth's specifications, *Technometrics* **1**, 31-48.

1959b R. B. Blackman and J. W. Tukey. *The Measurement of Power Spectra from the Point of View of Communications Engineering*. Dover, New York. [Table of contents, preface and glossary in *CWJWT*, I (1984b), 257-277.]

1959c The estimation of (power) spectra and related quantities. *On Numerical Approximation*. (R. E. Langer, ed.) 389-411. University Wisconsin Press, Madison, WI. [Also in *CWJWT*, I (1984b), 279-307.]

1959d Discussion of the papers of Messrs. Satterthwaite and Budne, *Technometrics* **1**, 166-174.

1959e Appendix 9: Equalization and pulse shaping techniques applied to the determination of initial sense of Rayleigh waves. *The Need for Fundamental Research in Seismology*. 60-129. Report of the Panel on Seismic Improvement, Washington, U. S. State Department. [Also in *CWJWT*, I (1984b), 309-357.]

1959f An introduction to the measurement of spectra. *Probability and Statistics, The Harald Crameŕ Volume*. (Ulf Grenander, ed.) 300-303. Almqvist and Wiksell, Stockholm. [Also in *CWJWT*, I (1984b), 359-395.]

1959g Little pieces of mixed factorials, (abstract 627) *Biometrics* **15**, 641-642.

1959h R. P. Abelson and J. W. Tukey. Efficient conversion of non-metric information into metric information. *1959 Proceedings of the Social Statistics Section*. 226-230. American Statistical Association, Washington, DC. [Also in *CWJWT*, VI (1990a), 393-406.]

1960a Where do we go from here? *J. Amer. Statist. Assoc.* **55**, 80-93. [Also in *CWJWT*, III (1986c), 107-126.]

1960b B. F. Green, Jr. and J. W. Tukey. Complex analyses of variance: General problems, *Psychometrika* **25**, 127-152. [Also in *CWJWT*, VII (1992a), 75-106.]

1960c S. Siegel and J. W. Tukey. A nonparametric sum of ranks procedure for relative spread in unpaired samples, *J. Amer. Statist. Assoc.* **55**, 429-445.

1960d Discussion of the papers of Messrs. Anscombe and Daniel, *Technometrics* **2**, 160-165.

1960e Conclusions vs. decisions, *Technometrics* **2**, 423-433. [Also in *CWJWT*, III (1986c), 127-141.]

1960f A survey of sampling from contaminated distributions. *Contributions to Probability and Statistics: Essays in Honor of Harold Hotelling*. (I. Olkin et al., eds.) Chapter 39, 448-485. Stanford University Press, Stanford, CA.

1961a Discussion, emphasizing the connection between analysis of variance and spectrum analysis, *Technometrics* **3**, 191-219. [Also in *CWJWT*, I (1984b), 397-435.]

1961b Curves as parameters, and touch estimation. *Proc. 4th Berkeley Symp. Math. Stat. Prob.* **1**, 681-694. University of California Press, Berkeley, CA. [Also in *CWJWT*, I (1984b), 437-454.]

1961c Statistical and quantitative methodology. *Trends in Social Science*. (D. P. Ray, ed.) 84-136. Philosophical Library, New York. [Also in *CWJWT*, III (1986c), 143-185.]

1962a The future of data analysis, *Ann. Math. Statist.* **33**, 1-67 and 812. [Also in *CWJWT*, III (1986c), 391-484.]

1962b Keeping research in contact with literature: citation indices and beyond, *Trans. I.R.E. Professional Group on Engineering Writing and Speech* **5**, 78-82.

1962c Keeping research in contact with literature: citation indices and beyond, *J. Chem. Doc.* **2**, 34-37.

1962d (with many others). Strengthening the Behavioral Sciences, The Behavioral Science Subpanel of the President's Science Advisory Committee, *Science* **136**, 233-241.

1962e (with others). Meeting manpower needs in science and technology. Report No. 1: Graduate training in engineering, mathematics and physical sciences. (A report of the President's Science Advisory Committee.) The White House, Washington, D.C., 45 pages.

1963a B. P. Bogert, M. J. R. Healy and J. W. Tukey. The quefrency alanysis of time series for echoes: cepstrum, pseudo-autocovariance, cross-cepstrum and saphe-cracking, *Proceedings of the Symposium on Time Series Analysis*. (M. Rosenblatt, ed.) Chapter 15, 209-243. Wiley, New York. [Also in *CWJWT*, I (1984b), 455-493.]

1963b F. J. Anscombe and J. W. Tukey. The examination and analysis of residuals, *Technometrics* **5**, 141-160.

1963c R. P. Abelson and J. W. Tukey. Efficient utilization of non-numerical information in quantitative analysis: General theory and the case of simple order, *Ann. Math. Statist.* **34**, 1347-1369. [Also in *CWJWT*, VI (1990a), 445-476.]

1963d D. H. McLaughlin and J. W. Tukey. Less vulnerable confidence and significance procedures for location based on a single sample: Trimming/Winsorization 1, *Sankhyā, Series A* **25**, 331-352.

1963e What can data analysis and statistics offer today? *Ocean Wave Spectra: Proceedings of a Conference*. (National Academy of Sciences, ed.) 347-350. Prentice-Hall, Englewood Cliffs, NJ. [Also in *CWJWT*, I (1984b), 495-502.]

1963f A citation index for statistics and probability, *Bulletin of the Int. Stat. Inst.* **40**, 747-756.

1963g A tagging system for journal articles and other citable items: a status report. Annual Report for 1963 under National Science Foundation Grant NSF-GN-297 (from the Office of Science Information Service).

1964a R. D. Luce and J. W. Tukey. Simultaneous conjoint measurement: A new type of fundamental measurement, *J. Math. Psych.* **1**, 1-27. [Also in *CWJWT*, III (1986c), 485-516.]

1964b Samuel Stanley Wilks. *Year Book of the American Philosophical Society, 1964*. 147-154.

1965a J. W. Cooley and J. W. Tukey. An algorithm for the machine calculation of complex Fourier series, *Math. of Computa.* **19**, 297-301. [Also in CWJWT, II (1985a), 651-658.]

1965b Data analysis and the frontiers of geophysics, *Science* **148**, 1283-1289. [Also in *CWJWT*, II (1985a), 659-675.]

1965c Mathematics. *Listen to Leaders in Science.* (A. Love and J. S. Childers, eds.) Chapter 15, 217-232. Tupper and Love, Atlanta, (David McKay, New York). Reprinted in AT&T Bell Laboratories booklet entitled *Dimensions in Communications: Engineering, Science, Mathematics.*

1965d Which part of the sample contains the information? *Proc. Nat. Acad. Sci. U.S.A.* **53**, 127-134. [Also in *CWJWT*, VI (1990a), 477-487.]

1965e T. E. Kurtz, R. F. Link, J. W. Tukey and D. L. Wallace. Short-cut multiple comparisons for balanced single and double classifications: Part I, Results, *Technometrics* **7**, 95-161. [Also in *CWJWT*, VIII (1992d), 353-430.]

1965f The technical tools of statistics, *American Statistician* **19**, 23-28.

1965g J. W. Tukey and M. B. Wilk. Data analysis and statistics: Techniques and approaches. *Proceedings of the Symposium on Information Processing in Sight Sensory Systems.* (P. W. Nye, ed.) 7-27. California Institute of Technology, Pasadena. [Also in *CWJWT*, V (1988a), 1-22.]

1965h E. S. Pearson and J. W. Tukey. Approximate means and standard deviations based on distances between percentage points of frequency curves, *Biometrika* **52**, 533-546.

1965i T. E. Kurtz, R. F. Link, J. W. Tukey and D. L. Wallace. Short-cut multiple comparisons for balanced single and double classifications: Part 2. Derivations and approximations, *Biometrika* **52**, 485-498. [Also in *CWJWT*, VIII (1992d), 431-452.]

1965j F. F. Stephan, J. W. Tukey and others. Memorial to Samuel S. Wilks, *J. Amer. Statist. Assoc.* **60**, 939-966.

1965k T. E. Kurtz, R. F. Link, J. W. Tukey and D. L. Wallace. Authors' reply to Anscombe's comments, *Technometrics* **7**, 169. [Also in *CWJWT*, VIII (1992d), 453-454.]

1965l The future of processes of data analysis. *Proceedings of the Tenth Conference on Design of Experiments in Army Research, Development and Testing.* 691-729. U.S. Army Research Office, Durham, NC. [Also in *CWJWT*, IV (1986e), 517-547.]

1965m The inevitable collision between computation and data analysis. *Proceedings of the IBM Scientific Computing Symposium on Statistics October 21-23, 1963.* 141-152. International Business Machines Corp., Armonk, NY.

1965n J. W. Tukey, M. Alexander, H. S. Bennett, N. C. Brady, J. C. Calhoun, Jr., J. C. Geyer, A. J. Haagen-Smit, N. Hackerman, J. B. Hartgering, D. Pimentel, R. Revelle, L. H. Roddis, W. H. Stewart, and J. L. Whittenberger. *Restoring the Quality of Our Environment, Report of the Environmental Pollution Panel, President's Science Advisory Committee.* Government Printing Office, Washington, DC.

1966a Statement on Pollution. *Federal Research and Development Programs, Hearings before a Sub-Committee of the Committee on Government Operations, House of Representatives, 89th Congress, Second Session, January 1966.* 121-126. U.S. Government Printing Office, Washington, DC.

1966b T. E. Kurtz, R. F. Link, J. W. Tukey and D. L. Wallace. Correlation of ranges of correlated deviates, *Biometrika* **53**, 191-197.

1966c A practicing statistician looks at the transactions, *IEEE Trans. on Information Theory* **IT-12**, 87-91.

1966d (with many others). Summary of the National Halothane Study, *J. Amer. Med. Assoc.* **197**, 775-788.

1966e Use of numerical spectrum analysis in geophysics, *Bull. Int. Inst. Stat.* **41**, 267-307. [Also in *CWJWT*, II (1985a), 677-738.]

1966f J. W. Tukey and J. L. Buckley. Earth — man's polluted spaceship. (The Call of the Vanishing Wild — 6.) *The Christian Science Monitor*, Boston, MA. October 24, 1966, 9.

1966g J. W. Tukey and M. B. Wilk. Data analysis and statistics: An expository overview, *AFIPS Conference Proceedings, 1966 Fall Joint Computer Conference* **29**, 695-709. AFIPS Press, Reston, VA. [Also in *CWJWT*, IV (1986e), 549-578.]

1966h R. A. Haubrich and J. W. Tukey. Spectrum analysis of geophysical data. *Proceedings of the IBM Scientific Computing Symposium on Environmental Sciences*. 115-128. International Business Machines Corp., Armonk, NY. [Also in *CWJWT*, II (1985a), 739-754.]

1966i M. L. Tompkins and J. W. Tukey. Permuted (circularly-shifted) indexes to abbreviations: a mechanically prepared aid to serial identification, *Proceedings of the 29th Annual Meeting of ADI*, 347-355. American Documentation Institute, Washington, DC.

1967a What can mathematicians do for the Federal Government? *Amer. Math. Monthly*, Part II, 101-109.

1967b C. Bingham, M. D. Godfrey and J. W. Tukey. Modern techniques of power spectrum estimation, *IEEE Trans. on Audio and Electroacoustics* **AU-15**, 56-66. [Also in *CWJWT*, II (1985a), 781-810.]

1967c Discussion on Anscombe's paper [Topics in the investigation of linear relations fitted by the method of least squares], *J. Roy. Stat. Soc., Series B* **29**, 47-48.

1967d Discussion of Session I, *SIAM Review* **9**, 308-310.

1967e A statistician's comment. *Electronic Handling of Information: Testing and Evaluation*. (A. Kent, O. E. Taulbee, J. Belzer and G. D. Goldstein, eds.) 41-47. Thompson Book Co., Washington, DC. [Also in *CWJWT*, IV (1986e), 579-585.]

1967f (with others). Computers in undergraduate education: mathematics, physics, statistics, and chemistry. Proceedings of a Conference sponsored by The National Science Foundation, University of Maryland. December 8-9, 1967, 73 pages.

1968a W. J. Dixon and J. W. Tukey. Approximate behavior of the distribution of Winsorized t (Trimming/Winsorization 2), *Technometrics* **10**, 83-98.

1968b An introduction to the calculations of numerical spectrum analysis. *Spectral Analysis of Time Series*. (B. Harris, ed.) 25-46. Wiley, New York. [Also in *CWJWT*, II (1985a), 811-835.]

1968c The true purpose of transformation, (abstract) *Biometrics* **24**, 1041.

1968d Proceedings of the workshop: Practical applications of the frequency approach to EEG analysis. *Advances in EEG Analysis*, Suppl. 27 to *Electroencephalography and Clinical Neurophysiology*. 10-11.

1968e F. Mosteller and J. W. Tukey. Data Analysis, Including Statistics. *Handbook of Social Psychology*. (G. Lindzey and E. Aronson, eds.) 2nd Edition, Vol. 2, Chapter 10, 80-203. Addison-Wesley, Reading, MA. [Sections A, B, D, E and F in *CWJWT*, IV (1986e), 601-720. Section C in *CWJWT*, V (1988a), 123-148.]

1968f Is statistics a computing science? *The Future of Statistics*. (D. G. Watts, ed.) 19-38. Academic Press, New York.

1969a W. M. Gentleman, J. P. Gilbert and J. W. Tukey. The smear-and-sweep analysis. *The National Halothane Study*, a study of the possible association between halothane anesthesia and postoperative hepatic necrosis. Chapter 4, 287-315. National Institutes of Health/National Institute of General Medical Sciences, Bethesda, MD.

1969b A further inquiry into institutional differences by means of superstandardization (a regression adjustment beyond standardization). *The National Halothane Study*, a study of the possible association between halothane anesthesia and the postoperative hepatic necrosis. Appendix 2, 358-369. National Institutes of Health/National Institute of General Medical Sciences, Bethesda, MD.

1969c W. M. Gentleman and J. W. Tukey. Should we superstandardize agent comparisons? — General considerations and apparent conclusions. *The National Halothane Study*, a study of the possible association between halothane anesthesia and postoperative hepatic necrosis. Appendix 4, 372-380. National Institutes of Health/National Institute of General Medical Sciences, Bethesda, MD.

1969d W. M. Gentleman and J. W. Tukey. Development of certain formulas. *The National Halothane Study*, a study of the possible association between halothane anesthesia and postoperative hepatic necrosis. Appendix 5, 381-386. National Institutes of Health/National Institute of General Medical Sciences, Bethesda, MD.

1969e W. M. Gentleman and J. W. Tukey. Assessment of various structural regression coefficients. *The National Halothane Study*, a study of the possible association between halothane anesthesia and postoperative hepatic necrosis. Appendix 6, 387-391. National Institutes of Health/National Institute of General Medical Sciences, Bethesda, MD.

1969f Analyzing data: Sanctification or detective work, *Amer. Psychologist* **24**, 83-91. [Also in *CWJWT*, IV (1986e), 721-739.]

1969g J. L. Dolby, H. L. Resnikoff and J. W. Tukey. A ruly code for serials, *Proc. of the Amer. Soc. Info. Sci.* **6**, 113-124.

1970a (with others). *1969-1970 Science: National results and illustrations of group comparisons, July 1970.* Report I. National Assessment of Educational Progress, Ann Arbor, MI.

1970b Some further inputs. *Geostatistics: a colloquium.* (D. F. Merriam, ed.) 163-174. Plenum Press, New York. [Also in *CWJWT*, IV (1986e), 741-751.]

1970c *Exploratory Data Analysis.* Volume I, limited preliminary edition. Addison-Wesley, Reading, MA.

1970d *Exploratory Data Analysis.* Volume II, limited preliminary edition. Addison-Wesley, Reading, MA.

1970e Some graphic and semigraphic displays, (abstract) *Technometrics* **12**, 205. [See also 1972a.]

1970f (with many others). *Cleaner Air for the Nation; Report.* President's Task Force on Air Pollution, U.S. Government Printing Office, Washington, DC.

1971a *Exploratory Data Analysis.* Volume III, limited preliminary edition. Addison-Wesley, Reading, MA.

1971b (with many others). *Federal Statistics.* Volumes I, II. U.S. Government Printing Office, Washington, DC.

1971c R. J. Epp, J. W. Tukey and G. S. Watson. Testing unit vectors for correlation, *J. Geophys. Res.* **76**, 8480-8483. [Also in *CWJWT*, VI (1990a), 489-496.]

1971d (with others). A conversation on population, environment and human well-being. (Reported by F. F. Darling and R. F. Dasmann). Belmont House, Elkridge, MD. May 20 and 21, 1971. The Conservation Foundation, Washington, D.C., 66 pages.

1972a Some graphic and semigraphic displays. *Statistical Papers in Honor of George W. Snedecor*. (T. A. Bancroft, ed.) 293-316. Iowa State University Press, Ames, IA. [Also in *CWJWT*, V (1988a), 37-62.]

1972b How computing and statistics affect each other. *The Babbage Memorial Meeting: Report of Proceedings*. 19-37. The British Computer Society and the Royal Statistical Society, London.

1972c No paper.

1972d Data analysis, computation and mathematics, *Quart. App. Math.* **30**, 51-65. (Special Issue: Symposium on the Future of Applied Mathematics) [Also in *CWJWT*, IV (1986e), 753-775.]

1972e D. F. Andrews, P. J. Bickel, F. R. Hampel, P. J. Huber, W. H. Rogers and J. W. Tukey. *Robust Estimates of Location: Survey and Advances*. Princeton University Press, Princeton, NJ.

1972f K. David, T. Dobzhansky, R. W. Gerard, H. B. Glass, E. R. Hilgard, J. V. Neel, H. A. Simon and J. W. Tukey. Recommendations with respect to the behavioral and social aspects of human genetics, *Proc. Nat. Acad. Sci. USA* **69**, 1-3.

1972g W. H. Rogers and J. W. Tukey. Understanding some long-tailed symmetrical distributions, *Statistica Neerlandica* **26**, 211-226.

1973a D. F. Andrews and J. W. Tukey. Algorithm AS 61--Six-line plots, *J. Roy. Stat. Soc., Series C* **22**, 265-269.

1973b D. F. Andrews and J. W. Tukey. Teletypewriter plots for data analysis can be fast: 6-line plots, including probability plots, *J. Roy. Stat. Soc., Series C* **22**, 192-202. [Also in *CWJWT*, V (1988a), 93-108.]

1973c Lags in statistical technology. *Proc. of the First Canadian Conf. in Applied Statistics*. (C. S. Carter et al., eds.) 96-104. American Statistical Assoc., Montreal, Canada Chapter. [Also in *CWJWT*, IV (1986e), 777-791.]

1973d The zig-zagging climb from initial observation to successful improvement. *Frontiers of Educational Measurement and Information Systems — 1973*. (W. E. Coffman, ed.) 113-120. Houghton Mifflin, Boston, MA.

1973e J. L. Dolby and J. W. Tukey. *The Statistics CumIndex*. The R&D Press, Los Altos, CA.

1973f *Index to Statistics and Probability: Citation Index*. Vol. 2 of the Information Access Series. The R&D Press, Los Altos, CA.

1973g I. C. Ross and J. W. Tukey. *Index to Statistics and Probability: Locations and Authors*. Vol. 5 of the Information Access Series. The R&D Press, Los Altos, CA.

1973h (with others, as Chairman). *Chemicals and Health*. Report of the President's Science Advisory Committee, September. U.S. Government Printing Office, Washington, DC.

1973i Exploratory data analysis as part of a large whole. *Proceedings of the Eighteenth Conference on the Design of Experiments in Army Research, Development and Testing, Part I.* 1-10. U.S. Army Research Office, Durham, NC. [Also in *CWJWT*, IV (1986e), 793-804.]

1974a J. H. Friedman and J. W. Tukey. A projection pursuit algorithm for exploratory data analysis, *IEEE Transactions on Computers* **C-23**, 881-890. [Also in *CWJWT*, V (1987b), 149-170.]

1974b Statement on growth and its implications for the future. *Hearings Before the Subcommittee on Fisheries and Wildlife Conservation and the Environment of the House of Representatives.* Serial No. 93-35, 54-56. U.S. Government Printing Office, Washington, DC.

1974c A. E. Beaton and J. W. Tukey. The fitting of power series, meaning polynomials, illustrated on band-spectroscopic data, *Technometrics* **16**, 147-185.

1974d Introduction to today's data analysis. *Critical Evaluation of Chemical and Physical Structural Information.* (D. R. Lide, Jr. and M. A. Paul, eds.) 3-14. National Academy of Sciences, Washington, DC.

1974e A. E. Beaton and J. W. Tukey. Comments on the fitting of power series. *Critical Evaluation of Chemical and Physical Structural Information.* (D. R. Lide, Jr. and M. A. Paul, eds.) 15-35. National Academy of Sciences, Washington, DC.

1974f Data analysis for molecular structure: today and tomorrow. *Critical Evaluation of Chemical and Physical Structural Information.* (D. R. Lide, Jr. and M. A. Paul, eds.) 48-58. National Academy of Sciences, Washington, DC.

1974g Named and faceless values: An initial exploration in memory of Prasanta C. Mahalanobis, *Sankhyā*, Series A, **36**, 125-176. [Also in *CWJWT*, VI (1990a), 497-552.]

1974h A further analysis of the first phase of the Princeton Robustness Study; examples of less standard two-way table analysis. *Exploring Data Analysis: The Computer Revolution in Statistics.* (W. J. Dixon and W. L. Nicholson, eds.) Chapter 5, 229-294. University of California Press, Berkeley and Los Angeles, CA.

1974i M. A. Fisherkeller, J. H. Friedman and J. W. Tukey. PRIM-9: An interactive multidimensional data display and analysis system. A.E.C. Scientific Computer Information Exchange Meeting, May 2-3. [*Note:* This item is a **movie**, not a paper.]

1974j M. A. Fisherkeller, J. H. Friedman and J. W. Tukey. PRIM-9: An interactive multidimensional data display and analysis system. SLAC-PUB-1408. Stanford Linear Accelerator Center, Stanford, CA. [Also in *CWJWT*, V (1988a), 307-327 and *Dynamic Graphics for Statistics.* (W. S. Cleveland and M. E. McGill, eds.) 91-109. Wadsworth Advanced Books & Software, Pacific Grove, CA.]

1975a Instead of Gauss-Markov least squares, what? *Applied Statistics.* (R. P. Gupta, ed.) 351-372. North-Holland, Amsterdam.

1975b N. Mantel, N. R. Bohidar, C. C. Brown, J. L. Ciminera and J. W. Tukey. An improved Mantel-Bryan procedure for "safety" testing of carcinogens, *Cancer Research* **35**, 865-872.

1975c D. R. McNeil and J. W. Tukey. Higher-order diagnosis of two-way tables, illustrated on two sets of demographic empirical distributions, *Biometrics* **31**, 487-510.

1975d I. C. Ross and J. W. Tukey. *Index to Statistics and Probability: Permuted Titles, A-Microbiology.* Information Access Series. Vol. 3. R&D Press, Los Altos, CA.

1975e I. C. Ross and J. W. Tukey. *Index to Statistics and Probability: Permuted Titles, Microclimatic-Z*. Information Access Series. Vol. 4. R&D Press, Los Altos, CA.

1975f Mathematics and the picturing of data. *Proceedings of the International Congress of Mathematicians* **2**, 523-531. Canadian Mathematical Congress, Montreal, Canada.

1975g M. A. Fisherkeller, J. H. Friedman and J. W. Tukey. PRIM-9: An interactive multidimensional data display and analysis system. *Data: Its Use, Organization and Management*. 140-145. Association for Computing Machinery, New York.

1976a Usable Resistant/Robust Techniques of Analysis. *Proceedings of the First ERDA Statistical Symposium*. (W. L. Nicholson and J. L. Harris, eds.) 1-31. Battelle Pacific Northwest Laboratories, Richland, WA.

1976b Methodology and the Statistician's Responsibility for Both Accuracy and Relevance, *Statistical Reporter*, No. 76-13, 253-262.

1976c No paper.

1976d Robust/Resistant Smoothing. *Selected Examples of ERDA Research in the Mathematical and Computer Sciences*. 80-82. ERDA 76-118. National Technical Information Service, Springfield, VA.

1976e Discussion of: Role of Statistics in National Health Policy Decisions, *American Journal of Epidemiology* **104**, 380-385.

1976f Discussion of paper by Lave and Seskin [Does air pollution cause mortality?]. *Proceedings of the 4th Symposium on Statistics and the Environment* 37-41. American Statistical Association, Washington, DC.

1976g J. P. Gilbert, F. Mosteller and J. W. Tukey. Steady social progress requires quantitative evaluation to be searching. *The Evaluation of Social Programs*. (Clark C. Abt, ed.) 295-312. Sage Publications, Beverly Hills, CA.

1976h (with others). *Halocarbons: Environmental Effects of Fluoromethane Release*. Committee on Impacts of Stratospheric Change, National Academy of Sciences, Washington, DC.

1977a *Exploratory Data Analysis*. First Edition. Addison-Wesley, Reading, MA. [Table of contents in *CWJWT*, V (1988a), 31-35.]

1977b F. Mosteller and J. W. Tukey. *Data Analysis and Regression: A Second Course in Statistics*. Addison-Wesley, Reading, MA.

1977c Modern statistical techniques in data analysis. *The Proceedings of the Fifth Biennial International Codata Conference*. (B. Dreyfus, ed.) 9-18. Pergamon Press, Oxford and New York.

1977d Some thoughts on clinical trials, especially problems of multiplicity, *Science* **198**, 679-684.

1977e (with others). *Response to Ozone Protection, Sections of the Clean Air Amendments 1977. An Interim Report*. 1-61. Committee on Impacts of Stratospheric Change, National Academy of Sciences, Washington, DC.

1977f Statistical considerations important in analyzing measured changes in water-wall tube thickness. *Proceedings of the Second Stationary Source Combustion Symposium* **V**, Addendum. 79-92. U.S. Environmental Protection Agency, National Technical Information Service, Springfield, VA.

1977g Discussion of Session I and III. *Proceedings of Second ERDA Statistical Symposium.* 128-141 and 262-273. National Technical Information Service, Springfield, VA.

1978a R. McGill, J. W. Tukey and W. A. Larsen. Variations of box plots, *American Statistician* **32**, 12-16. [Also in *CWJWT*, V (1988a), 63-77.]

1978b Statistics at Princeton. *A Princeton Companion.* (A. Leitch, ed.) 449-451. Princeton University Press, Princeton, NJ.

1978c J. L. McCarthy and J. W. Tukey. Exploratory analysis of aggregate voting behavior: Presidential elections in New Hampshire, 1896-1972, *Social Science History* **2**, 292-331.

1978d (with 95 others: The Anturane Reinfarction Trial Research Group.) Sulfinpyrazone in the prevention of cardiac death after myocardial infarction, *The New England Journal of Medicine* **298**, 289-295.

1978e The ninther, a technique for low-effort robust (resistant) location in large samples. *Contributions to Survey Sampling and Applied Statistics in Honor of H. O. Hartley.* (H. A. David, ed.) 251-258. Academic Press, New York.

1978f R. L. Hooper, C. A. Bennett, H. J. C. Kouts, F. C. Leone and J. W. Tukey. Panel Discussion: *Proceedings of the 1977 DOE Statistical Symposium.* (D. A. Gardiner and T. Truett, eds.) 175-183. National Technical Information Service, Springfield, VA.

1978g D. R. Brillinger, L. V. Jones and J. W. Tukey. *The Management of Weather Resources II: The Role of Statistics in Weather Resources Management.* U.S. Government Printing Office, Washington, DC.

1978h (with 156 others). The Twin Rivers program on energy conservation in housing: Highlights and conclusions, *Energy and Buildings* **1**, No. 3, 207-242. Elsevier Sequois S.A., Lausanne.

1978i B. S. Atal, J.-J. Chang, M. V. Mathews and J. W. Tukey. Inversion of articulatory-to-acoustic transformation in the vocal tract by a computer-sorting technique, *J. Acoust. Soc. Am.* **63**, 1535-1555.

1978j Some thoughts on clinical trials, especially problems of multiplicity. *Evaluation Studies Review Annual.* (T. D. Cook and Associates, eds.) Vol. 3, 327-332. Sage, Beverly Hills, CA. [Reprint of 1977d.]

1978k A data analyst's comments on a variety of points and issues. *Event-Related Brain Potentials in Man.* 139-154. Academic Press, New York. [Also in *CWJWT*, II (1985a), 915-934.]

1978l Comment on a paper by H. L. Gray et al. [A new approach to ARMA modeling], *Commun. Statist. Simula. Computa.* **B7**, 79-84. [Also in *CWJWT*, II (1985a), 935-937.]

1978m Discussion of paper by Herbert Schimmel [Evidence for possible acute health effects of ambient air pollution from time series analysis: methodological questions and some new results based on New York City daily mortality, 1963-1976], *Bulletin of the New York Academy of Medicine* **54**, 1111-1112.

1978n Comments on a paper by C. W. J. Granger [Seasonality: causation, interpretation, and implications]. *Seasonal Analysis of Economic Time Series.* (A. Zellner, ed.) 50-53. U.S. Government Printing Office, Washington, DC. [Also in *CWJWT*, II (1985a), 939.]

1979a Re-expression and 'plotting', (abstract) *The Institute of Mathematical Statistics Bulletin* **8**, 274.

1979b See **1980h**.

1979c Study of robustness by simulation: particularly improvement by adjustment and combination. *Robustness in Statistics.* (R. L. Launer and G. N. Wilkinson, eds.) 75-102. Academic Press, New York.

1979d Robust techniques for the user. *Robustness in Statistics.* (R. L. Launer and G. N. Wilkinson, eds.) 103-106. Academic Press, New York.

1979e Methodology, and the statistician's responsibility for BOTH accuracy AND relevance, *J. Amer. Statist. Assoc.* **74**, 786-793. [An enlarged version of 1976b.]

1979f Comment on Emanuel Parzen [Nonparametric statistical data modeling], *J. Amer. Statist. Assoc.* **74**, 121-122. [Also in *CWJWT*, IV (1986e), 805-809.]

1979g Statistical Mapping: What should not be plotted. *Proceedings of the 1976 Workshop on Automated Cartography.* 18-26. DHEW Publication No. (PHS) 79-1254. [Also in *CWJWT*, V (1988a), 109-121.]

1980a We need both exploratory and confirmatory, *American Statistician* **34**, 23-25. [Also in *CWJWT*, IV (1986e), 811-817.]

1980b No paper.

1980c Methodological comments focused on opportunities. *Multivariate Techniques in Communication Research.* (P. R. Monge and J. Cappella, eds.) Chapter 16, 489-528. Academic Press, New York. [Also in *CWJWT*, IV (1986e), 819-867.]

1980d See **1982j**.

1980e See **1981e**.

1980f Styles of data analysis, and their implications for statistical computing. *COMPSTAT 1980: Proceedings in Computational Statistics.* (M. M. Barrett and D. Wishart, eds.) 21-42. Physica-Verlag, New York.

1980g Can we predict where 'time series' should go next? *Directions in Time Series.* (D. R. Brillinger and G. C. Tiao, eds.) 1-31. Institute of Mathematical Statistics, Hayward, CA. [Also in *CWJWT*, II (1985a), 941-980.]

1980h J. H. Friedman, J. W. Tukey and P. A. Tukey. Approaches to analysis of data that concentrate near intermediate-dimensional manifolds. *Data Analysis and Informatics.* (E. Diday, L. Lebart, J. T. Pagès and R. Tomassone, eds.) 3-13. North-Holland, Amsterdam. [Also in *CWJWT*, V (1988a), 289-305.]

1981a D. C. Hoaglin, B. Iglewicz and J. W. Tukey. Small-sample performance of a resistant rule for outlier detection. *1980 Proceedings of the Statistical Computing Section.* 148-152. American Statistical Association, Washington, DC.

1981b See **1982m**.

1981c See **1982n**.

1981d *Analiz Rezul'tatov Nablyudeniyi. (The Analysis of the Results of Observations.* Russian edition of *Exploratory Data Analysis).* Translated by Ph.D. candidates A. F. Kusznira, A. L. Petrosyana and E. L. Reznikova, under the direction of B. F. Pisarenko. Mir Press, Moskva, USSR.

1981e P. A. Tukey and J. W. Tukey. Graphic display of data sets in 3 or more dimensions. *Interpreting Multivariate Data.* (V. Barnett, ed.) Chapters 10, 11 and 12, 189-275. Wiley, Chichester. [Lecture format in *CWJWT*, V (1988a), 189-288.]

1982a F. Mosteller and J. W. Tukey. Combination of results of stated precision: I. The optimistic case, *Utilitas Mathematica* **21A**, 155-178. [Also in *CWJWT*, VI (1990a), 591-611.]

1982b Another look at the future. *Computer Science and Statistics: Proceedings of the 14th Symposium on the Interface.* (K. W. Heiner, R. S. Sacher and J. W. Wilkinson, eds.) 2-8. Springer-Verlag, New York. [Also in *CWJWT*, V (1988a), 403-418.]

1982c J. W. Tukey and P. A. Tukey. Some graphics for studying four-dimensional data. *Computer Science and Statistics: Proceedings of the 14th Symposium on the Interface.* (K. W. Heiner, R. S. Sacher and J. W. Wilkinson, eds.) 60-66. Springer-Verlag, New York. [Also in *CWJWT*, V (1988a), 171-188.]

1982d No paper.

1982e See **1984k**.

1982f Discussion [The Role of Statistical Graduate Training]. *Teaching of Statistics and Statistical Consulting.* (J. S. Rustagi and D. A. Wolfe, eds.) 379-389. Academic Press, New York. [Also in *CWJWT*, IV (1986e), 881-889.]

1982g See **1983b**.

1982h N. Mantel, J. W. Tukey, J. L. Ciminera and J. F. Heyse. Tumorigenicity assays, including use of the jackknife, *Biom. J.* **24**, 579-596.

1982i A. Seheult and J. W. Tukey. Some resistant procedures for analyzing 2^n factorial experiments. *Utilitas Mathematica* **21B**, 57-98.

1982j C. L. Mallows and J. W. Tukey. An overview of techniques of data analysis, emphasizing its exploratory aspects. *Some Recent Advances in Statistics.* (J. Tiago de Oliveira et al., eds.) 111-172. Academic Press, London. [Also in *CWJWT*, IV (1986e), 891-967.]

1982k F. Mosteller and J. W. Tukey. *Analiz Dannykh I Regressiya*, Russian edition of *Data Analysis and Regression: A Second Course in Statistics.* Translated from English by Yu. N. Blagovshchenskogo, editing and preface by Yu. P. Adleva. Volume I and II. Financy I Statistika Press, Moskva, USSR.

1982l C. L. Mallows and J. W. Tukey. An overview of techniques of data analysis, emphasizing its exploratory aspects. *Some Recent Advances in Statistics.* (J. Tiago de Oliveira and B. Epstein, eds.) (limited national edition), 111-172. Lisboa, Publicacoes do II Centenario da Academia das Ciencias de Lisboa [another edition of 1982j].

1982m Introduction to styles of data analysis techniques. *Modern Data Analysis.* (R. L. Launer and A. F. Siegel, eds.) 1-11. Academic Press, New York. [Also in *CWJWT*, IV (1986e), 969-983.]

1982n The use of smelting in guiding re-expression. *Modern Data Analysis.* (R. L. Launer and A. F. Siegel, eds.) 83-102. Academic Press, New York.

1982o Discussion of the paper by A. C. Atkinson [Regression diagnostics, transformations and constructed variables]. *J. Roy. Stat. Soc., Series B* **44**, 31.

1983a M. B. Breckenridge and J. W. Tukey. The relationship of empirical analysis to more narrowly modeled analysis. *Age, Time, and Fertility: Applications of Exploratory Data Analysis.* Appendix A, 274-281. Academic Press, New York.

1983b (D. C. Hoaglin, F. Mosteller and J. W. Tukey, eds.) *Understanding Robust and Exploratory Data Analysis.* Wiley, New York.

1983c H. I. Braun and J. W. Tukey. Multiple comparisons through orderly partitions: the maximum subrange procedure. *Principals of Modern Psychological Measurement: A Festschrift for Frederic M. Lord*. 55-65. Lawrence Erlbaum Associates, Inc., Hillsdale, NJ. [Also in *CWJWT*, VIII (1992d), 455-467.]

1983d D. C. Hoaglin, F. Mosteller and J. W. Tukey. Introduction to more refined estimators. *Understanding Robust and Exploratory Data Analysis*. (D. C. Hoaglin, F. Mosteller and J. W. Tukey, eds.) Chapter 9, 283-296. Wiley, New York.

1984a Styles of spectrum analysis. *A Celebration in Geophysics and Oceanography — 1982, in Honor of Walter Munk*. Reference Series 84-5, 100-103. Scripps Institute of Oceanography, La Jolla, CA. [Also in *CWJWT*, II (1985a), 1143-1153.]

1984b *The Collected Works of John W. Tukey, Volume I: Time Series, 1949 - 1964*. (D. R. Brillinger, ed.) Wadsworth Advanced Books & Software, Monterey, CA.

1984c Comments on paper by F. P. Agterberg [Use of spatial analysis in mineral resource evaluation], *Mathematical Geology* **16**, 591-594.

1984d Data analysis: history and prospects. *Statistics: An Appraisal*. (H. A. David and H. T. David, eds.) 183-202. Iowa State University Press, Ames, IA. [Also in *CWJWT*, IV (1986e), 985-1001.]

1984e See **1985l**.

1984f Acceptance Speech. As recipient of Deming Award, *Newsletter*. Metropolitan Section, American Society for Quality Control. (January)

1984g Discussion of the paper by Drs. Spiegelhalter and Knill-Jones [Statistical and knowledge-based approaches to clinical decision-support systems, with an application to gastroenterology], *J. Roy. Stat. Soc., Series A* **147**, 62-64.

1984h J. W. Tukey and R. W. Hamming. Measuring noise color. *The Collected Works of John W. Tukey, Volume I: Time Series, 1949 - 1964*. 1-127. Wadsworth Advanced Books & Software, Monterey, CA.

1984i The spectral representation and transformation properties of the higher moments of stationary time series. *The Collected Works of John W. Tukey, Volume I: Time Series, 1949 - 1964*. 165-184. Wadsworth Advanced Books & Software, Monterey, CA.

1984j Mathematics 596: An introduction to the frequency analysis of time series. *The Collected Works of John W. Tukey, Volume I: Time Series, 1949 - 1964*. 503-650. Wadsworth Advanced Books & Software, Monterey, CA.

1984k F. Mosteller and J. W. Tukey. Combination of results of stated precision: II. A more realistic case. *W. G. Cochran's Impact on Statistics*. (P. S. R. S. Rao and J. Sedransk, eds.) 223-252. Wiley, New York. [Also in *CWJWT*, VI (1990a), 613-644.]

1985a *The Collected Works of John W. Tukey, Volume II: Time Series, 1965 - 1984*. (D. R. Brillinger, ed.) Wadsworth Advanced Books & Software, Monterey, CA.

1985b D. C. Hoaglin, F. Mosteller and J. W. Tukey. *Exploring Data Tables, Trends, and Shapes*. (D. C. Hoaglin, F. Mosteller and J. W. Tukey, eds.) Wiley, New York.

1985c Comment on Hahn's paper [More intelligent statistical software and statistical expert systems: Future directions], *American Statistician* **39**, 12-14.

1985d Comment on Ericksen/Kadane [Estimating the population in a census year: 1980 and beyond], *J. Amer. Statist. Assoc.* **80**, 127-128.

1985e Fourier methods in the frequency analysis of data, and souvenir sheets. *The Collected Works of John W. Tukey, Volume II: Time Series, 1965 - 1984*. 755-780. Wadsworth Advanced Books & Software, Monterey, CA.

1985f Nonlinear (nonsuperposable) methods for smoothing data. *The Collected Works of John W. Tukey, Volume II: Time Series, 1965 - 1984*. 837-856. Wadsworth Advanced Books & Software, Monterey, CA.

1985g First Scott lecture. *The Collected Works of John W. Tukey, Volume II: Time Series, 1965 - 1984*. 857-884. Wadsworth Advanced Books & Software, Monterey, CA.

1985h Second Scott lecture. *The Collected Works of John W. Tukey, Volume II: Time Series, 1965 - 1984*. 885-914. Wadsworth Advanced Books & Software, Monterey, CA.

1985i When should which spectrum approach be used? *The Collected Works of John W. Tukey, Volume II: Time Series, 1965 - 1984*. 981-1000. Wadsworth Advanced Books & Software, Monterey, CA.

1985j D. R. Brillinger and J. W. Tukey. Spectrum analysis in the presence of noise: some issues and examples. *The Collected Works of John W. Tukey, Volume II: Time Series, 1965 - 1984*. 1001-1141. Wadsworth Advanced Books & Software, Monterey, CA.

1985k J. W. Tukey and P. A. Tukey. Computer graphics and exploratory data analysis: An introduction. *Computer Graphics '85, Conference Proceedings, Vol. III*. 773-785. National Computer Graphics Assoc., Fairfax, VA. [Also in *CWJWT*, V (1988a), 419-436.]

1985l Improving crucial randomized experiments — especially in weather modification — by double randomization and rank combination. *Proceedings of the Berkeley Conference in Honor of Jerzy Neyman and Jack Kiefer*. (L. M. Le Cam and R. A. Olshen, eds.) Volume I, 79-108. Wadsworth Inc., Belmont, CA.

1985m Discussion of Peter J. Huber [Projection pursuit], *Ann. Statist.* **13**, 517-518.

1985n D. C. Hoaglin and J. W. Tukey. Checking the shape of discrete distributions. *Exploring Data Tables, Trends, and Shapes*. (D. C. Hoaglin, F. Mosteller and J. W. Tukey, eds.) Chapter 9, 345-416. Wiley, New York.

1985o The variance of slopes of lines fitted to groups: an analysis of the Johnstone and Velleman Monte Carlo results, *J. Amer. Statist. Assoc.* **80**, 1055-1059.

1985p J. W. Tukey, J. L. Ciminera, and J. F. Heyse. Testing the statistical certainty of a response to increasing doses of a drug. *Biometrics* **41**, 295-301.

1985q Discussion of the paper by C. Chatfield [The initial examination of data], *J. Roy. Stat. Soc., Series A* **148**, 246-247.

1986a Sunset Salvo, *American Statistician* **40**, 72-76. [Also in *CWJWT*, IV (1986e), 1003-1016.]

1986b An alphabet for statisticians' expert systems. *Artificial Intelligence and Statistics*. (W. A. Gale, ed.) 401-409. Addison-Wesley, Reading, MA.

1986c *The Collected Works of John W. Tukey, Volume III: Philosophy and Principles of Data Analysis, 1949 - 1964*. (L. V. Jones, ed.) Wadsworth Advanced Books & Software, Monterey, CA.

1986d Data analysis and behavioral science or learning to bear the quantitative man's burden by shunning badmandments. *The Collected Works of John W. Tukey, Volume III: Philosophy and Principles of Data Analysis, 1949 - 1964*. 187-389. Wadsworth Advanced Books & Software, Monterey, CA.

1986e *The Collected Works of John W. Tukey, Volume IV: Philosophy and Principles of Data Analysis, 1965 - 1986.* (L. V. Jones, ed.) Wadsworth Advanced Books & Software, Monterey, CA.

1986f What have statisticians been forgetting? *The Collected Works of John W. Tukey, Volume IV: Philosophy and Principles of Data Analysis, 1965 - 1986.* 587-599. Wadsworth Advanced Books & Software, Monterey, CA.

1986g Choosing techniques for the analysis of data. *The Collected Works of John W. Tukey, Volume IV: Philosophy and Principles of Data Analysis, 1965 - 1986.* 869-874. Wadsworth Advanced Books & Software, Monterey, CA.

1986h Do derivations come from heaven? *The Collected Works of John W. Tukey, Volume IV: Philosophy and Principles of Data Analysis, 1965 - 1986.* 875-880. Wadsworth Advanced Books & Software, Monterey, CA.

1986i No paper.

1986j See **1987a**.

1986k The interface with computing: in the small or in the large *Computer Science and Statistics: Proceedings of the 18th Symposium on the Interface.* (T. J. Boardman, ed.) 3-7. American Statistical Assoc., Washington, DC.

1986l See **1987h**.

1986m D. C. Hoaglin, B. Iglewicz and J. W. Tukey. Performance of some resistant rules for outlier labeling, *J. Amer. Statist. Assoc.* **81**, 991-999.

1986n See **1988h**.

1986o Discussion of Session VIII-b, *Proceedings of the Second Annual Research Conference.* 408-410. Bureau of the Census, Washington, DC.

1986p Discussion of paper by H. Wainer [The SAT as a social indicator: a pretty bad idea]. *Drawing Inferences from Self-Selected Samples.* (H. Wainer, ed.) 24-25. Springer-Verlag, New York.

1986q Discussion of paper by B. Singer [Self-selection and performance-based ratings: a case study in program evaluation.] *Drawing Inferences from Self-Selected Samples.* (H. Wainer, ed.) 52-55. Springer-Verlag, New York.

1986r Discussion of paper by J. Heckman [Alternative methods for evaluating the impact of intervention]. *Drawing Inferences from Self-Selected Samples.* (H. Wainer, ed.) 58-60. Springer-Verlag, New York.

1986s Discussion of paper by R. Glynn, N. Laird and D. Rubin [Mixture modeling versus selection modeling with nonignorable nonresponse]. *Drawing Inferences from Self-Selected Samples.* (H. Wainer, ed.) 143. Springer-Verlag, New York.

1986t Comments on paper by J. Heckman and R. Robb [Alternative methods for solving the problem of selection bias in evaluating the impact of treatments on outcomes]. *Drawing Inferences from Self-Selected Samples.* (H. Wainer, ed.) 108-110. Springer-Verlag, New York.

1986u Discussion of paper by D. R. Brillinger [The natural variability of vital rates and associated statistics], *Biometrics* **42**, 729-732.

1987a Configural Polysampling [1985 John von Neumann Lecture], *SIAM Review* **29**, 1-20.

1987b See **1988a**.

1987c See **1988b**.

1987d See **1988c**.

1987e See **1988d**.

1987f See **1988e**.

1987g Discussion of paper by Terry P. Speed [What is an analysis of variance?], *Annals of Statistics* **15**, 936-937.

1987h E. G. Johnson and J. W. Tukey. Graphical exploratory analysis of variance illustrated on a splitting of the Johnson and Tsao data. *Design, Data, and Analysis by Some Friends of Cuthbert Daniel.* (C. Mallows, ed.) 171-244. Wiley, New York.

1987i Discussion of paper by M. C. Jones and R. Sibson [What is projection pursuit?], *Journal of the Royal Statistical Society* **150**, 33.

1987j Lopressor Intervention Trial Research Group. The Lopressor Intervention Trial: Multicentre study of metoprolol in survivors of acute myocardial infarction, *European Heart Journal* **8**, 1056-1064.

1987k Comment on paper by R. A. Becker, W. S. Cleveland and A. R. Wilks [Dynamic graphics for data analysis], *Stat. Sci.* **2**, 383-385.

1988a *The Collected Works of John W. Tukey, Volume V: Graphics, 1965 - 1985.* (W. S. Cleveland, ed.) Wadsworth Advanced Books & Software, Pacific Grove, CA.

1988b Data analysis and statistics: Techniques and approaches. [Part] *The Collected Works of John W. Tukey, Volume V: Graphics, 1965 - 1985.* 1-22. Wadsworth Advanced Books & Software, Pacific Grove, CA.

1988c Notch plots for counted rates. [Part] *The Collected Works of John W. Tukey, Volume V: Graphics, 1965 - 1985.* 79-92. Wadsworth Advanced Books & Software, Pacific Grove, CA.

1988d Control and stash philosophy for two-handed, flexible, and immediate control of graphic display. *The Collected Works of John W. Tukey, Volume V: Graphics, 1965 - 1985.* 329-382. Wadsworth Advanced Books & Software, Pacific Grove, CA. [Also in *Dynamic Graphics for Statistics.* (W. S. Cleveland and M. E. McGill, eds.) 133-178. Wadsworth Advanced Books & Software, Pacific Grove, CA.]

1988e Thoughts on the evolution of dynamic graphics for data-modification display. *The Collected Works of John W. Tukey, Volume V: Graphics, 1965 - 1985.* 383-401. Wadsworth Advanced Books & Software, Pacific Grove, CA.

1988f Frederick Mosteller and John W. Tukey: A Conversation, *Stat. Sci.* **3**, 136-144.

1988g J. D. Emerson, D. C. Hoaglin, J. W. Tukey and G. Y. Wong. Exploring Some Adverb-Adjective Data, *Chance* **1(3)**. 42-48.

1988h K. Kafadar and J. W. Tukey. A Bidec *t* Table, *J. Amer. Statist. Assoc.* **83**, 532-539.

1989a See **1990a**.

1989b See **1990b**.

1989c See **1990c**.

1989d See **1990d**.

1989e See **1990e**.

1989f See **1990f**.

1989g See **1990g**.

1989h See **1990h**.

1989i See **1990i**.

1989j See **1990j**.

1989k See **1990k**.

1989l See **1990l**.

1989m See **1990m**.

1989n See **1990n**.

1989o See **1990o**.

1989p See **1990p**.

1989q See **1990q**.

1989r J. L. Ciminera and J. W. Tukey. Control-charting automated laboratory instruments when many successive differences may be zero, *J. Qual. Technol.* **21**, 7-15.

1989s J. L. Ciminera and J. W. Tukey. Use of augmented box plots for reviewing measurement results collected over time, *J. Qual. Technol.* **21**, 187-198.

1989t See **1989s**.

1989u See **1989y**.

1989v See **1990r**.

1989w SPES in the years ahead, *Proc. of Amer. Statist. Assoc. Sesquicentennial 1988-1989 meetings*, Washington, D.C. 175-182.

1989x E. Ericksen, J. B. Kadane and J. W. Tukey. Adjusting the 1980 Census of Housing and Population, *J. Amer. Statist. Assoc.* **84**, 927-944.

1989y J. L. Ciminera, J. F. Heyse and J. W. Tukey. Techniques for the graphical analysis of possibly skewed measurements, *J. Qual. Technol.* **21**, 223-231.

1990a *The Collected Works of John W. Tukey, Volume VI: More Mathematical, 1938 - 1984*. (C. L. Mallows, ed.) Wadsworth Advanced Books & Software, Pacific Grove, CA.

1990b The finite case of the "Problem of the Nile." *The Collected Works of John W. Tukey, Volume VI: More Mathematical, 1938 - 1984*. 35-40. Wadsworth Advanced Books & Software, Pacific Grove, CA.

1990c The present state of fiducial probability. *The Collected Works of John W. Tukey, Volume VI: More Mathematical, 1938 - 1984*. 55-118. Wadsworth Advanced Books & Software, Pacific Grove, CA.

1990d Handouts for the Wald lectures 1958. *The Collected Works of John W. Tukey, Volume VI: More Mathematical, 1938 - 1984*. 119-148. Wadsworth Advanced Books & Software, Pacific Grove, CA.

1990e Souvenir sheets for "The criticism of transformations." *The Collected Works of John W. Tukey, Volume VI: More Mathematical, 1938 - 1984*. 157-165. Wadsworth Advanced Books & Software, Pacific Grove, CA.

1990f The practical relationship between the common transformations of percentages or fractions and of amounts. *The Collected Works of John W. Tukey, Volume VI: More Mathematical, 1938 - 1984*. 211-219. Wadsworth Advanced Books & Software, Pacific Grove, CA.

1990g Introduction to guided re-expression. *The Collected Works of John W. Tukey, Volume VI: More Mathematical, 1938 - 1984*. 221-236. Wadsworth Advanced Books & Software, Pacific Grove, CA.

1990h Determination of linear relations between systematic parts. *The Collected Works of John W. Tukey, Volume VI: More Mathematical, 1938 - 1984*. 317-329. Wadsworth Advanced Books & Software, Pacific Grove, CA.

1990i Standard confidence points. *The Collected Works of John W. Tukey, Volume VI: More Mathematical, 1938 - 1984*. 331-365. Wadsworth Advanced Books & Software, Pacific Grove, CA.

1990j Some general principles and approximations for sequentially studentized procedures. *The Collected Works of John W. Tukey, Volume VI: More Mathematical, 1938 - 1984*. 367-371. Wadsworth Advanced Books & Software, Pacific Grove, CA.

1990k One degree of freedom or several? Parsimony in detection of an effect. *The Collected Works of John W. Tukey, Volume VI: More Mathematical, 1938 - 1984*. 407-419. Wadsworth Advanced Books & Software, Pacific Grove, CA.

1990l A note on least squares and unbiased estimates. *The Collected Works of John W. Tukey, Volume VI: More Mathematical, 1938 - 1984*. 421-427. Wadsworth Advanced Books & Software, Pacific Grove, CA.

1990m Better coordinates for combinations of order statistics. *The Collected Works of John W. Tukey, Volume VI: More Mathematical, 1938 - 1984*. 429-433. Wadsworth Advanced Books & Software, Pacific Grove, CA.

1990n Souvenir sheet for "Use of control classifications: adjustment for inadequacy of broad classes." *The Collected Works of John W. Tukey, Volume VI: More Mathematical, 1938 - 1984*. 435-443. Wadsworth Advanced Books & Software, Pacific Grove, CA.

1990o NkNS forests. *The Collected Works of John W. Tukey, Volume VI: More Mathematical, 1938 - 1984*. 553-556. Wadsworth Advanced Books & Software, Pacific Grove, CA.

1990p Some further multivariate suggestions. *The Collected Works of John W. Tukey, Volume VI: More Mathematical, 1938 - 1984*. 557-583. Wadsworth Advanced Books & Software, Pacific Grove, CA.

1990q Steps toward a universal univariate distribution analyzer. *The Collected Works of John W. Tukey, Volume VI: More Mathematical, 1938 - 1984*. 585-590. Wadsworth Advanced Books & Software, Pacific Grove, CA.

1990r Data based graphics: visual display in the decades to come, *Statistical Science* **5**, 327-339. [Also in *Proc. of Amer. Statist. Assoc. Sesquicentennial 1988-1989 meetings*, Washington, D.C. 366-381.]

1990s S. Morgenthaler and J. W. Tukey. The next future of data analysis. *Data Analysis, Learning Symbolic and Numeric Knowledge*. (E. Diday, ed.) Nova Science Publishers, New York.

1990t (with others). Proposed guidelines for statistical adjustment of the 1990 Census. Hearing before the SubCommittee on Census and Population of the Committee on Post Office and Civil Service, House of Representatives, 101st Congress, Second Session, January 30, 1990, Serial No. 101-43, U. S. Government Printing Office, Washington, D.C., 177-190.

1990u (with others). 1990 Census coverage evaluation operations. Hearing before the SubCommittee on Census and Population of the Committee on Post Office and Civil Service, House of Representatives, 101st Congress, Second Session, September 11, 1990, Serial No. 101-79, U.S. Government Printing Office, Washington, D.C., 1-142.

1990v Foreword *for* The effect of changes in national assessment: Disentangling the NAEP 1985-86 reading anomaly. (A. E. Beaton and R. Zwick). Report No. 17-TR-21, National Assessment of Educational Progress, Educational Testing Service, Rosedale Road, Princeton, NJ 08541-0001.

1991a See **1992a**.

1991b See **1992b**.

1991c (S. Morgenthaler and J. W. Tukey, eds.) *Configural Polysampling: A Route to Practical Robustness*. Wiley, New York.

1991d S. Morgenthaler and J. W. Tukey. Chapter 1: Background. *Configural Polysampling: A Route to Practical Robustness*. (S. Morgenthaler and J. W. Tukey, eds.) 1-8. Wiley, New York.

1991e S. Morgenthaler and J. W. Tukey. Chapter 3: Key ideas and outline. *Configural Polysampling: A Route to Practical Robustness*. (S. Morgenthaler and J. W. Tukey, eds.) 21-36. Wiley, New York.

1991f S. Morgenthaler and J. W. Tukey. Chapter 7: Point estimation of location: technical choices. *Configural Polysampling: A Route to Practical Robustness*. (S. Morgenthaler and J. W. Tukey, eds.) 87-108. Wiley, New York.

1991g Chapter 13: Appendix on a circular approach to cubature. *Configural Polysampling: A Route to Practical Robustness*. (S. Morgenthaler and J. W. Tukey, eds.) 213-221. Wiley, New York.

1991h (D. C. Hoaglin, F. Mosteller and J. W. Tukey, eds.) *Fundamentals of Exploratory Analysis of Variance*. Wiley, New York, 430 pages.

1991i F. Mosteller, A. Parunak and J. W. Tukey. Mean squares, F. tests, and estimates of variance. *Fundamentals of Exploratory Analysis of Variance*. (D. C. Hoaglin, F. Mosteller and J. W. Tukey, eds.) Chapter 7, 146-164. Wiley, New York.

1991j D. C. Hoaglin and J. W. Tukey. Qualitative and quantitative confidence. *Fundamentals of Exploratory Analysis of Variance*. (D. C. Hoaglin, F. Mosteller and J. W. Tukey, eds.). Chapter 12, 336-364. Wiley, New York.

1991k F. Mosteller, J. W. Tukey and C. S. Youtz. Assessing changes. *Fundamentals of Exploratory Analysis of Variance*. (D. C. Hoaglin, F. Mosteller and J. W. Tukey, eds.). Chapter 11, 295-335. Wiley, New York.

1991l F. Mosteller and J. W. Tukey. Purposes of analyzing data that comes in a form inviting us to apply tools from the analysis of variance. *Fundamentals of Exploratory Analysis of Variance*. (D. C. Hoaglin, F. Mosteller and J. W. Tukey, eds.). Chapter 2, 24-39. Wiley, New York.

1991m D. C. Hoaglin, F. Mosteller and J. W. Tukey. Concepts and examples in analysis of variance (ANOVA). *Fundamentals of Exploratory Analysis of Variance.* (D. C. Hoaglin, F. Mosteller and J. W. Tukey, eds.). Chapter 1, 1-23. Wiley, New York.

1991n The philosophy of multiple comparisons, (1989 Miller Lecture presented at Stanford University), *Statistical Science* **6**(1), 98-116.

1991o Use of many covariates in clinical trials, *Int. Stat. Rev.* **59**(2), 123-137.

1991p Consumer datesware. *Directions in Robust Statistics and Diagnostics, Part II, The IMA Volumes in Mathematics and its Applications.* **34**. (W. Stahel and S. Weisberg, eds.) 297-308. Springer-Verlag, New York.

1991q Graphical displays for alternate regression fits. *Directions in Robust Statistics and Diagnostics, Part II, The IMA Volumes in Mathematics and its Applications.* **34**. (W. Stahel and S. Weisberg, eds.) 309-325. Springer-Verlag, New York.

1991r (with others). The experience and legacy of NAPAP (Report of the Oversight Review Board of the National Acid Precipitation Assessment Program), U.S. Government Printing Office, Washington, D.C., 40 pages.

1991s A. Sivak and J. W. Tukey. Comments on predicting EPA's forthcoming CO standards in light of new clinical evidence, *Risk Analysis* **11**, 575-576.

1991t See **1992q**.

1991u (with others). Oversight hearing to review the progress of coverage evaluation procedures. *Hearing before the SubCommittee on Census and Population Committee on Post Office and Civil Service, House of Representatives, 102nd Congress, First Session, June 27, 1991,* **Serial No. 102-19,** U. S. Government Printing Office, Washington, D. C. 1-95.

1992a *The Collected Works of John W. Tukey, Volume VII: Factorial & ANOVA, 1949 - 1962.* (D. R. Cox, ed.) Wadsworth Advanced Books & Software, Pacific Grove, CA.

1992b W. T. Federer, D. S. Robson and J. W. Tukey. Handout for meeting of experimental station statisticians "Tests for non-additivity." *The Collected Works of John W. Tukey, Volume VII: Factorial & ANOVA, 1949 - 1962.* 19-73. Wadsworth Advanced Books & Software, Pacific Grove, CA.

1992c Exbrids: Simple nearly symmetrizing re-expressions for exponentially distributed quantities [Festschrift in Honor of Geoffrey S. Watson]. *The Art of Statistical Science.* [K. S. Mardia, ed.) 183-187. John Wiley & Sons, Ltd., West Sussex, England.

1992d See **1993a**.

1992e See **1993b**.

1992f See **1993c**.

1992g See **1993d**.

1992h See **1993e**.

1992i See **1993f**.

1992j See **1993g**.

1992k (with Hansen, K. M.). Tuning a major part of a clustering algorithm, *Int. Stat. Rev.* **59**(1), 21-43.

1992l (with Kafadar, K.). An approach to U. S. cancer death rates involving urbanness and geographic contiguity: 1. A simple adjustment for urbanness, ***Int. Stat. Rev.*** **60**(3). To appear.

1992m Where should multiple comparisons go next?, ***Proceedings Charles W. Dunnett Symposium*** (29 May - 1 June 1991, McMaster University, Hamilton, Ontario, Canada). Chapter 12, 187-207. Marcel Dekker Publishing Company.

1992n (with Heyse, J. and Ciminera, J.). Tests for qualitative treatment-by-center interaction using a 'pushback' procedure, ***Statistics in Medicine***. To appear.

1992o (with Ciminera, Joseph L., Heyse, Joseph F., Nguyen, Ha H.). Evaluation of multicenter clinical trial data using adaptations of the Mosteller-Tukey procedure, ***Statistics in Medicine***. To appear.

1992p No paper.

1992q Discussion of paper by Bradley Efron [Jackknife-after-bootstrap standard errors and influence functions], *J. Roy. Stat. Soc., Series B* **54**, 120-121.

1993a *The Collected Works of John W. Tukey, Volume VIII: Multiple Comparisons, 1948 - 1983.* (H. I. Braun, ed.) Chapman and Hall, New York, NY.

1993b The problem of multiple comparisons. *The Collected Works of John W. Tukey, Volume VIII: Multiple Comparisons, 1948 - 1983*. 1-300. Chapman and Hall, New York, NY.

1993c Equalizing unequal accuracies of means. *The Collected Works of John W. Tukey, Volume VIII: Multiple Comparisons, 1948 - 1983*. 301-313. Chapman and Hall, New York, NY.

1993d Reminder sheets for "Allowances for various types of error rates." *The Collected Works of John W. Tukey, Volume VIII: Multiple Comparisons, 1948 - 1983*. 335-339. Chapman and Hall, New York, NY.

1993e Reminder sheets for "Multiple comparisons." *The Collected Works of John W. Tukey, Volume VIII: Multiple Comparisons, 1948 - 1983*. 341-345. Chapman and Hall, New York, NY.

1993f Souvenir sheets for "Confidence procedures are better." *The Collected Works of John W. Tukey, Volume VIII: Multiple Comparisons, 1948 - 1983*. 347-351. Chapman and Hall, New York, NY.

1993g Reminder sheets for "Discussion of paper on multiple comparisons by Henry Scheffé." *The Collected Works of John W. Tukey, Volume VIII: Multiple Comparisons, 1948 - 1983*. 469-475. Chapman and Hall, New York, NY.

1993h (with Cohen, M. and Dalal, S.). Robust, smoothly — heterogeneous variance regression, *Applied Statistics* **42**(1). To appear in 1993 volume.

INDEX TO BIBLIOGRAPHY OF JOHN W. TUKEY 1938-1993

A. Index by Subject

(entries are identifications in list; "47bcdj" is "1947b, 1947c, 1947d, and 1947j")

B. Index by Co-Authors

(entries are identifications in list; "47bcdj" is "1947b, 1947c, 1947d, and 1947j")

C. Index by Conventional Journal

(entries are identifications in list; "47bcdj" is "1947b, 1947c, 1947d, and 1947j"; entries give "volume/identification")

AMSX	Annals of Math'l. Statistics 15/44a, 16/45a, 17/46abc, 18/47bcdj, 19/48ab, 20/49ek, 21/50dh, 27/56adf, 28/57abcdf, 29/58bcg, 33/62a, 34/63c
APST	Amer. Psychologist 24/69f
ARTS	Amer. Rev. of Tuberculosis 62/50e
ASLJ	Astronomical J. 114/51e
ASTN	Amer. Statistician 3/49b, 19/65f, 32/78a, 34/80a, 39/85c, 40/86a
ASTX	Applied Statistics 22/73ab
ASTX1	Annals of Statistics 13/85m, 15/87g
BIIS	Bull. Inst. Int. Statistique 40/63f, 41b/66e
BMLJ	Biomedical Journal 24/82h
BMSP	Proc. Berkeley Symp. Math. Stat. Prob. 4/61b
BMTA	Biometrika 42/55c, 43/56g, 44/57g, 52/65hi, 53/66b
BMTX	Biometrics 5/49h, 7/51a, 10/53d, 11/55f, 15/59g, 24/68c, 37/75c, 41/85p, 42/86u
BNYM	Bull. New York Acad. of Medicine 54/78m
BSTJ	Bell System Tech'l J. 37/58de
CCAS	Proc. Xth Canadian Conf. in Applied Statistics 1/73c
CHNE	Chance 1/88g
CRSH	Cancer Research 37/75b
CSMR	Christian Science Monitor ??/66f
CSSC	Communications in Statistics — Simulation and Computation B7/78l
CSSP	Computer Science and Statistics: Proc. Xth Symposium on the Interface 14/82bc, 18/86k
DMLJ	Duke Math'l. J. 9/42b
EBLS	Energy and Buildings 1/78h
ECNA	Econometrica 20/52b
EHTJ	European Heart Journal 8/87j
ESRA	Evaluation Studies Review Annual 3/78j
HBLY	Human Biology 20/48c, 21/49a, 22/50g
IAMR	SIAM Rev. 9/67d, 29/87a
IMSB	Inst. of Math. Statistics Bull. 8/79a
IQCL	Industrial Quality Control 6/49ij, 6/50bc
ISLR	Internat. Statist. Rev. 59/91o, 59/92k, 60/92l
JAMA	J. of the Amer. Medical Assoc. 197/66d
JASA	J. of the Amer. Statistical Assoc. 41/46de, 42/47hi, 44/49glm, 45/50f, 47/52ac, 48/53b, 49/54ad, 50/55d, 55/60ac, 60/65j, 74/79ef, 80/85do, 81/86m, 83/88h, 84/89x

JASA1	J. of the Acoustical Society of Amer. 63/78i
JCDN	J. of Chemical Documentation 2/62c
JCPY	J. of Clinical Psychology 6/50a
JCRH	J. of Cycle Research 6/57h
JGRH	J. of Geophysical Research 76/72c
JIAS	J. of the Interamer. Statistical Institute 14/56h
JMPY	J. of Math'l Psychology 1/64a
JORA	J. of the Operations Research Soc. of Amer. 3/55e
JQTY	J. of Quality Technology 21/89rstuy
JRSA	J. of the Royal Statistical Soc., Series A 147/84g, 150/87i
JRSB	J. of the Royal Statistical Soc., Series B 29/67c, 54/92q
JRSC	J. of the Royal Statistical Soc., Series C (see Applied Statistics)
JSPY	J. of Sedimentary Petrology 26/56e
MCMN	Math. of Computation 19/65a
MGLY	Mathematical Geology 16/84c
MSBK	Matematicheskikh Sbornik 4/38b
NEJM	New England J. of Medicine 298/78d
PBIC	Proc. Xth Biennial International Codata Conf. 5/77c
PCPS	Proc. of the Cambridge Philosophical Soc. 53/57e
PICM	Proc. of the International Congress of Math'ns. 50/52d, 74/75f
PMTE	Portugaliae Mathematica 3/42a
PNAS	Proc. of the National Academy of Sciences 25/39a, 53/65d, 69/72f
PSCA	Psychometrika 23/58a, 25/60b
PSCS	Proc. Stat. Comp. Sect. ASA 80/81a
PSSC	Proc. of the Xth Stationary Source Combustion Symp. 2/77f
PSSE	Proc. Xth Symp. of Statistics and the Environment 4/76f
QAMX	Quarterly of Applied Math. 5/47ag, 30/72d
QCCP	Quality Control Conference Papers, 51/51d
RANS	Risk Analysis 11/91s
SMDE	Statistics in Medicine -/92no
SCNE	Science 109/49cd, 122/55a, 131/62d, 148/65b, 198/77d
SNAA	Sankhyā, Series A 25/63d, 36/74g
SNRA	Statistica Neerlandica 26/72g
SRPR	Statistical Reporter (U.S.) 76/76b
SSCE	Statistical Science 2/87k, 3/88f, 5/90r, 6/91n

D. Index for Miscellaneous Appearances

(entries are identifications in list; "47bcdj" is "1947b, 1947c 1947d, and 1947j")

E. Index by Editors

(entries are identifications in list; "47bcdj" is "1947b, 1947c, 1947d, and 1947j")

F. Acknowledgment

The efforts of David C. Hoaglin in reviewing and correcting this bibliography is deeply appreciated.

Foreword to the Multiple Comparisons Volume

John W. Tukey

Henry Braun's descriptive introduction, which follows this foreword, places the various items in contexts appropriate to their epochs (1949-54, except 1948 for item 2, 1965 for items 7, 8, and 9, and 1983 for item 10). So this foreword can focus on history and background.

My connection with "multiple comparisons" began with an invitation, by the Baltimore Chapter of ASQC, to come to Baltimore and give a talk on "industrial uses of the range." After I had accepted, I recognized an obligation to look around for some such uses. With help from Richard F. Link and David L. Wallace, who were students at Princeton at the time, I put together the requested talk, which included the use of the Studentized range to examine individual comparisons among multiple results.

Questions led to techniques; techniques and competing techniques (Scheffé, Duncan) then led to deeper analysis and the development of philosophy. Work in this area absorbed much of my attention at this time, as can be inferred from the relatively small numbers of entries in my bibliography for 1951, 1952, 1953 (4 each), 1954 (5), and 1955 (6).

The major first 300 pages of the present volume — "The Problem of Multiple Comparisons" — should have been issued, in book form, in the middle fifties. One reason this did not happen was the only piece of bad advice I ever had from Walter Shewhart! He told me it was unwise to put a book out until one was sure that it contained the last word of one's thinking. Even for author's who will write a single book — or like Walter, one book as such, and an organized account of a set of lectures — this was *not* a wise general principle, especially in a fast-moving field. I respected and trusted Walter, as I should have almost always, so much that completing and issuing "The Problem of Multiple Comparisons" slipped lower and lower in my "sin pile."

Strangely enough, after about 40 years, I feel there is still merit in giving it more general circulation (a moderate number of copies have been asked for and supplied through the years).

Besides the philosophical issues that Henry Braun points to, he also notices "effective use of tables to reduce size and/or improve accessibility." While the rise of the computer has made it possible to be far less pedestrian about critical values that depend upon 2 or more parameters, the examples of empirical search for empirical descriptions are still highly relevant to the further development of statistical theory — theory as a whole, including formula manipulation, data analysis of tables, results of experimental sampling, and all the manifold ways in which data-analysis relevant results flow from assumptions, conventional or unconventional.

Item 1 antedates the general push on multiple comparisons. Items 2 to 6, and 11 belong to the push. Items 7 and 8 represent the use of early computers to construct tables so data could be analyzed without computers. Had they been done 20 (or even 40) years earlier, they might have had a much larger impact, during the decades where the computer had not even reached its childhood. As it was, they were barely in time to receive moderate use. They may provide an opportunity for someone to rework the basic ideas and effective approximations for use in a modern computer. The advantages of range-based assessment of variability are by no means confined to simplifying hand calculation. There can be advantages of behavior of the procedures as well.

Henry Braun, perhaps with a co-author's modesty, has not stressed one aspect of item 10 which I find of great importance:

- Recognition of the usefulness of — and need for — a *variety* of procedures reflecting the *varied strength* of different experiments (particularly, perhaps, when that strength grows with the development and strengthening of any particular subject-matter field).

It would be wrong, I argue strongly, to try to find a single multiple comparisons procedure for general use. There can be, very occasionally, "a man for all seasons"; but we do not dare to seek "a single procedure for all experiments."

Princeton, NJ
September 1992

Introduction to the Multiple Comparisons Volume

Henry I. Braun

As readers of earlier volumes in this series are aware, John Tukey has made important and often seminal contributions to each of the areas of statistical thought and practice that have been covered. All will be relieved to learn that with this volume the run remains unbroken. But more here than elsewhere, perhaps, John also succeeded in setting the questions that have guided research and subsequent practice for some thirty-five years.

In a relatively short period of two or three years in the early 1950's, John brought to the attention of the wider statistical community the intriguing problems inherent in this area and provided a useful framework for thinking about them. With the "appearance" of the never-published "The Problem of Multiple Comparisons" he succeeded in making substantial advances and demonstrating the usefulness of his schema.

That there is any agreement about the terms of the debate in this area is largely due to the force and cogency of John's arguments. And let it be said that "Multiple Comparisons" is an area of statistical practice that has engendered more than its share of controversy and confusion. The confusion often begins with the title, which affords a perfect example of synecdoche — the use of a less inclusive term for a more inclusive one. "Simultaneous Inference" is a more appropriate label for this branch of statistics, with multiple comparisons properly regarded as a particular set of techniques. Nonetheless, historical usage decrees that the two titles will be treated interchangeably for some time to come.

But more than labels are at stake. Simultaneous Inference is very much concerned with good statistical practice: How do we properly make multiple inferences from a set of data in the context of a scientific investigation? As John often pointed out, the answer must depend on the nature of the investigation the kinds of questions asked and the nature of the consequences following from the inferences. These issues do not permit a neat mathematical argument. They require hard thinking about substantive matters — on which reasonable people may agree to disagree. Before getting started, an approach must be selected: multiple significance statements, multiple confidence statements or a multiple decision approach.

As always, John's style has been to ask questions and suggest answers that are sensitive to practical concerns.

Another battle revolved around the issue of whether and how to control error rates. By now, most statisticians would agree that the interpretability of a family of statements requires the overall error rate to be bounded at levels ordinarily attached to single statements. But in the early 1950's this was not so clear, especially with David Duncan advocating what may be termed a more liberal approach. The more conservative approach, strongly argued by John, eventually prevailed as people recognized the untoward consequences of the alternative.

A related issue is, to use John's term, error-rate budgeting. This is concerned with how the allowed level of the test can be apportioned among the different kinds of statements to be made. For example, in the analysis of a factorial experiment we may want to make inferences about main effects, two-factor interactions, three-factor interactions, and so on. The relative importance we accord to the different kinds of statements should influence the proportion of the error rate that we assign to the classes. In the simplest setting, the one factor experiment, we may place more interest on pairwise comparisons among treatments than on contrasts or general linear combinations. These considerations lead to choice of technique, by which the budgeting question is implicitly decided.

John also created a comprehensive taxonomy of simultaneous inference problems (multiple comparisons, multiple determinations, etc.) together with a guide to appropriate procedures. This is particularly useful for practitioners who often don't have the leisure to sort out the various techniques and determine which are particularly suitable for the problem at hand. Of course, this area is not without controversy, as anyone who has followed the studentized range versus F-statistic argument. However, the reader of papers 6 and 11 will find that John's view is a balanced one and his recommendations are driven by his interpretation of the needs of users.

I should like to note that Professor James Dolby originally agreed to serve as the editor of this volume. His untimely death prevented him from doing so. I was honored to be asked to take his place and can only hope that my efforts justify the confidence placed in me. I should especially like to thank my colleagues Bruce Kaplan, Kathleen Sheehan and Min-Hwei Wang, who provided invaluable assistance in preparing the manuscript of "The Problem of Multiple Comparisons" for publication. Their painstaking work has greatly enhanced the readability of the text through elimination of various errors and ambiguities. The statistical world owes them a vote of thanks.

Comments On The Individual Papers

Henry I. Braun

1. THE PROBLEM OF MULTIPLE COMPARISONS [1953U6]

The Problem of Multiple Comparisons is a landmark treatment of a broad range of issues in that area of statistics usually called simultaneous inference. Dealing with both theoretical and practical issues, it already displays a number of features that modern readers will recognize as characteristic of John's "later" style. These include the creative use of language, effective use of tables (employing data-analytic strategies to reduce size and/or improve accessibility) and a grasp of the interplay between statistical ideas and substantive problems. Above all, there is the profound insight or intuitive leap that repays the careful reader many fold.

As I am not an historian, I will not venture to speculate how this book came to influence the field. Suffice it to say, that nearly forty years after it was written, it is not a mere historical curiosity. There is discussion here of philosophy and approach to practice which still bear careful attention and thought. (While you may not agree with everything John says, you certainly have to work hard to develop plausible counter-arguments.) It provides an intellectual challenge on nearly every page.

The text itself is grouped into six parts (labelled A through F) that are divided into chapters, further divided into sections. The numbering of chapters and sections is continuous through the text; there are altogether 36 chapters and 225 sections. An outline and summary begins in Chapter 6, Section 30. The interested or casual reader should begin there before delving further into the text. While the book should be read linearly, browsing can be fun. The more philosophical sections can often be read without regard to the preceding material.

2. EQUALIZING UNEQUAL ACCURACIES OF MEANS [1948U4]

The principal focus of this paper is to show how an inference problem involving observed means with variances in known ratios can be transformed into one involving means with equal variances. The purpose, of course, is to facilitate the application of techniques developed for the latter structure. The example given involves an application of the straggler test for separating group means.

3. COMPARING INDIVIDUAL MEANS IN THE ANALYSIS OF VARIANCE [1949f]

This is probably Tukey's earliest sustained contribution to the problem of multiple comparisons. Motivated by the important practical problem of separating treatment means into distinct groups after the conventional analysis of variance, he introduces the gap-straggler-variance test. The problem of controlling the frequency of false positives is recognized but disposed of only with a conjecture on error bounds. The paper is noteworthy in that it presents a significance-based method, rather than the confidence interval based methods Tukey later preferred. Procedures later proposed by Ryan (1959) and Welsch (1977) represent outgrowths of some of the thinking found here.

4. REMINDER SHEETS FOR "ALLOWANCES FOR VARIOUS TYPES OF ERROR RATES" [1952U2]

In this short handout, Tukey sets out his position on three important issues in simultaneous inference; namely, the kind of statements to be made, the setting of error rates and the budgeting thereof. He clearly distinguishes his position from that of David Duncan, especially in the matter of controlling the error rate over an experiment. He describes the least significant difference (*LSD*) procedure and its relatives, representing the allowance as a product of factors arising from different sources.

5. REMINDER SHEETS FOR "MULTIPLE COMPARISONS" [1952U3]

This handout, presents a partial taxonomy of simultaneous inference problems and approaches to controlling error rates. Tukey introduces the phrase *WSD*, wholly significant differences and provides a useful table for implementing the procedure.

6. SOUVENIR SHEETS FOR "CONFIDENCE PROCEDURES ARE BETTER" [1954U5]

In this set of reminder sheets, Tukey states his stand on the main issues in the philosophy of multiple comparisons practice. He unequivocally takes exception to the position of David Duncan, an earlier speaker on the same program.

7. SHORT-CUT MULTIPLE COMPARISONS FOR BALANCED SINGLE AND DOUBLE CLASSIFICATIONS: PART 1, RESULTS [1965e]

This rather lengthy paper describes a number of short-cut multiple comparisons procedures employing various combinations of the range to form a suitable denominator for a test statistic. While the procedures are perhaps not as useful now, the discussion on the philosophy underlying practice are still valuable. The questions raised in 1952U2 are dealt with at length, especially the choice of significance versus confidence. It is interesting that considerable space is devoted to justifying the bounding of experimentwise error rates at conventional levels.

8. SHORT-CUT MULTIPLE COMPARISONS FOR BALANCED SINGLE AND DOUBLE CLASSIFICATIONS: PART 2 DERIVATIONS AND APPROXIMATIONS [1965i]

Tukey and his collaborators provide the mathematical derivations and approximations to support the procedures advocated in 1965e.

9. AUTHORS' REPLY TO ANSCOMBE'S COMMENTS [1965k]

There were two comments on 1965e, one by T. E. Jackson of Eastman Kodak and one by Frank Anscombe. The authors replied only to the latter who questioned the use of the error rate terminology and the general applicability of simultaneous confidence interval procedures. Anscombe suggested that most interesting practical problems should be formulated in a decision-theoretic framework. The principal argument in the reply is that in most practical settings the appropriate action space is not clear so that techniques which convey the uncertainty attached to the different comparisons of interest should be generally useful.

10. MULTIPLE COMPARISONS THROUGH ORDERLY PARTITIONS: THE MAXIMUM SUBRANGE PROCEDURE [1983c]

This paper marks a return to consideration of simultaneous inference procedures involving statements of significance unaccompanied by confidence intervals. One line of research in the literature had been concerned with controlling experimentwise error rates not only for the null hypothesis of all treatment means equal but also over a broad range of hypotheses that could be represented as a partition of the treatment means into two or more blocks of equal means. Most classic techniques failed in this more ambitious task. A number of complex procedures had been proposed and the maximum subrange procedure introduced here was meant to provide a computationally simpler and more understandable alternative. An interesting feature of these procedures was that they all employed "interference" — that the inference about one of pair of means was affected by the configuration and location of the estimates of the other means.

11. REMINDER SHEETS FOR "DISCUSSION OF PAPER ON MULTIPLE COMPARISONS BY HENRY SCHEFFÉ" | [circa 1951]

This set of reminder sheets clearly explicates the differences in Tukey's approach to multiple comparisons from that of Henry Scheffé. The application of simultaneous inference ideas to complex ANOVA designs is particularly worthwhile.

REFERENCES

Ryan, T. (1959). *Multiple Comparisons in Psychological Research,* Psychological Bulletin **56**, 26-47.

Welsch, R. E. (1977). Stepwise multiple comparison procedures, *J. Am. Stat. Assoc.* **72**, 566-575.

1

The Problem Of Multiple Comparisons

John W. Tukey

(Contents summarized at page 63ff.)

1. INTRODUCTION

Having measured

(1) the yields of six varieties of wheat; *or*

(2) the efficiencies of six types of catalyst; *or*

(3) the results of six methods of teaching a high-school subject; *or*

(4) the average income of families reading each of six national magazines; *or*

(5) the frequencies of certain blood groups among the residents of six areas; *or*

(6) the weather four days after each of six objectively defined types of weather map;

The Collected Works of John W. Tukey (1993) Chapman and Hall, New York, NY.

Dedicated to the memory of Charles P. Winsor

Prepared in connection with research sponsored by the Office of Naval Research.

Previously unpublished manuscript (1953).

and knowing something of his accuracy of measurement, the naive, unspoiled investigator wishes, at the very least, to compare each of the six determinations with every other, and to consider what he has learned about the differences between many, if not all, pairs. Until now, this natural, and the writer believes wise, desire has been in conflict with the desire of an increasing number of investigators to attach some definite statistical meaning to their statements. An investigator with both desires has been likely to be told that what he wishes to do involves non-independent tests of significance, with the implication that such things are not spoken of in polite society.

We present in this paper a rather complete account of how to attain both desires, not only with regard to making all simple comparisons (i.e. differences of pairs of determinations) but also with regard to examining all contrasts between determinations or even all linear combinations of determinations.

An account of how to do this follows this introduction. (The order of presentation has been chosen to put first those matters of interest to the most readers.) Some may wish to begin instead with the discussion of what is being done in Part B (p. 71) or with the derivation of distributions and factors in Part C (p. 110).

The basic probability considerations are classical, and the novelty of the approach lies in the way they are assembled. (Some of the basic ideas were discovered independently by H. K. Nandi [49], by Cornfield, Halperin and Greenhouse [14] and by the writer [74,75,76,77].) A fuller account of the historical background is given in Chapter 12 (p. 104).

A more complete summary and table of contents is given in Chapter 6 (p. 63).

2. CLASSICAL NOTATIONS

The six determinations of the example in the introduction (the reader is invited to think of that specific example closest to his own interest) may each be a total (or an arithmetic mean) of 3 or 7 or 10 individual measurements. The investigator's knowledge of the accuracy of his measurements may come from the scatter among these individual measurements. If this is the case, and if we had focussed our attention on *just two* determinations, then familiar statistical procedures can be applied to give the desired sort of answer.

These procedures would very probably be based on the distribution of Student's t, which was the beginning of modern "small sample" statistics and remains one of its most important foundations. In the simplest situations they would be based on the assumption that we could treat the individual measurements whose arithmetic mean is a given determination as a random sample from an infinite population upon whose average value our interest is focussed. We shall call this average value of the infinite population the *determinand* corresponding to the given determination. (It is the "true" value of the determination.) Our only interest in the determinations lies in what they tell us about the determinands, and for the case of only two determinations we would classically assert that

$$\text{determinand}_A - \text{determinand}_B$$

lies between

$$(\text{determination}_A - \text{determination}_B) \pm t\, s\, \sqrt{2}$$

where t is a critical value of "Student's t" and s is the square-root of an estimate of the variance of a single determination.

Like all statements based on finite data this conclusion may have been wrong. The goal of the statistical analysis has not been to be always right — it has been to control the probability of error. We may fix this probability, usually called the *significance level*, or the complement of the *confidence coefficient*, at any desired level by choosing a corresponding probability level in determining the critical value of t. We shall wish to deal with the generalizations of this probability in terms of *error rates*, since this will give us greater flexibility without loss of simplicity. In the case at hand, the error rate is the long-run value of

$$\frac{\text{number of erroneous statements}}{\text{number of comparisons attempted}},$$

when the assumption that we can treat the sets of individual measurements as a sample is justified. It is natural to be more specific by calling this error rate an *error rate per comparison*.

The critical value of Student's t depends on both the probability level we select, and on a measure of the accuracy with which our estimate s^2, which is subject in its turn to sampling variation, reflects the actual variance of a determination. The customary measure is a number of *degrees of freedom*, where many degrees of freedom imply high accuracy. We shall abbreviate this measure of stability by the letters *DDF* which refer to "denominator degree of freedom."

Finally, we shall refer to the $\pm t\, s \sqrt{2}$ affixed to the difference of the two determinations as an *allowance*, since it allows for the effects of those sources of variability which manifest themselves in differences among individual measurements.

The concepts we have discussed in this section are all classical, all that is new are words and notations, including: "determination," "determinand," "error rate," "*DDF*," "allowance." These novelties have been introduced to smooth our way in the more general situation in which we are really interested.

3. NEW NOTATIONS

We are actually concerned with a *family* of m (in the examples, 6) determinations, rather than with only a pair. There are $\frac{1}{2} m(m-1)$ different pairs of determinations among the m determination of such a family and hence an equal number of *simple comparisons* (a simple comparison = the numerical difference between the determinations of a pair). We could analyze each of these pairs separately by the method indicated in the last section which would indeed control the error rate per comparison, but if we fix the error rate at 5% per comparison we will make an average of 5% of 45, namely $2\frac{1}{4}$, erroneous statements about the comparisons within one family. To most statisticians and most investigators, $2\frac{1}{4}$ errors per family seems too high. If we agree, and wish to make statements about individual comparisons, then we need additional kinds of error rates.

Two definitions which come naturally to mind compare with the one already introduced as follows:

$$\text{error rate per comparison} = \frac{\text{erroneous statements}}{\text{comparisons}},$$

$$\text{error rate per family} = \frac{\text{erroneous statements}}{\text{families}},$$

$$\text{error rate familywise} = \frac{\text{erroneous families}}{\text{families}}.$$

Here, "number of" is implied in every numerator and every denominator, and a family is erroneous if an erroneous statement is made about one or more of its comparisons.

"Error rate per family" and "error rate familywise" are numerically quite similar whenever they are interestingly small, say <10%. Either could be used in practice, but since "error rate familywise" has theoretical advantages we shall use it.

Classically, the only sort of procedure available to us was an overall test to see if the determinands were "significantly different" or not. The level of significance of such a test is its error rate familywise when all determinands are equal. Thus the standard of security is not being lowered if we make the change from a "test of significance at a 5% significance level" to a "system of allowances with an error rate of 5% familywise" (that is to say, to a procedure yielding "simultaneous confidence statements with a joint confidence coefficient of 95%"). There are many advantages to such a change, as is illustrated in many of the numerical examples to follow and discussed generally in Part D (p. 161).

4. BATCHES VS. FAMILIES

Sometimes our determinations are mainly of individual rather than comparative interest — sometimes we wish to set limits principally on individual determinands, rather than differences or contrasts. In such a case we identify the weaker relation among determinations (or among the corresponding determinands) as the relationship among a *batch* of determinands rather than among a family.

Here, too, there are different possibilities of defining an error rate. The leading contenders are:

$$\text{error rate per determination} = \frac{\text{erroneous statements}}{\text{determinations}},$$

$$\text{error rate per batch} = \frac{\text{erroenous statements}}{\text{batches}},$$

$$\text{error rate batchwise} = \frac{\text{erroneous batches}}{\text{batches}}.$$

Again, "number of" may be prefixed to all numerators and denominators. This time, "statements" are intervals for single determinands and "erroneous batches" are batches about one or more of whose determinands an erroneous statement has been made.

Again error rates "per X" and "X-wise" (here X = batch, earlier X = family) are practically almost indistinguishable. Again, we prefer to use "X-wise."

We have subordinated batches to families in this introduction, and shall continue to do so in Part A, since this reflects their relative importance and frequency of occurrence in actual use. When we come to Parts B and C, we shall place batches before families, since the theoretical considerations for batches are slightly simpler to discuss and serve as an easy and convenient introduction to families.

PART A
USER´S ACCOUNT

CHAPTER 1: ALLOWANCES

5. INTRODUCTION

This part provides a summary of the use of allowances, with brief examples. It is a "How to do it" or, as some might say, "cookbook" account, giving directions and suggestions without reasons or argument. These reasons and arguments are not omitted, being merely postponed to later parts. For accounts of the essentials of what is being done (Part B), the derivation of the tables given (Part C), a general discussion of reasons for choices among procedures (Part D), a muster and comparison of available and foreseeable systems of allowances (Part E), or a muster and comparison of selfconsistent significance procedures (Part F), the reader may wish to begin with the corresponding part. A more detailed account of the contents of these various parts is to be found in Chapter 6 at the end of the present part.

We shall usually assume that the user has available:

(1) the values $x_1, x_2, \ldots, x_m$ of m determinations, one corresponding to each of m determinands;

(2) knowledge that the variances of these m determinations are really equal;

(3) an estimate, s^2, of this common variance;

(4) an equivalent number of degrees of freedom, *DDF*, which can reasonably be assigned to s^2.

We shall discuss the case where (2) is relaxed in Chapter 4, and particular cases where the estimate of the common variance is obtained by short-cut methods in Chapters 2 and 3.

6. FAMILIES OF DETERMINATIONS

It is convenient to make the computation in two steps. (We could reduce this to one step, at the cost of much more extensive tables.) The first step is to compute the least significant difference, *LSD*, from

$$LSD = [LSD/s](s)$$

and the second step is to compute the wholly significant difference, *WSD*, from

$$WSD = [WSD/LSD](LSD),$$

where $[LSD/s]$ and $[WSD/LSD]$ are both given to working accuracy in Table 1.

For an error rate of 5% familywise, allowances are to be calculated as follows:

(1) for a simple comparison (difference of 2 determinations) the allowance is equal to the *WSD*;

(2) for any linear combination of determinations (including all contrasts) the allowance is the product of the *WSD* and the norm of the linear combination.

If the determinations are $x_1, x_2, \ldots, x_m$, and the linear combination is

$$x = c_1 x_1 + c_2 x_2 + \ldots + c_m x_m$$

then its norm is the greater of the two numbers:

sum of positive c's and $-$(sum of negative c's)

which can be symbolically expressed in the form

$$\text{norm}\{c_i\} = \max\{\Sigma^+ c_i, -\Sigma^- c_i\}$$

where Σ^+ indicates summation of those c_i which are positive, and Σ^- that of those which are negative.

Table 1

Factors Required in Finding LSD's and WSD's

DDF	$60/DDF$	$[LSD/s]$	m	$[WSD/LSD]$	m	$[WSD/LSD]$
1		18.0	2*	$1.00+0.0/DDF$	21	$1.82+2.5/DDF$
2		6.10	2	$1.02+0.2/DDF$	22	$1.83+2.6/DDF$
3		4.50	3	$1.20+0.4/DDF$	23	$1.85+2.7/DDF$
4		3.92	4	$1.32+0.6/DDF$	24	$1.86+2.8/DDF$
5		3.64	5	$1.40+0.8/DDF$	25	$1.87+2.8/DDF$
6		3.46	6	$1.46+1.0/DDF$	26	$1.88+2.9/DDF$
7		3.34	7	$1.51+1.2/DDF$	27	$1.89+3.0/DDF$
8		3.26	8	$1.55+1.4/DDF$	28	$1.90+3.1/DDF$
9		3.20	9	$1.59+1.5/DDF$	29	$1.91+3.1/DDF$
10	6	3.15	10	$1.62+1.6/DDF$	30	$1.91+3.2/DDF$
	5	3.08	11	$1.65+1.7/DDF$	35	$1.95+3.4/DDF$
	4	3.01	12	$1.67+1.8/DDF$	40	$1.99+3.7/DDF$
	3	2.95	13	$1.70+1.9/DDF$	45	$2.01+3.9/DDF$
	2	2.89	14	$1.72+2.0/DDF$	50	$2.04+4.1/DDF$
	1	2.83	15	$1.74+2.1/DDF$	60	$2.08+4.4/DDF$
	0	2.77	16	$1.75+2.2/DDF$	80	$2.15+5.0/DDF$
Values of $[LSD/s]$ are $\sqrt{2}$ times upper 5% points of Student's $\lvert t \rvert$			17	$1.77+2.3/DDF$	100	$2.19+5.4/DDF$
			18	$1.78+2.4/DDF$	200	$2.37+6.9/DDF$
			19	$1.80+2.4/DDF$	500	$2.61+9.1/DDF$
			20	$1.81+2.5/DDF$	∞	$\infty+\infty/DDF$

* = For comparison of the 2 determinations only

Values of $[WSD/s]$ are upper 5% points of the studentized range

An example of the calculation and application of such allowances follows:

Roberts and Jessup [111, p. 14] have reported the results of bomb calorimeter measurements on the heat of combustion of alpha-methylstyrene with results for $-\Delta U_c$ (in absolute joules per gram of CO_2) which can be summarized as follows:

monomer (alpha-methylstyrene): $x_{118} = 12709.3$

polymer 5400 (fraction I): $x_{5400} = 12615.3$

polymer 3750 (fraction II): $x_{3750} = 12613.1$

polymer 2200 (fraction III): $x_{2200} = 12609.6$

polymer 1300 (fraction IV): $x_{1300} = 12601.6$

$s^2 = 1.13,$

$DDF = 25,$

thus we find, using Table 1 twice

$$s = \sqrt{1.13}$$

$$[LSD/s] = 2.91$$

$$LSD = 3.09$$

$$[WSD/LSD] = 1.40 + \frac{0.8}{25} = 1.43$$

$$WSD = 4.42.$$

We are thus entitled (with 5% error rate familywise) to accept all such statements as

(1) the excess of polymer 5400 polymer 1300 in heat of combustion is

$$12615.3 - 12601.6 \pm 4.4 = 13.7 \pm 4.4 = \text{between } 9.3 \text{ and } 18.1;$$

(2) the excess of polymer 3750 over polymer 2200 in heat of combustion is

$$12613.1 - 12609.6 \pm 4.4 = 3.5 \pm 4.4 = \text{between } -0.9 \text{ and } 7.9;$$

(3) the excess of monomer over polymer 5400 in heat of combustion is

$$12709.3 - 12615.3 \pm 4.4 = 94.0 \pm 4.4 = \text{between } 89.6 \text{ and } 98.4;$$

(4) the excess of monomer over the mean of the four polymer fractions in heat of combustion is

$$12{,}709.3 - \frac{12615.3 + 12613.1 + 12609.6 + 12601.6}{4} \pm 4.4$$

$$= 99.4 \pm 4.4 = \text{between } 95.0 \text{ and } 103.8;$$

(5) the heat of combustion of monomer is

$$12709.3 \pm 4.4 = \text{between } 12704.9 \text{ and } 12713.7;$$

(6) the heat of combustion of polymer 3750 is

$$12613.1 \pm 4.4 = \text{between } 12608.7 \text{ and } 12617.5;$$

(7) and so on — .

We have a family of determinations, and a system of allowances for all simple comparisons, contrasts and other linear combinations.

In their paper, Roberts and Jessup present a simple quantitative model for the physical chemistry of polymerization which implies that the heats of combustion should take the form

$$a \text{ (for monomer)}$$

$$a + b + \frac{c}{M} \text{ (for polymer} - M)$$

where M is the molecular weight and a, b, c are constants to be estimated. (Their results are stated for another quantity at another reference temperature, but an examination of the conversion factors and correction constants used, shows that they must also apply to the original data.)

Thus we would expect, if fluctuations of measurement were negligible,

$$x_{5400} = 1.18x_{3750} - .18x_{1300}$$

and so would be led to compute

$$x_{5400} - 1.18x_{3750} + .18x_{1300}$$

which is a contrast of norm 1.18 and value

$$-.13 \pm 1.18\,(4.42) = -.13 \pm 5.2 = \text{between } -5.1 \text{ and } +5.3$$

which might reasonably vanish.

We also find

$$x_{3750} - .69x_{5400} - .31x_{2200} = -0.4 \pm 4.4$$

$$x_{3750} - .85x_{5400} - .15x_{1300} = -0.1 \pm 4.4$$

$$x_{2200} - .52x_{5400} - .48x_{1300} = 0.9 \pm 4.4$$

$$x_{2200} - .61x_{3750} - .39x_{1300} = 1.0 \pm 4.4$$

$$x_{1300} + 1.07x_{5400} - 2.07x_{2200} = -1.9 \pm 2.07\,(4.42)$$

$$= -1.9 \pm 9.1.$$

Clearly no single point shows any tendency to deviate from the line indicated by the others (the only surprising aspect is the small size of the deviations).

It should be noted that Table 1 has two entries for [*WSD*/*LSD*] for the case of 2 determinations. The smaller is to be used when *only* the difference between the two determinations is of interest. The larger is to be used when other linear combinations also matter. (Further discussion in Chapter 10).

If the far more dangerous basis of a 5% error rate per comparison is to be used, the appropriate allowance is the *LSD*. A cautionary footnote is required, as is discussed in Section 41.

The case of unequal variance (in known ratio) is discussed in Section 22.

Graphical presentation is discussed in Section 21.

Simple procedures for altering the error rate from 5% to any reasonable desired value are given in Sections 23 and 24.

7. BATCHES OF DETERMINATIONS

It is again convenient to compute the allowance in two steps. The individual allowance, *IA*, is computed from

$$IA = [IA/s]s,$$

and the parallel allowance, *PA*, is computed from

$$PA = [PA/IA](IA),$$

where $[IA/s]$ and $[PA/IA]$ are tabled to working accuracy in Table 2.

For an error rate of 5% per determination, the *IA* is used as the allowance.

For an error rate of 5% batchwise, the *PA* is used as the allowance. This is the more conservative procedure, and should probably be regarded as the normal procedure.

When we are using the *PA*, we can also set an allowance for $x = \Sigma c_i x_i$ in the form

$$\Sigma c_i x_i \pm (\Sigma \mid c_i \mid)(PA)$$

(when necessary we denote $\Sigma \mid c_i \mid$ as $\text{norm}_p \{c_i\}$).

An example of the calculation and application of the procedure follows:

We return to data of Jessup and Roberts [111, p. 14] analyzed above as an example of a family of determinations and proceed to treat it as a batch of determinations. Again

$x_{118} = 12709.3$ (monomer)

$x_{5400} = 12615.3$ (polymer, fraction I)

$x_{3750} = 12613.1$ (polymer, fraction II)

$x_{2200} = 12609.6$ (polymer, fraction III)

$x_{1300} = 12601.6$ (polymer, fraction IV)

$s^2 = 1.13$

$DDF = 25$

Table 2

Factors required in finding IA's and PA's

DDF	$60/DDF$	$[IA/s]$	m	$[PA/IA]$	m	$[PA/IA]$
1		12.71	1	$1.00+0.0/DDF$	21	$1.55+1.9/DDF$
2		4.30	2	$1.14+0.3/DDF$	22	$1.56+1.9/DDF$
3		3.18	3	$1.22+0.5/DDF$	23	$1.56+1.9/DDF$
4		2.78	4	$1.27+0.6/DDF$	24	$1.57+1.9/DDF$
5		2.57	5	$1.31+0.8/DDF$	25	$1.57+2.0/DDF$
6		2.45	6	$1.34+0.9/DDF$	26	$1.57+2.0/DDF$
7		2.36	7	$1.37+1.0/DDF$	27	$1.58+2.1/DDF$
8		2.31	8	$1.39+1.1/DDF$	28	$1.59+2.1/DDF$
9		2.26	9	$1.41+1.2/DDF$	29	$1.59+2.2/DDF$
10	6	2.23	10	$1.43+1.3/DDF$	30	$1.60+2.2/DDF$
	5	2.18	11	$1.45+1.3/DDF$	35	$1.62+2.3/DDF$
	4	2.13	12	$1.46+1.4/DDF$	40	$1.64+2.5/DDF$
	3	2.09	13	$1.47+1.5/DDF$	45	$1.67+2.5/DDF$
	2	2.04	14	$1.48+1.6/DDF$	50	$1.68+2.6/DDF$
	1	2.00	15	$1.50+1.6/DDF$	60	$1.71+2.8/DDF$
	0	1.96	16	$1.51+1.6/DDF$	80	$1.75+3.0/DDF$
			17	$1.51+1.7/DDF$	100	$1.78+3.3/DDF$
Values of $[IA/s]$			18	$1.52+1.8/DDF$	200	$1.87+4.1/DDF$
are upper 5% points			19	$1.53+1.8/DDF$	500	$1.99+5.3/DDF$
of student's $\|t\|$			20	$1.54+1.8/DDF$	∞	$\infty+\infty/DDF$

Values of $[PA/s]$ are upper 5% points of the studentized maximum modulus

and now, using Table 2 twice,

$$s = 1.06$$

$$[IA/s] = 2.06$$

$$IA = 2.18$$

$$[PA/IA] = 1.31 + \frac{0.8}{25} = 1.34$$

$$PA = 2.92.$$

We are thus entitled to accept (with 5% error rate batchwise) the statements as:

(1) the heat of combustion of monomer is

$$12709.3 \pm 2.9 = \text{between } 12706.4 \text{ and } 12712.2;$$

(2) the heat of combustion of fraction I is

$$12615.3 \pm 2.9 = \text{between } 12612.4 \text{ and } 12618.2;$$

(3) the heat of combustion of fraction II is

$$12613.1 \pm 2.9 = \text{between } 12610.2 \text{ and } 12616.0;$$

(4) the heat of combustion of fraction III is

$$12609.6 \pm 2.9 = \text{between } 12606.7 \text{ and } 12612.5;$$

(5) the heat of combustion of fraction IV is

$$12601.6 \pm 2.9 = \text{between } 12598.7 \text{ and } 12604.5;$$

and their logical consequences, which include:

(6) the excess of fraction I over fraction IV in heat of combustion is

$$12615.3 \pm 2.9 - 12601.6 \pm 2.9 = 13.7 \pm 5.8 = \text{between } 7.9 \text{ and } 19.5;$$

(7) the excess of monomer over fraction I in heat of combustion is

$$12709.3 \pm 2.9 - 12615.3 \pm 2.9 = 94.0 \pm 5.8 = \text{between } 88.2 \text{ and } 99.8;$$

(8) the excess of monomer over the mean of the four polymer fractions in heat of combustion is

$$12709.3 \pm 2.9 - \frac{12615.3 \pm 2.9 + 12613.1 \pm 2.9 + 12609.6 \pm 2.9 + 2601.6 \pm 2.9}{4}$$

$= 99.4 \pm 5.8 =$ between 93.6 and 105.2.

Notice that:

(1) it is possible to make statements about comparisons, other contrasts, etc. within a batch;

(2) statements about contrasts are sharper within a family than within a batch (±4.4 rather than ±5.8 in the example);

(3) statements about individual determinations are sharper within a batch than within a family (±2.9 rather than ±4.4 in the example);

(4) it is *not* legitimate to treat a set of determinations as a family when considering contrasts and as a batch when considering individuals.

It should further be recorded that if we wished to accept an error rate of 5% per determination, we could substitute the individual allowance of ±2.2 for the parallel allowance of ±2.9.

As a further example, consider an experiment on the response of potato yield to fertilizer, carried out at Wimblington, England in 1934 and reported by Yates [115, p. 9]. The experiment was a 2^3 in nitrogen (N), potassium (K) and manure (D) in 4 replications. We shall be concerned with the batch of 7 factorial effects formed from the observations, namely

$$x_N = 333$$

$$x_K = 2271$$

$$x_{NK} = 105$$

$$x_D = 2987$$

$$x_{ND} = 161$$

$$x_{KD} = -669$$

$$x_{NKD} = -63$$

for which

$$s^2 = 2776$$

$$DDF = 21$$

so that

$$s = 53$$

$$[IA / s] = 2.09$$

$$IA = 111$$

$$[PA/IA] = 1.37 + \frac{1.0}{21} = 1.42$$

$$PA = 157$$

and thus we may conclude with 5% error rate familywise that, for example,

(1) the main effect of potassium is

$$2271 \pm 157 = \text{between } 2114 \text{ and } 2428;$$

(2) the main effect of manure exceeds that of potassium by

$$2987 \pm 157 - (2271 \pm 157) = 716 \pm 314 = \text{between } 402 \text{ and } 1030;$$

(3) moreover

$$x_N + x_{ND} = 333 \pm 157 + 161 \pm 157 = 494 \pm 314 = \text{between } 180 \text{ and } 808;$$

which is of interest, since it is proportional to the effect of nitrogen at the upper level of manure.

Simple procedures for altering the error rate from 5% to any reasonable desired value are given in Chapter 4 (Section 23).

CHAPTER 2: SINGLE CLASSIFICATIONS

8. INTRODUCTION

Having set forth the essentials of the general procedure for making multiple comparisons, it is now time for us to treat the most important special cases. There are two classes of cases where the arrangement of observations clearly indicated in what ways the estimate of common variance, s^2, should be calculated. These are balanced single and double classification situations. (More complex patterns of observation frequently imply choices which have to be settled with the aid of additional information as to what should be considered the common variance of the determinations. This is the familiar problem of "what is the appropriate error term." We leave a discussion of such situations to another account.)

We shall treat the balanced single classification situation in four forms:

(1) the classical form, but concluding with the use of allowances rather than an F-test of significance;

(2) a condensed form, the same in result, but with arithmetic minimized;

(3) a short-cut form, with a rougher and readier estimate of variance;

(4) a half-cut form, intermediate between the last two.

The first two give equivalent results, so that (1) is to be used only when (i) the tables for (2) are not available, or when (ii) consistency of calculation with that for more complex designs is strikingly important. The choice between (3), (4) or one of the first two is discussed further in Chapter 5 (Section 29).

9. THE CLASSICAL FORM

We have a balanced single classification situation when each of the m determinations x_i is the arithmetic mean (or, at our choice, the total) of an equal number, r, of observations $y_{i1}, y_{i2}, \ldots, y_{ir}$, where no one of the observations of the ith determination is more closely related to any one of the observations of another determination, say the hth, than any other, and where our computation starts from the rm observations, y_{ij}. We classically calculate as follows:

$$
\begin{aligned}
SS_i &= \text{sum of squares of observations in } i\text{th determination} \\
&= y_{i1}^2 + y_{i2}^2 + \ldots + y_{ir}^2, \\
T_i &= \text{total of observations in } i\text{th determination} \\
&= y_{i1} + y_{i2} + \ldots + y_{ir}, \\
SSD_i &= \text{sum of squares of deviations in the } i\text{th group} \\
&= SS_i - (T_i^2) / r, \\
TSSD &= \text{total sum of squares of deviations} \\
&= SS_1 + SS_2 + \ldots + SS_m - (T_1^2 + T_2^2 + \ldots + T_m^2) / r, \\
DDF &= m(r-1), \\
MSD &= \text{mean square deviation (estimated variance of an observation)} \\
&= TSSD / m(r-1) = TSSD / DDF, \\
s^2 &= \text{estimated variance of a determination} = \begin{cases} MSD/r \text{ for means,} \\ r(MSD) \text{ for totals.} \end{cases}
\end{aligned}
$$

Having found s^2 and DDF, we are ready to go ahead with the procedures of the last chapter.

An example of the computation follows:

Dockray, Meadows and Smith [108, p. 702] give the percentage of recoverable fiber in two samples of card strips from each of three companies in the cotton waste trade as measured with a Shirley analyzer. The observations rounded to the nearest %, and the resulting computations are as follows:

	Company B	Company C	Company E
	90	91	92
	74	93	90
SS_i	13576	16930	16564
T_i	164	184	182
$T_i^2/2$	13448	16928	16562
SSD_i	128	2	2

hence

$$TSSD = 128 + 2 + 2 = 132$$

$$DDF = 3$$

$$MSD = {}^{132}/_{3} = 44$$

$$s^2 \text{ (means)} = 22$$

$$s^2 \text{ (total)} = 88$$

and, from Table 1,

$$[LSD/s\,] = 4.50$$

$$[WSD/LSD\,] = 1.20 + \frac{0.4}{3} = 1.33$$

$$[WSD/s\,] = 6.00$$

so that, finally,

$$s \text{ (means)} = 4.70$$

$$s \text{ (totals)} = 9.40$$

$$WSD \text{ (means)} = 28.2$$

$$WSD \text{ (totals)} = 56.4$$

and we may conclude, with 5% error rate familywise, that, for example:

(1) the total percentage recoverable fiber for company C exceeds that of company B by

$$184 - 164 \pm 56 = 20 \pm 56 = \text{between } -36 \text{ and } +76$$

in the total of 2 samples (and hence by between −18 and +38 for one sample);

(2) the average percentage recoverable fiber for company E exceeds that for company C by

$$91 - 92 \pm 28 = -1 \pm 28 = \text{between } -29 \text{ and } +27.$$

The data suggests doubt of the assumption of equal variability, but they do not disprove it. (After all, in two tosses of each of three coins with identical probabilities of heads falling between 25% and 41%, the most likely pattern of heads and tails is HT, TT, TT, which shows as much apparent variation of variance.) Information on the characteristics of the cotton waste trade might throw light on the question.

10. THE CONDENSED FORM

If we observe that, for totals,

$$WSD = [WSD/LSD\,]\,[LSD/s\,]s$$

$$= [WSD/LSD\,]\,[LSD/s\,]\left(r\,\frac{TSSD}{m\,(r-1)}\right)^{1/2},$$

$$= [WSD/LSD\,]\,[LSD/s\,]\left(\frac{r}{m\,(r-1)}\right)^{1/2}(TSSD)^{1/2},$$

we see that we can tabulate a single factor for the ratio of WSD to $\sqrt{TSSD}$. This is done in Table 3 for an error rate of 5% familywise.

An example of the condensed calculation is the following, where we have made use of

$$TSSD = (SS_1 + \ldots + SS_m) - (T_1^2 + \ldots + T_m^2)\,/r$$

to reduce the number of divisions to one:

Wilson [114, p. 220], quoted by Clark [105, p. 169] gives length/width ratios for 300 pollen mother cells (100 each from the top, bottom and middle fractions of the anthers) for each of 3 species of *Musa*. The data are given in broad groups and are here analyzed in (linearly) coded form (1 = (1.0 to 1.4), . . ., 5 = (3.0 to 3.4)). We find:

Table 3

Values of [(*WSD* for totals)/$\sqrt{TSSD}$] for a Balanced Single Classification and an Error Rate of 5% Familywise

m = number of determinations	r = number of replications: 2	3	4	5	6	7	8	9	10	12	15	20	30	60	∞
2	6.09	3.40	2.83	2.58	2.44	2.35	2.29	2.25	2.21	2.16	2.12	2.08	2.04	2.00	1.96
3	4.81	3.07	2.63	2.44	2.32	2.24	2.20	2.16	2.13	2.08	2.05	2.01	1.98	1.95	1.92
4	4.08	2.78	2.42	2.26	2.17	2.11	2.06	2.03	2.01	1.96	1.94	1.90	1.87	1.84	1.82
5	3.56	2.56	2.19	2.12	2.03	1.99	1.95	1.91	1.89	1.86	1.83	1.80	1.77	1.74	1.71
6	3.22	2.38	2.12	1.99	1.93	1.89	1.84	1.82	1.80	1.77	1.74	1.71	1.69	1.67	1.65
7	2.98	2.24	2.01	1.89	1.83	1.78	1.76	1.74	1.72	1.69	1.66	1.64	1.62	1.60	1.58
8	2.80	2.12	1.92	1.81	1.75	1.71	1.69	1.66	1.64	1.62	1.60	1.58	1.56	1.54	1.52
9	2.64	2.03	1.83	1.74	1.69	1.65	1.62	1.60	1.58	1.56	1.54	1.52	1.50	1.48	1.46
10	2.51	1.94	1.76	1.67	1.63	1.58	1.57	1.55	1.53	1.51	1.48	1.47	1.45	1.43	1.41
11	2.39	1.87	1.70	1.62	1.57	1.53	1.51	1.49	1.48	1.46	1.44	1.42	1.41	1.39	1.37
12	2.29	1.80	1.65	1.57	1.53	1.49	1.47	1.45	1.43	1.41	1.40	1.38	1.36	1.34	1.33
13	2.21	1.75	1.60	1.52	1.48	1.45	1.43	1.41	1.40	1.38	1.36	1.35	1.33	1.31	1.30
14	2.13	1.70	1.56	1.48	1.44	1.41	1.39	1.37	1.36	1.35	1.33	1.31	1.30	1.28	1.27
15	2.06	1.65	1.52	1.45	1.41	1.38	1.36	1.34	1.33	1.31	1.30	1.28	1.27	1.25	1.24
16	2.00	1.61	1.48	1.42	1.37	1.34	1.32	1.31	1.30	1.28	1.27	1.25	1.24	1.22	1.21
17	1.94	1.56	1.44	1.38	1.34	1.32	1.30	1.28	1.27	1.25	1.24	1.23	1.21	1.19	1.18
18	1.89	1.53	1.41	1.35	1.31	1.29	1.27	1.25	1.25	1.23	1.22	1.20	1.19	1.17	1.16
19	1.85	1.50	1.38	1.32	1.28	1.26	1.24	1.23	1.22	1.20	1.19	1.18	1.16	1.15	1.14
20	1.81	1.47	1.35	1.30	1.26	1.24	1.22	1.21	1.20	1.19	1.17	1.16	1.14	1.13	1.12
							60/r	=	6	5	4	3	2	1	0

Species:	ZA	MR	Br	total
code 1	190	58	130	—
code 2	91	141	142	—
code 3	18	67	21	—
code 4	1	25	7	—
code 5	0	9	0	—
T_i	430	686	505	—
T_i^2	184900	470596	255025	910521
SS_i	732	1850	999	3581
TSSD	—	—	—	545.9
$\sqrt{TSSD}$				23.4
factor				1.92
WSD for totals				44.9

Thus we conclude that, in terms of coded totals for 300 cells:

(1) MR exceeds ZA in length/width ratio by

$$686 - 430 \pm 45 = 256 \pm 45 = \text{between } 211 \text{ and } 301$$

(2) Br exceeds ZA in length/width ratio by

$$505 - 430 \pm 45 = 75 \pm 45 = \text{between } 30 \text{ and } 120$$

and so on. (These results may be expressed in terms of mean length/width ratio (not coded) by division by 600.) Thus we have only to:

(1) sum the squares of all observations;

(2) total the observations in each group;

(3) square these totals and sum these squares;

(4) divide this latter sum by the number of observations per group and subtract the result from the sum of squares from step (1) to form the *TSSD*;

(5) take the square root of the *TSSD* and multiply it by the factor from Table 3 to find the *WSD* for totals;

(6) in case we desire to work in terms of means rather than totals, to divide both the group totals and the *WSD* for totals by the number of observations per group in order to find group means and the *WSD* for totals.

This is relatively simple, but still involves a certain amount of squaring and square-rooting.

11. THE SHORT-CUT FORM

In most cases it will be adequate to follow a simpler procedure, namely:

(1) to form the total and range for each group;

(2) to sum the ranges to form a total range;

Table 4

Values of [(*WSD* for totals)/(total range)] for a Balanced Single Classification and an Error Rate of 5% Familywise.

number of observations per group, range or determination	number of groups = number of ranges = number of determinations																		
	2	**3**	**4**	**5**	**6**	**7**	**8**	**9**	**10**	**11**	**12**	**13**	**14**	**15**	**16**	**17**	**18**	**19**	**20**
2	3.43	2.37	1.78	1.40	1.16	1.00	.87	.78	.70	.66	.63	.58	.50	.47	.44	.42	.40	.38	.36
3	1.91	1.44	1.13	.94	.80	.70	.62	.56	.51	.47	.43	.40	.38	.36	.33	.32	.30	.29	.27
4	1.63	1.25	1.01	.84	.72	.63	.57	.51	.47	.43	.40	.37	.35	.33	.31	.29	.28	.27	.25
5	1.53	1.19	.96	.81	.70	.61	.55	.50	.45	.42	.39	.36	.34	.32	.30	.29	.27	.26	.25
6	1.50	1.18	.95	.80	.69	.61	.55	.49	.45	.42	.39	.36	.34	.32	.30	.29	.27	.26	.25
7	1.49	1.17	.95	.80	.69	.61	.55	.50	.45	.42	.39	.36	.34	.32	.30	.29	.28	.26	.25
8	1.49	1.17	.96	.81	.70	.62	.55	.50	.46	.42	.39	.37	.35	.33	.31	.29	.26	.27	.25
9	1.50	1.18	.97	.82	.71	.62	.56	.51	.47	.43	.40	.37	.35	.33	.31	.30	.28	.27	.26
10	1.52	1.20	.98	.83	.72	.63	.57	.52	.47	.44	.41	.38	.35	.34	.32	.30	.29	.27	.26
11	1.54	1.21	.99	.84	.73	.64	.58	.52	.48	.44	.41	.38	.36	.34	.32	.31	.29	.28	.27
12	1.56	1.23	1.00	.85	.74	.65	.59	.53	.49	.45	.42	.39	.37	.35	.33	.31	.30	.28	.27
13	1.58	1.25	1.02	.86	.75	.66	.59	.54	.49	.46	.42	.40	.37	.35	.33	.32	.30	.29	.27
14	1.60	1.26	1.03	.87	.76	.67	.60	.55	.50	.46	.43	.40	.38	.36	.34	.32	.31	.29	.28
15	1.62	1.28	1.05	.89	.77	.68	.61	.56	.51	.47	.44	.41	.38	.36	.34	.33	.31	.30	.28
16	1.64	1.30	1.06	.90	.78	.69	.62	.56	.52	.48	.44	.41	.39	.37	.35	.33	.31	.30	.29
17	1.66	1.31	1.08	.91	.79	.70	.63	.57	.52	.48	.45	.42	.39	.37	.35	.33	.32	.30	.29
18	1.68	1.33	1.09	.92	.80	.71	.64	.58	.53	.49	.46	.43	.40	.38	.36	.34	.32	.31	.30
19	1.70	1.34	1.10	.93	.81	.72	.65	.59	.54	.50	.46	.43	.40	.38	.36	.34	.33	.31	.30
20	1.72	1.36	1.11	.95	.82	.73	.65	.59	.54	.50	.47	.44	.41	.39	.37	.35	.33	.32	.30

(3) to multiply this total range by a factor from Table 4, to obtain a *WSD* for totals; and

(4) in case we desire to work in terms of means rather than totals, to divide group totals and the *WSD* for totals by the number of observations per group, in order to find group means and a *WSD* for means.

The resulting *WSD* will not be identical with that obtained from the long process, but its properties are similar on the average. When we need a distinction, which will be rarely, we shall identify this short-cut *WSD* as the $W_s SD$. An example of the computation is as follows:

Davies [107, p. 74 of 2nd edition] has reported the yields of napthalene black 12B, expressed in grams of standard color, obtained in each of 5 replicate measurements on each of 6 samples of H-acid (a well-known dye intermediate). The observations, totals and ranges are as follows:

Sample:	1	2	3	4	5	6
Color (gms):	1440	1490	1510	1440	1515	1445
	1440	1495	1550	1445	1515	1450
	1520	1540	1560	1465	1625	1455
	1545	1555	1595	1545	1630	1480
	1580	1560	1605	1595	1635	1520
Total:	7525	7640	7820	7490	7920	7350
Range:	140	70	95	155	120	75

The sum of the ranges is

$$140 + 70 + 75 + 155 + 120 + 75 = 655$$

from Table 4 we find a factor of .70, so that

$$WSD(\text{totals}) = (.70)\,(655) = 459.$$

Thus we may accept, with an error rate of 5% familywise, such statements as:

(1) the average yield of napthalene black 12B for H-acid like sample 5 exceeds that from H-acid like sample 6 by

$$7920 - 7350 \pm 459 = 570 \pm 459 = \text{between } 111 \text{ and } 1029$$

in a total of 5 determinations;

(2) the average yield of napthalene black 12B from H-acid like sample 1 exceeds that from H-acid like sample 2 by

$$7525 - 7640 \pm 459 = -115 \pm 459 = \text{between } -574 \text{ and } +344.$$

12. THE HALF-CUT FORM

In cases where moderate inhomogeneity of variance or non-normality of distribution is suspected, it may be advisable to calculate a modified total range from

$$\text{modified total range} = [(m - \frac{1}{r}(1 - \frac{1}{m}))\,(\text{sum of squared ranges})]^{1/2}$$

and multiply this by the factor from Table 4 in order to obtain a *WSD* for totals. The only loss in so doing is the additional effort of m squarings and one square-rooting. When we need to identify such a half-cut *WSD* we denote it as W_hSD. An example of the calculation follows:

Johnson [110, p. 212] has reported the mental age in months of 6 samples of 6 pupils each, randomly chosen from the same grade in 6 different urban schools. The observations, totals and ranges are as follows:

School	1	2	3	4	5	6
	158	156	160	159	164	153
	157	155	158	155	163	148
	153	154	156	148	162	145
	151	153	155	147	160	144
	144	151	150	146	154	144
	143	149	145	145	151	136
Totals	906	918	924	900	954	870
Ranges	15	7	15	14	13	17
(Ranges)^2	225	49	225	196	169	289

Thus

$$\text{Modified total range} = [(6 - {}^{1}/_{6}(1 - {}^{1}/_{6}))\,(225 + 49 + 225 + 196 + 169 + 289)]^{1/2}$$

$$= \sqrt{5765} = 75.93,$$

factor = .69,

WSD (totals) = 52.

This example illustrates

(1) the computational procedure for the half-cut form; and

(2) the fact that the half-cut form is not usually necessary;

for the actual total range is 81, and the effect of the added computation has been to change this by 6%.

13. DESIGN

With a method of analyzing data occuring in a certain class of patterns there comes a desire to design patterns so as to give predetermined accuracy. For the classical analysis of a single classification such procedures have been established ([33] . . . , but note the correction in [45]). It will clearly be convenient to have such a design procedure to accompany the *WSD*.

We suppose that the investigator possesses:

(1) an estimate S_w^2 of the variance corresponding to an individual measurement;

(2) an equivalent number of degrees of freedom, DDF^*, which may reasonably be assigned to S_w^2;

(3) a bound, B, below which he desires the *WSD* for the experiment to fall; and

(4) a level of confidence, which we will suppose to be either 50% or 90%, with which he wishes the *WSD* to fall below this bound.

As emphasized by Harris, Horvitz and Mood [33], the estimate, S_w^2, and its DDF^* may be pure guesses. Even if they are, the design procedure is a rational and reasonable way of taking such (somewhat nebulous) information into account in planning an experiment or observational program.

The number of determinations, m, we shall assume fixed. Thus we have only to choose r. It is shown in Chapter 18 that we must choose r to satisfy

$$r \geq \frac{S_w^2}{B^2} [WSD/LSD]^2 [LSD/s]^2 (F^*)$$

where F^* is the 50% or 90% point, respectively, of the F (or variance ratio) distribution with $DDF = m(r-1)$ and DDF^* degrees of freedom. Since the last three factors on the right involve r, this is a rather fussy inequality to handle directly.

However, an approximate table of the product of these factors is easy to give, and the following procedure will determine r relatively easily:

(1) Use Table 5 to obtain an approximate value, r', for r.

(2) Calculate the right-hand side of the inequality for r'.

(3) If the right-hand side lies between r' and $r'-1$ then r' is the desired value.

(4) If the right-hand side is less than $r'-1$, repeat the process with $r''=r'-1$.

(5) If the right-hand side is greater than r' and less than $r'+1$, then $r=r'+1$ is the desired value.

(6) If the right-hand side is greater than $r'+1$, repeat the process with $r''=r'+1$.

An average of about one and a half cycles will be required.

An example of the computation is the following:

Let us suppose a further experiment is to be conducted along the lines of the H-acid example at the end of Section 11, where 18 samples of H-acid are to be compared, and where a *WSD* for means of no more than 50 is required with 90% confidence.

We can argue backward from the *WSD* (totals) found in that example, treating it as if it had been obtained by the conventional procedure (an not by the short-cut). We have, from Table 1, using $DDF = 20$,

$$[LSD/s] = 2.92$$

$$[WSD/LSD] = 1.46 + \frac{1.0}{20} = 1.51$$

Table 5

Approximate Values of Factor Multiplying S_w^2/B^2 to Give the Right-Hand-Side of the Design Inequality for a 5% Error Rate Familywise in a Single Classification

90% confidence			
	r = 2	**r = 3**	**r = 4 or more**
$m = 2$	$86 + 220/DDF^*$	$30 + 100/DDF^*$	$8 + 55/r + 80/DDF^*$
$m = 3$	$72 + 220/DDF^*$	$33 + 130/DDF^*$	$11 + 55/r + 120/DDF^*$
$m = 5$	$59 + 220/DDF^*$	$35 + 150/DDF^*$	$15 + 50/r + 160/DDF^*$
$m = 10$	$50 + 230/DDF^*$	$36 + 170/DDF^*$	$20 + 45/r + 200/DDF^*$
$m = 20$	$47 + 250/DDF^*$	$37 + 250/DDF^*$	$26 + 30/r + 250/DDF^*$

50% confidence			
	r = 2	**r = 3**	**r = 4 or more**
$m = 2$	26	13	$8 + 11/r$
$m = 3$	25	17	$11 + 14/r$
$m = 5$	27	20	$15 + 13/r$
$m = 10$	30	24	$20 + 11/r$
$m = 20$	32	28	$25 + 10/r$

whence from *WSD* (totals) = 459

$$s \text{ (totals)} = 103$$

$$s^2 \text{ (totals)} = 10609$$

$$S_w^2 = \frac{s^2 \text{ (totals)}}{S} = 2122,$$

and we may reasonably assign this $DDF^* = 15$, since it was obtained in a somewhat roundabout and approximate way.

For the new experiment then,

$$m = 18$$

$$S_w^2 = 2100$$

$$DDF^* = 15$$

$$B = 50$$

$$\text{confidence} = 90\%$$

and so Table 5 yields, as the approximate design inequality

$$r' \geq \frac{2100}{2500}\left(25 + \frac{36}{r'} + \frac{230}{15}\right) = 33.8 + \frac{30.2}{r'}$$

whose smallest integer solution is $r' = 35$.

To obtain the exact design inequality, we assume $r' = 35$ to find

$$DDF = 18(34) = 612$$

$$[LSD/s] = 2.77$$

$$[WSD/LSD] = 1.78 + \frac{2.4}{612} = 1.78$$

$$F^* = 1.76 \qquad \text{(90\% of } F \text{ on 612 and 15)}$$

$$(1.78)^2\ (2.77)^2\ (1.76) = 42.7$$

and thus the right-hand side of the exact design inequality is

$$\frac{2100}{2500}(42.7) = 35.9$$

and hence, by (5) above,

$$r = 36.$$

Thus 36 tests on each sample of H-acid will be required.

Notice that, for most practical purposes, the 35 replications found for r' would probably be sufficient. The approximate design inequality frequently provides a satisfactory approximate design procedure.

CHAPTER 3: DOUBLE CLASSIFICATIONS

14. INTRODUCTION

Even more common than the balanced single classification situation is the balanced double classification situation, where each observation falls in a cell determined by the intersection of two classifications. As examples we might mention

(1) the yields of six varieties of wheat in each of five locations; *or*

(2) the efficiencies of six types of catalyst in converting each of five feed stocks; *or*

(3) the results of six methods of teaching a high-school subject in schools of each of five sizes or in each of five states; *or*

(4) the average income of families reading each of six national magazines in each of five metropolitan areas; *or*

(5) the frequencies of certain blood groups among the residents of six different areas whose surnames belong to each of five different groups; *or*

(6) the weather four days after each of six objectively defined types of weather maps in each of five times of year.

In any of these examples, there is one classification into six categories and one into five. There are $6 \times 5 = 30$ cells, and one observation in each cell. This observation may be a percentage, total or mean, but we resolutely turn our face for the present against any inquiry into the parts from which individual observations are composed, leaving cases where further analysis may be desired to another account.

It is clearly appropriate to inquire of a set of observations arranged in this way about the determinations corresponding to either classification or indeed, about those corresponding to both. There are 2 families of determinations here, one for each classification.

15. THE CLASSICAL FORM

If we write y_{ij} for the observation common to category i for the first classification and category j of the second classification, where $1 \leq i \leq m$ and $1 \leq j \leq r$, we have the conventional calculation:

$$y_{i+} = y_{i1} + y_{i2} + \ldots y_{ir},$$

$$y_{+j} = y_{1j} + y_{2j} + \ldots y_{mj},$$

$$y_{++} = y_{1+} + y_{2+} + \ldots + y_{m+} = y_{+1} + y_{+2} + \ldots + y_{+r},$$

$$SSD = (y_{11}^2 + y_{12}^2 + \ldots + y_{mr}^2) - \frac{1}{r}(y_{1+}^2 + y_{2+}^2 + \ldots + y_{m+}^2)$$

$$- \frac{1}{m}(y_{+1}^2 + y_{+2}^2 + \ldots + y_{+r}^2) + \frac{1}{rm} y_{++}^2,$$

$$DDF = (m-1)(r-1),$$

$$MSD = SSD / \left[(m-1)(r-1)\right] = SSD/DDF,$$

$$s^2 \text{ (for } y_{i+}) = r(MSD),$$

$$s^2 \text{ (for } y_{+j}) = m(MSD),$$

$$s^2 \text{ (for } y_{i.} = \frac{1}{r} y_{i+}) = \frac{1}{r}(MSD),$$

$$s^2 \text{ (for } y_{.j}) = \frac{1}{m}(MSD),$$

and then

$$WSD \text{ (for any)} = [WSD/LSD\,]\,[LSD/s\,] \text{ (} s \text{ for same).}$$

An example of the computation follows:

Tippett [112], also quoted by Freeman [109 p. 65], has described an experiment on yarn strength (measured as the product of count and lea strength in pounds) where 3 roller weightings (A, B, C) were combined with 4 roving sets (1,2,3,4). The observations and marginal totals were:

		WEIGHTS (i)			
		A	B	C	Total
SETS (j)	1	1577	1535	1592	4704
	2	1690	1640	1652	4982
	3	1800	1783	1810	5393
	4	1642	1621	1663	4926
	Total	6709	6579	6717	20005

Our calculations will be appreciably simplified if we code by subtracting 1500 from each observation, whence we have

	A	B	C	Total
1	77	35	92	204
2	190	140	152	482
3	300	283	310	893
4	142	121	163	426
Total	709	579	717	2005

so that, in coded terms

$$y_{A.} = 177.25$$

$$y_{B.} = 144.75$$

$$y_{C.} = 179.25$$

$$SSD = (77^2 + 190^2 + \ldots + 163^2) - (709^2 + 579^2 + 717^2)/4$$

$$- (204^2 + 482^2 + 893^2 + 426^2)/3 + 2005^2/12 = 1362.7$$

$$DDF = 6$$

$$MSD = 227.1$$

$$s^2 \text{ (for } y_{i+}) = 4(227.1) = 908.4$$

$$s^2 \text{ (for } y_{+j}) = 3(227.1) = 681.3$$

$$s^2 \text{ (for } y_{i.}) = {}^1\!/\!_4(227.1) = 56.8$$

$$s^2\ (y_{.j}) = {}^1\!/\!_3(227.1) = 75.7$$

and, referring to Table 1,

$$[LSD/s\,] = 3.46$$

$$[WSD/LSD\,] \text{ (for roller wts)} = 1.20 + \frac{0.4}{6} = 1.27$$

$$[WSD/s\,] \text{ (for roller wts)} = 4.40$$

$$[WSD/LSD] \text{ (for roving sets)} = 1.32 + \frac{0.6}{6} = 1.42$$

$$[WSD/s] \text{ (for roving sets)} = 4.91$$

so that

$$WSD \text{ (for roller wt. totals)} = 132.6$$

$$WSD \text{ (for roller wt. means)} = 33.1$$

$$WSD \text{ (for roving set totals)} = 128.1$$

$$WSD \text{ (for roving set means)} = 42.7.$$

Thus we conclude that, for example;

(1) roller weight C exceeds roller weight B in the total of 4 yarn strengths by

$$6717 - 6579 \pm 133 = 138 \pm 133 = \text{between 5 and 271;}$$

(2) roller weight A exceeds roller weight B in mean yarn strength by

$$1677.25 - 1644.75 \pm 33.1 = 32.5 \pm 33.1 = \text{between } -0.6 \text{ and } 65.6;$$

(3) roving set 3 exceeds roving set 1 in mean yarn strength by

$$1797.7 - 1568.0 \pm 42.7 = 229.7 \pm 42.7 = \text{between 187.0 and 272.4.}$$

We are clearly in a situation similar to that for the single classification, and can consider streamlining matters.

16. THE CONDENSED FORM

The computationally simplest version runs as follows:

(1) total the observations by rows and columns, find a grand total from both as a check;

(2) form four quantities of the general form:

(sum of squares of totals)/(observations per total)

where

(a) totals = individual observations;

(b) totals = column totals;

(c) totals = row totals;

(d) totals = grand total;

(3) find the *SSD* from these four results; adding (a) to (d) and then subtracting (b) and (c), and find its square root;

(4) find the *WSD* (for column totals) and the *WSD* (for row totals) from the factors in Table 6 and the $\sqrt{SSD}$ just obtained.

Except for many squarings and one square-rooting this is reasonably simple.

As an example, consider observations by Chapanis, Rouse, and Schachter [104, p. 431] also quoted by Clark [105, p. 168-9] on the number of correct responses made by 5 subjects reading the Luckiesh-Moss low-contrast test chart under various experimental conditions. The observations and marginal totals are as follows: (The experiment was actually a latin square, with days as the third classification, which we omit here.)

	Loud sound	Weak sound	Heavy pressure	Light pressure	Control	Total
A	21	22	20	22	22	107
B	22	16	23	19	23	103
C	14	14	23	24	20	95
D	29	24	24	24	28	129
E	16	15	14	15	13	73
Total	102	91	104	104	106	507

Table 6

Values of [*WSD* for totals)/$\sqrt{SSD}$] for Balanced Double Classifications and an Error Rate of 5% Familywise

r = number of measurements per determination

m = number of determinations	2	3	4	5	6	7	8	9	10	12	15	20	30	60	∞
2	26.4	7.40	5.21	4.39	3.96	3.74	3.57	3.46	3.35	3.24	3.14	3.04	2.94	2.85	2.77
3	8.28	4.33	3.54	3.20	3.60	2.88	2.80	2.74	2.68	2.62	2.56	2.50	2.44	2.39	2.34
4	5.57	3.46	2.94	2.72	2.58	2.50	2.44	2.38	2.34	2.30	2.26	2.22	2.18	2.14	2.10
5	4.45	2.99	2.60	2.41	2.32	2.26	2.22	2.18	2.14	2.10	2.06	2.02	1.99	1.96	1.93
6	3.78	2.69	2.36	2.22	2.13	2.08	2.04	2.01	1.98	1.95	1.92	1.89	1.86	1.83	1.80
7	3.40	2.48	2.20	2.07	1.99	1.95	1.91	1.88	1.86	1.83	1.80	1.78	1.75	1.73	1.71
8	3.09	2.30	2.08	1.96	1.88	1.84	1.81	1.79	1.77	1.74	1.71	1.68	1.66	1.64	1.62
9	2.88	2.17	1.97	1.86	1.79	1.76	1.73	1.70	1.68	1.66	1.64	1.61	1.59	1.57	1.55
10	2.71	2.07	1.88	1.78	1.72	1.68	1.65	1.63	1.61	1.59	1.57	1.55	1.53	1.51	1.49
11	2.55	1.98	1.80	1.70	1.65	1.62	1.60	1.58	1.56	1.54	1.52	1.50	1.48	1.46	1.44
12	2.46	1.90	1.73	1.64	1.59	1.56	1.54	1.52	1.50	1.48	1.46	1.44	1.43	1.41	1.39
13	2.34	1.83	1.67	1.59	1.54	1.52	1.50	1.48	1.46	1.44	1.42	1.40	1.38	1.36	1.35
14	2.24	1.77	1.62	1.54	1.50	1.48	1.46	1.44	1.42	1.40	1.38	1.36	1.34	1.32	1.31
15	2.16	1.71	1.57	1.50	1.46	1.44	1.42	1.40	1.38	1.36	1.34	1.33	1.31	1.29	1.28
16	2.08	1.67	1.53	1.46	1.42	1.39	1.37	1.35	1.34	1.32	1.31	1.29	1.28	1.26	1.25
17	2.02	1.63	1.49	1.42	1.38	1.36	1.34	1.32	1.31	1.30	1.28	1.26	1.25	1.23	1.22
18	1.96	1.58	1.45	1.39	1.35	1.33	1.31	1.29	1.28	1.27	1.25	1.24	1.22	1.20	1.19
19	1.91	1.55	1.42	1.36	1.33	1.30	1.28	1.27	1.26	1.24	1.23	1.21	1.20	1.18	1.17
20	1.86	1.51	1.38	1.33	1.30	1.27	1.25	1.24	1.23	1.22	1.20	1.19	1.17	1.16	1.16
									60/*r* = 6	5	4	3	2	1	0

The four quantities are:

$$(21^2 + 22^2 + 14^2 + \ldots + 13^2)/1 = 10773$$

$$(102^2 + 91^2 + 104^2 + 104^2 + 106^2)/5 = 10310.6$$

$$(107^2 + 103^2 + 95^2 + 129^2 + 73^2)/5 = 10610.6$$

$$507^2/25 = 10281.96$$

so that:

$$TSSD = 10773 - 10310.6 - 10610.6 + 10281.96 = 133.76$$

$$\sqrt{TSSD} = 11.6$$

factor (either totals) = 2.41

WSD (either totals) = 28.

We clearly conclude that

(1) there need not be consistent differences due to experimental conditions;

(2) such differences, for the conditions explored to not exceed

$$|106 - 91| + 28 = 43$$

for a total over 5 subjects, and thus differences in means among these 5 conditions do not exceed 8.8 (clearly we do not know too much about possible effects of experimental conditions!);

(3) difference among subjects exist, of magnitude at least

$$129 - 73 - 28 = 28$$

for a total over 5 conditions, and thus differences in means among these 5 amount to at least 5.6.

In cases where computing machines are not used, the computation of *TSSD* can be simplified by subtracting convenient functions of rows and columns. If we subtract 20, 20, 20, 25 and 15 from the five rows we obtain

					(T)
1	2	0	2	2	7
2	−4	3	−1	3	3
−6	−6	3	4	0	−5
4	−1	−1	−1	3	4
1	0	−1	0	−2	−2
2	−9	4	4	6	7

and if we then subtract 0, −2, 1, 1, 1 from the five columns we obtain

					(T)
1	4	−1	1	1	6
2	−2	2	−2	2	2
−6	−4	2	3	−1	−6
4	1	−2	−2	2	3
1	2	−2	−1	−3	−3
2	1	−1	−1	1	2

for which the four quantities are, as may be checked by mental arithmetic,

$$1^2 + 2 + (-6) + \ldots + 2^2 + (-3)^2 = 154$$

$$(2^2 + 1^2 + (-1)^2 + (-1)^2 + 1^2)/5 = 1.6$$

$$(6^2 + 2^2 + (-6)^2 + 3^2 + (-3)^2)/5 = 18.8$$

$$2^2/25 = 0.16$$

and the *TSSD* is, again,

$$154 - 1.6 - 18.8 + 0.16 = 133.76.$$

17. THE SHORT-CUT FORM

Again we present a short-cut based on ranges, although it is not quite as simple as for single classifications. Moreover it is not as unique, depending on the order of the "columns" and on which classification has been chosen as providing "columns." (We write "columns" in quotes since we shall describe the calculation in these terms, recognizing that either actual columns or actual rows may be chosen as "columns," the other becoming "rows.") The step by step procedure is as follows:

(1) calculate "column" totals, "row" totals and the two check grand totals;

(2) repeat the "left-hand column" to the "right" of the "right-hand column" and write in differences in the direction of "rows";

(3) form ranges of ***differences*** in the direction of "columns" and sum to form a total range of differences;

(4) multiply this by the two factors from Table 7 to obtain the two *WSD*'s for totals;

(5) when desired, divide the column totals ***and*** their *WSD* by the number of columns to obtain row means ***and*** their *WSD*.

An example of the computation follows:

Warren [113], also quoted by Tippett [112, p. 127 of 2nd ed., p. 183 of 4th ed.] gives the mean head breadths (in millimeters) of samples of small solider termites taken from each of five nests during each of five months. The observations, differences, totals, and ranges are as follows:

	Nest 668		Nest 670		Nest 672		Nest 674		Nest 675		(Nest 668)	Totals
Nov.	2.273		2.479		2.404		2.497		2.456		(2.273)	12.059
		+.206		−.075		+.043		+.009		−.183		(.000)
Jan.	2.332		2.603		2.457		2.388		2.626		(2.332)	12.406
		+.271		−.146		−.069		+.238		−.294		(.000)
Mar.	2.375		2.613		2.452		2.515		2.633		(2.375)	12.588
		+.238		−.161		+.063		+.118		−.258		(.000)
May	2.373		2.557		2.396		2.445		2.487		(2.373)	12.258
		+.184		−.161		+.049		+0.42		−.114		(.000)
Aug.	2.318		2.377		2.279		2.312		2.410		(2.318)	11.696
		+.059		−.098		+.033		+.098		.092		(.000)
Totals	11.671		12.629		11.988		12.107		12.612			
Ranges		.212		.086		.132		.229		.202		.861

Reference to Table 7 shows that

$$\text{factor for either} = .60$$

Table 7

Values of [*WSD* for totals)/(total range of differences)] for a Balanced Double Classification and an Error of 5% Familywise. (Upper entries for totals A, lower entries for totals B)

Number of ranges = Number of totals A

Number of differences per range = Number of totals *B*	2	3	4	5	6	7	8	9	10	11	12	13	14	15	16	17	18	19	20
	6.36	2.17	1.53	1.17	.96	.80	.68	.62	.54	.49	.45	.41	.38	.36	.34	.31	.30	.28	.27
2	6.36	1.95	1.40	1.12	.96	.84	.76	.70	.65	.61	.58	.55	.52	.50	.48	.46	.45	.44	.42
	1.95	1.17	.90	.73	.62	.54	.48	.42	.38	.35	.32	.30	.28	.26	.25	.23	.22	.21	.20
3	2.17	1.17	.91	.77	.68	.61	.56	.52	.49	.46	.44	.42	.40	.38	.37	.36	.35	.34	.33
	1.45	.96	.76	.63	.54	.47	.42	.38	.34	.32	.29	.27	.25	.24	.23	.21	.20	.19	.18
4	1.56	.94	.76	.66	.58	.53	.49	.45	.43	.40	.39	.37	.35	.34	.33	.32	.31	.30	.29
	1.27	.90	.72	.60	.51	.45	.40	.36	.33	.30	.28	.26	.25	.23	.22	.21	.20	.19	.18
5	1.29	.84	.69	.60	.53	.49	.45	.42	.40	.38	.36	.34	.33	.32	.31	.30	.29	.28	.27
	1.18	.87	.70	.58	.50	.44	.39	.36	.33	.30	.28	.26	.24	.23	.22	.21	.20	.19	.18
6	1.14	.78	.64	.56	.50	.46	.43	.40	.38	.36	.34	.33	.31	.30	.29	.28	.27	.27	.26
	1.16	.85	.69	.58	.50	.44	.39	.36	.33	.30	.28	.26	.24	.23	.22	.21	.20	.19	.18
7	1.07	.74	.61	.54	.48	.44	.41	.38	.36	.34	.33	.31	.30	.29	.28	.27	.26	.26	.25
	1.15	.85	.69	.58	.50	.44	.40	.36	.33	.30	.28	.26	.25	.23	.22	.21	.20	.19	.18
8	1.00	.71	.59	.52	.47	.43	.40	.37	.35	.33	.32	.30	.29	.28	.27	.26	.26	.25	.24
	1.14	.85	.69	.58	.50	.45	.40	.36	.33	.31	.28	.27	.25	.24	.22	.21	.20	.19	.18
9	.96	.68	.57	.50	.45	.42	.39	.36	.34	.33	.31	.30	.29	.28	.27	.26	.25	.24	.24
	1.15	.86	.70	.59	.51	.45	.40	.37	.34	.31	.29	.27	.25	.24	.23	.21	.20	.19	.19
10	.93	.66	.56	.49	.44	.41	.38	.36	.34	.32	.31	.29	.28	.27	.26	.25	.25	.24	.23

Number of ranges = Number of totals B

Number of differences per range = Number of totals *B*	2	3	4	5	6	7	8	9	10	11	12	13	14	15	16	17	18	19	20
	1.15	.87	.71	.60	.52	.46	.41	.37	.34	.31	.29	.27	.26	.24	.23	.22	.21	.20	.19
11	.91	.65	.55	.48	.43	.40	.37	.35	.33	.31	.30	.29	.28	.27	.26	.25	.24	.24	.23
	1.16	.88	.71	.60	.52	.46	.41	.38	.35	.32	.30	.28	.26	.25	.23	.22	.21	.20	.19
12	.88	.63	.54	.47	.43	.39	.37	.35	.33	.31	.30	.28	.27	.26	.25	.25	.24	.23	.23
	1.17	.89	.72	.61	.53	.47	.42	.38	.35	.32	.30	.28	.26	.25	.24	.22	.21	.20	.19
13	.86	.62	.53	.47	.42	.39	.36	.34	.32	.31	.29	.28	.27	.26	.25	.24	.24	.23	.22
	1.18	.90	.73	.62	.54	.48	.43	.39	.36	.33	.31	.29	.27	.25	.24	.23	.22	.21	.20
14	.85	.62	.52	.46	.42	.39	.36	.34	.32	.30	.29	.28	.27	.26	.25	.24	.23	.23	.22
	1.19	.91	.74	.63	.55	.48	.43	.39	.36	.33	.31	.29	.27	.26	.24	.23	.22	.21	.20
15	.84	.61	.52	.46	.41	.38	.36	.33	.32	.30	.29	.28	.27	.26	.25	.24	.23	.23	.22
	1.21	.92	.76	.64	.55	.49	.44	.40	.37	.34	.31	.29	.28	.26	.25	.23	.22	.21	.20
16	.82	.60	.51	.45	.41	.38	.35	.33	.31	.30	.29	.27	.26	.25	.25	.24	.23	.23	.22
	1.22	.93	.76	.65	.56	.50	.44	.40	.37	.34	.32	.30	.28	.26	.25	.24	.23	.22	.21
17	.81	.60	.51	.45	.41	.38	.35	.33	.31	.30	.28	.27	.26	.25	.24	.24	.23	.22	.22
	1.23	.94	.77	.65	.57	.50	.45	.41	.38	.35	.32	.30	.28	.27	.25	.24	.23	.22	.21
18	.81	.59	.50	.45	.40	.37	.35	.33	.31	.29	.28	.27	.26	.25	.24	.24	.23	.22	.22
	1.24	.95	.78	.66	.57	.51	.46	.42	.38	.35	.33	.31	.29	.27	.26	.24	.23	.22	.21
19	.80	.59	.50	.44	.40	.37	.35	.33	.31	.29	.28	.27	.26	.25	.24	.23	.23	.22	.22
	1.26	.97	.79	.67	.58	.51	.46	.42	.39	.36	.33	.31	.29	.27	.26	.25	.23	.22	.21
20	.79	.58	.50	.44	.40	.37	.34	.32	.31	.29	.28	.27	.26	.25	.24	.23	.23	.22	.21

so that

$$WSD \text{ (either set of totals)} = .517$$

and so we may conclude (with 5% error rate familywise, as always) that, for example:

(1) March head breadths exceed August head breadths by

$$12.588 - 11.696 \pm .517 = .892 \pm .517 = \text{between } .375 \text{ and } 1.409$$

in a total over 5 nests (and hence by between .075 and.282 millimeters in mean);

(2) nest 670 head breadths exceed nest 668 head breadths by

$$12.629 - 11.671 \pm .517 = .958 \pm .517 = \text{between } .441 \text{ and } 1.475$$

in a total over 5 months (and hence by between .088 and .295 millimeters in mean).

18. THE HALF-CUT FORM

Again it may be desired, with rather messy data, to use

$$\left[(m - \frac{1}{r}(1 - \frac{1}{m})] \text{ (sum of squared ranges of differences))}\right]^{1/2}$$

in place of the total range of differences. An example of the computation follows:

Curtis [106], quoted by Anderson and Bancroft [101, p. 244], gives mean incomes for samples of regular, former, and potential patrons at 10 farmer's markets in North Carolina. The observations, totals and differences are as follows

	Asheville		Asheboro		Charlotte		Durham		Greensboro		Raleigh
Regular	30	(9)	39	(2)	41	(−12)	29	(3)	32	(−3)	29
Former	30	(6)	36	(1)	37	(−8)	29	(−2)	27	(0)	27
Potential	28	(−1)	27	(1)	28	(−4)	24	(−2)	22	(7)	29
Total	88		102		106		82		81		85
Range		10		1		8		5		10	

(cont'd)

	Raleigh		Goldsboro		Rocky Mount		Franklin		J'ville		Asheville
Regular	29	(−4)	25	(2)	27	(3)	30	(−1)	29	(1)	30
Former	27	(−3)	24	(4)	28	(1)	29	(−1)	28	(2)	30
Potential	29	(−9)	20	(7)	27	(−3)	24	(3)	27	(1)	28
Total	85		69		82		83		84		88
Range		5		5		6		4		1	

The sum of squared ranges i.e. $10^2 + 1^2 + \ldots + 4^2 + 1^2 = 393$ so that

$$\text{modified total range} = [(10 - {}^1/_3\,(1 - {}^1/_{10}))\ 393]^{1/2} = \sqrt{3812} = 61.7$$

Now

factor for market totals = .38

factor for type (of patron) totals = .49

WSD (market totals) = (.38) (61.7) = 23.4

WSD (type totals) = (.49) (61.7) = 30.2

and since the type totals were

regular 311

former 295

potential 256

We conclude that, for example;

(1) the average income of potential patrons is less than that of either regular or former patrons by at least 295 − 256 − 30 = 9 in a total over 10 markets (and hence at least 0.9 in mean);

(2) the average income at Goldsboro is less than that at Asheboro by

$$102 - 69 \pm 23 = 33 \pm 23 = \text{between 10 and 56}$$

in a total of 3 types of patron (and hence between 3 and 19 in mean).

19. DESIGN

When only one family of determinations is of real interest, the design problem for double classifications mimics that for single classifications, the only change being that *DDF* is now equal to $(m-1)(r-1)$. This requires a new table of approximate values of the factor, which is given as Table 8.

Table 8

Approximate Values of the Factor Multiplying $S_w^2/8^2$ in the Right-Hand-Side of the Design Inequality for 50% and 90% Confidence and an Error Rate Familywise of 5%

90% confidence

	r = 2	r = 3	r = 4 or more
$m=2$	$880+1800/DDF^*$	$86+220/DDF^*$	$8+80/DDF^*+(137-830/DDF^*)/r$
$m=3$	$160+400/DDF^*$	$49+160/DDF^*$	$11+120/DDF^*+90/r$
$m=5$	$77+270/DDF^*$	$40+170/DDF^*$	$15+160/DDF^*+60/r$
$m=10$	$54+230/DDF^*$	$37+200/DDF^*$	$20+200/DDF^*+50/r$
$m=20$	$47+260/DDF^*$	$38+240/DDF^*$	$26+250/DDF^*+30/r$

50% confidence

	r = 2	r = 3	r = 4 or more
$m=2$	150	26	$8+32/r$
$m=3$	50	21	$11+23/r$
$m=5$	33	22	$15+16/r$
$m=10$	32	25	$20+13/r$
$m=20$	32	28	$25+10/r$

An example of the calculation is the following:

Let us consider the example of Section 15 concerning the effect of roller weight on yarn strength. Let us suppose that we wish to carry out further tests to study 8 roller weights and that we wish to have, with 90% confidence, a *WSD* of no more than 30 for mean strengths. How many replications (roving sets) are required?

Here

$$m = 8$$

$$S_w^2 = 227 \text{ (the old } MSD\text{)}$$

$$DDF^* = 6 \text{ (the old } DDF\text{)}$$

$$B = 30$$

$$\text{confidence} = 90\%$$

and since an r of more than 3 will obviously be required, the approximate factor will be

$$18 + \frac{184}{6} + \frac{54}{r} = 49 + \frac{54}{r}$$

so that the approximate design inequality becomes

$$r' \geq \frac{227}{(30)^2} (49 + \frac{54}{r'}) = 12.4 + \frac{13.6}{r'}$$

whence $r' = 14$.

Now for $r' = 14$, we have

$$DDF = (10 - 1)\ (14 - 1) = 117$$

$$[WSD/LSD\,] = 1.62 + \frac{1.6}{117} = 1.63$$

$$[LSD/s\,] = 2.80$$

$$F^* = 2.75 \qquad \text{(90\% point of } F \text{ on 117 and 6)}$$

factor = $(1.63)^2\ (2.80)^2\ (2.75) = 57.3$
so that the design inequality becomes

$$14 > \frac{227}{900}\ 57.3 = 14.45$$

which is not fulfilled. Since $14.45 < 14 + 1$ however, we know that $r = 15$ is the solution.

Hence 15 replications (15 roving sets) are required.

When a double classification pattern is combined with interest in both families of determinations, the question of design becomes simpler. For it then runs like this — "What accuracy must the individual measurements have, in order that there be 50% or 90% confidence that the final *WSD* for the family of determinations be less than B''? (If the other family is of prime interest, repeat the process with r and m interchanged.) Usually such extra accuracy is to be obtained by making n measurements, whose mean will constitute one of the individual measurements y_{ij}. This amounts to replacing S_w^2 by S_w^2/n, and the design inequality becomes

$$r \geq \frac{S_w^2}{nB^2} \text{ (factor approximated in Table 8)}$$

which may be written

$$n \geq \frac{S_w^2}{B^2} \frac{\text{(factor approximated in Table 8)}}{r} .$$

An example of the computation is the following:

Let us consider the example of Section 16 concerning the number of correct responses made on a particular low-contrast test chart. Let us suppose that we have available as many equivalent forms of the chart as we wish to use, but are restricted to 12 subjects and 7 experimental conditions. We wish to obtain a *WSD* for experimental condition means of 0.8 or less with 50% confidence. We believe that, under the circumstances of the forthcoming experiment, the variance of reading one form is 6.2 and assign this mental estimate a stability corresponding to $DDF = 12$.

Then

$$S_w^2 = 6.2$$

$$DDF^* = 12$$

$$B = 0.8$$

$$\text{confidence} = 50\%$$

so that

$$\text{approximate factor} = 17 + {}^{15}/_{12} = 18.3$$

$$\frac{S_w^2}{B^2} = \frac{6.2}{0.64} = 9.7$$

and the approximate design inequality

$$n \geq 9.7\ \frac{18.3}{12} = 14.8.$$

The exact factor is to be found from

$$m = 7$$

$$DDF = (7-1)\ (12-1) = 66$$

$$WSD/LSD = 1.51 + \frac{1.2}{66} = 1.53$$

$$LSD/s = 2.83$$

$$F^* = 1.04 \qquad \text{(50\% point of } F \text{ on 66 and 12)}$$

$$\text{factor} = (1.53)^2\ (2.83)^2\ (1.04) = 19.5$$

so that the exact design inequality is

$$n \geq 9.7\ \frac{19.5}{12} = 15.8$$

and the conclusion is that n should be at least 16.

Thus at least 16 equivalent forms of low-contrast test chart should be read by each of 12 subjects under each of 7 experimental conditions, if a *WSD* of not more than 0.8 is to be had with 50% confidence.

CHAPTER 4: EXTENSIONS

20. GRAPHICAL PRESENTATION

Having dealt with the specific applications of the general procedure to the two simplest standard examples, it is now time to treat various extensions of the general procedure.

Our first extension will be to graphical presentation. There are three cases to be distinguished:

(1) cases where only comparisons or contrasts among the determinations of the family are of interest;

(2) cases where both comparisons or contrasts, and individual values or other linear combinations which are not contrasts, are of interest;

(3) cases where individual values are of interest, but no comparisons or contrasts are to be considered (the pure batch case).

Cases of this third sort should usually not lead to graphical presentation, since the prime purpose of graphical presentation is comparison. Exceptions usually involve comparing the various determinations of a batch, each with its own theoretically derived standard. In these exceptional cases, the observed determination is normally plotted at the center of a bar of length 2(*WSD*).

In a purely comparative or contrast situation, as in (1) above, the appropriate graphical presentation involves:

(1) each determination represented by a bar of total length equal to the *WSD*, centered at the observed value;

(2) no vertical scale, except for differences, (since only comparison is involved);

(3) mental computation of limits by comparing the top of one bar with the bottom of another, and vice versa.

This is not conventional, but straightforward.

In the mixed situation of (2) above, the appropriate graphical presentation involves:

(1) each determination represented by a bar or total length equal to the *WSD*;

(2) two vertical scales, close together, displaced by the amount of the *WSD*;

(3) mental computation of limits by comparing the top of one bar with the bottom of another or with the higher-reading vertical scale or by comparing the bottom of one bar with the top of another or with the lower-reading vertical scale.

Examples are given in Sections 47 and 57 (pp. 82ff. 95ff).

21. KNOWN INHOMOGENEITY

So far we have assumed that the determinations have a common variance. The natural generalization of this is to assume that the investigator has:

(2′) knowledge that the variance of the different determinations are nearly in known ratios.

That is to say, that

$$\operatorname{var}\{x_i\} = g_i^2\sigma^2,$$

where the g_i are known, and s^2 is an estimate of σ^2. It is shown in Section 175 that it is natural to use

$$\pm (WSD)\sqrt{g_i^2 + g_j^2}$$

as the allowance for $x_i - x_j$ and

$$\pm (WSD)\operatorname{norm}_g\{x\},$$

where

$$\operatorname{norm}_g\{x\} = \Sigma\, a_j b_k \sqrt{g_j^2 + g_k^2}$$

as the allowance for

$$x = \Sigma\, c_i x_i = \Sigma\, a_j x_j - \Sigma\, b_k x_k$$

where the a_j are the positive c_i and the b_k are the positive $-c_i$. Thus if

$$x = 3x_1 - 2x_2 + x_3 - 2x_7 = 3x_1 + x_3 - (2x_2 + 2x_7)$$

then

$$\text{norm}_g\{x\} = 6\sqrt{g_1^2 + g_2^2} + 2\sqrt{g_2^2 + g_3^2} + 6\sqrt{g_1^2 + g_7^2} + 2\sqrt{g_3^2 + g_7^2}.$$

An example of the calculation is the following:

Brownlee [103, p. 55 of 4th British Ed.] gives the throughputs obtained from units of chemical engineering plant before failure from corrosion. The results for 3 different foundries run as follows:

Foundry	Mean life	g_i^2	g_i
A	53.0	9	3.00
B	78.3	5	2.24
C	79.7	15	3.88

$$s^2 = 11.7,$$

$$DDF = 16,$$

$$s = 3.42,$$

$$[LSD/s] = 3.00,$$

$$LSD = 10.3,$$

$$[WSD/LSD] = 1.20 + \frac{0.4}{16} = 1.225,$$

$$WSD = 12.6.$$

Notice carefully that this *WSD cannot*, we repeat *cannot*, be applied directly to simple comparisons. The norm of $x_B - x_A$ is $\sqrt{5+9} = 3.74$ and so we are to accept that, for example,

(1) the excess mean throughput of foundry B units over foundry A units is

$$78.3 - 53.0 \pm (3.74)(12.6) = 25.3 \pm 47.10$$

$$= \text{between } -21.8 \text{ and } +72.4.$$

The approximation in the use of this procedure, which is being examined by T. E. Kurtz, is apparently in the conservative direction in the sense that the actual error rate tends to be somewhat less than the nominal error rate.

23. ERROR-RATE SHIFTING

To date we have concentrated on an error rate of 5%, mainly familywise, to a lesser extent batchwise, occasionally per comparison or per determination. There is nothing sacred about 5%, and indeed it is sometimes convenient to use other error rates, either because they are standard or because the spirit moves us. We have two choices at the level of practical arithmetic, the first and classical one being to repeat tables 1 to 4, 6 and 7 for each of a number of error rates. Since no six error rates would suffice for all users, 50 or more tables would shortly be desired, bulky to handle and tedious to compute. The second choice, and the one analogous to the choice we have already made in connection with Tables 1 and 2, is to find some simple empirical approximations which will serve our needs. As the reader already expects, we have made the second choice.

Table 9 contains the value necessary to convert a *WSD* for an error rate of 5% familywise to one for any chosen error rate between 0.01% and 50%. (The approximation is also relatively good beyond 0.01%.) Examples of its use are the following:

In Section 16 (page 37) we discussed data of Chapanis, Rouse and Schaefer [104, p. 431] on the number of correct responses on the Luckiesh-Moss low-contrast test chart under various conditions. Here $m = 5$ and $DDF = 16$. Let us shift the error rate for the *WSD*'s from 5% familywise to 20% familywise. We have

$$5h\,\% = 20\%$$

$$h = 4$$

$$u = -\log_{10} h = -.60$$

$$C = .125 + .0104(4) = .166$$

$$D = .63$$

$$C + \frac{D}{DDF} = .166 + \frac{.63}{16} = .205$$

Table 9

Values of Auxilliary Functions for Shifting the Error Rate of a *WSD* to any Rate Between 50% and 0.01% (or Lower)

	A	B	C	D
2	$.480 - .046u$	$1.09 + .68u$	$.202 + .0264h$	.40
3	$.368 - .033u$	$1.27 + .61u$	$.163 + .0137h$	.50
4	$.320 - .027u$	$1.40 + .57u$	$.138 + .0124h$	.57
5	$.291 - .026u$	$1.50 + .55u$	$.125 + .0104h$	.63
6	$.272 - .024u$	$1.60 + .53u$	$.116 + .0093h$	.66
7	$.257 - .022u$	$1.70 + .51u$	$.109 + .0083h$	.70
8	$.242 - .020u$	$1.77 + .50u$	$.105 + .0078h$	.73
9	$.235 - .019u$	$1.84 + .49u$	$.102 + .0074h$	.74
10	$.228 - .018u$	$1.90 + .48u$	$.099 + .0070h$	.77
11	$.223 - .018u$	$1.96 + .47u$	$.096 + .0067h$	.80
12	$.218 - .018u$	$2.02 + .47u$	$.095 + .0064h$	.80
13	$.214 - .017u$	$2.07 + .46u$	$.094 + .0061h$	.81
14	$.209 - .017u$	$2.12 + .46u$	$.093 + .0058h$	.81
15	$.205 - .016u$	$2.17 + .45u$	$.093 + .0056h$	.82
16	$.202 - .016u$	$2.21 + .45u$	$.092 + .0054h$	.83
17	$.199 - .016u$	$2.25 + .44u$	$.091 + .0052h$	.84
18	$.196 - .016u$	$2.30 + .44u$	$.090 + .0050h$	.85
19	$.194 - .016u$	$2.35 + .43u$	$.089 + .0048h$	.86
20	$.187 - .016u$	$2.38 + .43u$	$.087 + .0046h$	.88
21	$.186 - .016u$	$2.41 + .43u$	$.086 + .0044h$	.90
22	$.185 - .015u$	$2.43 + .42u$	$.086 + .0044h$	.90
23	$.184 - .015u$	$2.46 + .42u$	$.085 + .0043h$	.91
24	$.183 - .014u$	$2.48 + .42u$	$.084 + .0043h$	.92
25	$.181 - .013u$	$2.50 + .42u$	$.083 + .0043h$	.93
26	$.180 - .014u$	$2.53 + .42u$	$.082 + .0042h$	.94
27	$.179 - .013u$	$2.55 + .41u$	$.081 + .0042h$	.94
28	$.178 - .013u$	$2.57 + .41u$	$.080 + .0042h$	.95
29	$.177 - .013u$	$2.59 + .41u$	$.079 + .0041h$	.96
30	$.175 - .015u$	$2.61 + .41u$	$.078 + .0040h$	.97
35	$.169 - .015u$	$2.85 + .40u$	$.075 + .0039h$	1.00
40	$.165 - .014u$	$2.99 + .40u$	$.073 + .0038h$	1.04
45	$.161 - .013u$	$3.12 + .39u$	$.071 + .0037h$	1.06
50	$.157 - .012u$	$3.25 + .39u$	$.069 + .0035h$	1.09
60	$.149 - .010u$	$3.47 + .38u$	$.067 + .0032h$	1.13
80	$.139 - .009u$	$3.93 + .37u$	$.089 + .0030h$	1.20
100	$.132 - .008u$	$4.34 + .37u$	$.058 + .0029h$	1.24
200	$.120 - .007u$	$6.30 + .36u$	$.036 + .0026h$	1.41
500	$.110 - .005u$	$11.82 + .36u$	$.012 + .0023h$	1.52
∞	? – ?	? + ?	$-.016 + .0020h$	1.79

with $u = \log_{10} r = -\log_{10} h$, new error rate $= (5/r)\% = 5h\%$

$$WSD_{(5/r)\%} = WSD_{5\%}\left[1 + u\left(A + \frac{B}{DDF}\right)\right], \text{ for } u \text{ positive } (r \geq 1)$$

$$\log_{10}(WSD_{5h\%}) = \log_{10}(WSD_{5\%}) + u\left(C + \frac{D}{DDF}\right), \text{ for } u < 0\ (h \geq 1)$$

$$\log_{10}(WSD_{20\%}) = \log_{10}(WSD_{5\%}) + (-.60)(.205) = \log_{10}(WSD_{5\%}) - .123$$

$$WSD_{20\%} = (WSD_{5\%})(10^{-.123}) = .752\ WSD_{5\%}.$$

Thus

$$WSD_{20\%} \text{ (either totals)} = .752(28.0) = 21.1.$$

A glance at the data shows that altering the error rate from 5% to 20% has altered the conclusions very little.

In Section 6 (page 10) we discussed Roberts and Jessup's [111] measurements of heats of combustion of alpha — methylstyrene and some of its polymers. Here $m = 5$ and $DDF = 25$. Let us shift the error rate for the WSD and the resulting allowances from 5% familywise to 0.2% familywise. We have

$$5/r\ \% = 0.2\%$$

$$r = 25$$

$$u = \log_{10} r = 1.40$$

$$A = .291 - .026\,(1.40) = .255$$

$$B = 1.50 + .55u = 2.27$$

$$A + \frac{B}{DDF} = .255 + \frac{2.27}{25} = .345$$

$$WSD_{0.2\%} = 1.484\ WSD_{5\%} = 1.484\,(4.42) = 6.56.$$

Thus the result of the error rate shifting is to change the WSD from 4.4 to 6.6. Again, this has little effect on the conclusions.

To shift the error rate for LSD or IA, use the top line in Table 9, as in the following example.

Returning to the second example of Section 7, where

$$m = 7$$

$$DDF = 21$$

$$IA = 110$$

we might wish to make statements with an individual error rate of 0.5%. Then

$$u = 1.0$$

$$A = .480 - .046\ (1.0) = .434$$

$$B = 1.09 + .68\ (1.0) = 1.77$$

$$A + \frac{B}{DDF} = .434 + \frac{1.77}{21} = .518$$

$$IA_{0.5\%} = 1.518\ IA_{5\%} = (1.518)\ (110) = 167.$$

Thus, for x_N we have

333 ± 167 = between 166 and 500 with an individual error rate of 0.5%.

To shift a PA from 5% to $p\% = (5/r)\%$, where $0.01 > p/m < 5$, use the relation

$$\log_{10}\left\{\frac{PA_{(5/r)\%}}{PA_{5\%}}\right\} = \frac{\left(1.25 + \frac{24.6}{DDF}\right)\left(3 + \frac{20}{DDF}\right)}{\left(3 + \frac{20}{DDF} + \log_{10} m\right)\left(3 + \frac{20}{DDF} + \log_{10} m + \delta\right)}\,\delta$$

where

$$\delta = \log_{10} r + .011\left(1 - \frac{1}{r}\right).$$

(It is felt that the need for shifting PA's will be very infrequent, and that no Table is needed.)

Two major comments about error rate shifting need to be made:

(1) it is very easy to get any reasonable error rate you wish; and

(2) as m increases the ratio of $WSD_{(5/r)\%}$ to $WSD_{5\%}$ becomes closer and closer to unity (in many cases, the fractional increase is surprisingly small).

23. ERROR-RATE SHIFTING SHORT-CUTS AND HALF-CUTS

Exactly the same tables and formulas may be used to shift W_sSD's and W_hSD's, providing the correct number of degrees of freedom are used. Tables 10 and 11 provide the ratios of equivalent number of degrees of freedom for short-cut to actual numbers of degrees of freedom for the conventional sums of square deviations.

Table 10

Values of [(equivalent *DDF*)/*m* (*r* – 1)] for Balanced Single Classification Short-Cuts

r = number of measurements per range

m	2	3	4	5	6	8	10	12	14	16	18	20
2	.95	.98	.97	.94	.94	.91	.87	.84	.81	.78	.75	.73
5	.92	.95	.95	.94	.92	.89	.86	.83	.80	.77	.75	.72
10	.91	.93	.93	.92	.91	.89	.84	.82	.79	.76	.74	.72
20	.90	.93	.93	.92	.91	.88	.84	.82	.79	.76	.74	.71
∞	.90	.93	.93	.92	.91	.88	.84	.81	.78	.75	.73	.70

(*m* = number of determinations = number of ranges)

Table 11

Values of [(equivalent *DDF*)/(*c* – 1)(*r* – 1)] for Balanced Double Classification Short-Cuts

r = number of differences per range

m	2	3	4	6	8	10	12	12	16	18	20
2	1.00	1.00	.98	.94	.92	.89	.86	.83	.80	.78	.75
3	.95	.95	.95	.95	.95	.95	.95	.94	.93	.92	.91
4	.83	.83	.83	.83	.85	.87	.90	.90	.89	.89	.89
5	.77	.77	.77	.79	.82	.85	.86	.85	.84	.83	.82
7	.72	.72	.72	.73	.75	.80	.81	.81	.76	.78	.77
10	.68	.68	.68	.69	.71	.75	.76	.76	.76	.73	.72
15	.65	.65	.65	.66	.68	.71	.73	.72	.71	.70	.69
20	.64	.64	.64	.65	.65	.66	.67	.67	.67	.67	.67

(*m* = number of ranges)

Examples of the calculation are the following:

Let us consider the H-acid procedure at the end of Section 11, and shift the error rate from 5% to 23%. We have

$$m = 6$$

$$h = 4.6$$

$$u = -\log_{10} h = -.66$$

and from Table 10

$$\text{equivalent } DDF = (.94)(20) = 18.8$$

hence from Table 9,

$$C = .116 + .0093\ (4.6) = .159$$

$$D = .66$$

$$C + \frac{D}{DDF} = .159 + \frac{.66}{18.8} = .194$$

$$\log_{10} WSD_{23\%} = (.194)\ (-.66) + \log_{10} WSD_{5\%} = -.128 + \log_{10} WSD_{5\%}$$

$$WSD_{23\%} = 10^{-.128}\ WSD_{5\%} = .745\ WSD_{5\%} = (.745)\ (459) = 342\ .$$

Let us consider the termite example at the end of Section 17, and shift the error rate from 5% to 0.005%. We have

$$m = 5$$

$$r = 1000$$

$$u = \log_{10} r = 3$$

and from Table 11

$$\text{equivalent } DDF = (.78)(16) = 12.5$$

while from Table 9

$$A = .291 - .026\,(3.0) = .213$$

$$B = 1.50 + .55u = 3.15$$

$$A + \frac{B}{DDF} = .213 + \frac{3.15}{12.5} = .465$$

$$WSD_{0.005\%} = (1 + (3.0)\,(.465))\; WSD_{5\%} = 2.395\; WSD_{5\%} = (2.395)\,(.517) = 1.24\,.$$

CHAPTER 5: CAUTIONS AND ALTERNATIVES

24. INTRODUCTION

We have introduced one general procedure, and two special procedures for single and double classifications. In the present chapter we shall introduce another general procedure, and hint at a third. Clearly we have a responsibility to guide the investigator in his choice of method.

We have tried to meet this responsibility in two widely differing ways. In Parts D, E and F we have marshalled the philosophical and quantitative reasons for a choice. The sophisticated investigator may wish to start there. In Section 31 we have given as direct advice as we know how. This we hope will be useful to many in making their choices. Such choices, in the writer's present view, should be based on the extent of non-normality of distribution and inhomogeneity of variance present in the observations, so that we must open this chapter with a brief discussion of these dangers.

25. NON-NORMALITY OF DISTRIBUTION

In our discussion of the single and double classifications, we have implicitly assumed, as it is the custom to assume either explicitly or implicitly, that we could carry out our calculation "as if" the individual fluctuations were normally distributed.

The mathematician might wish to have "as if" replaced by "assuming" since only when the fluctuations are *exactly* normally distributed can he or the writer prove that the quantities we compute have *exactly* the properties we have asked them to have. To the statistician "as if" means that for fluctuation distributions *somewhere near* normality the calculated quantities will have *somewhere near* the properties asked for. He seeks for an understanding of:

(1) how near to normality is needed to keep the properties reasonably close to their nominal behavior; and

(2) when the fluctuations are non-normal what sort of non-nominality can be expected in the properties.

We shall return to (1) at the close of this chapter, pointing out here that for allowances the effect of excessively frequent large deviations is almost always to somewhat increase the error rates above their nominal values, — when we think we are working to 5% error rate we may be working to 6%, 7%, 8%, or 10%.

26. INHOMOGENEITY OF VARIANCE

The other conventional assumption, which we made explicitly in Section 5 was that m determinations had "substantially the same variance."

Again the mathematician would like to assume that the variances were identical, while the statistician would be concerned with the effects of small and moderate deviations from equality.

Leaving the question of how soon important effects set in aside for the moment, differences in variance from determination to determination almost always (probably always) raise the error rate, which of course becomes serious if the error rate among the comparisons, which seems to the writer to be of far less consequence. We shall return to this redistribution in Section 67.

27. THE SUBSTITUTE SYSTEM OF ALLOWANCES

When we are faced by serious non-normality of distribution or inhomogeneity of variance, it *may* pay us to use a substitute system of allowances. (This substitute system is based on the F-distribution rather than on the studentized range, and was introduced by Scheffé [68-69]).

Instead of calculating a *WSD*, which serves as an allowance for comparisons, and then using

$$\pm (WSD)\ \text{norm}_W \{c_i\}$$

as an allowance for contrasts and linear combinations, where

$$\text{norm}_W \{c_i\} = \max(\Sigma^+ c_i, -\Sigma^- c_i)$$

(note that since we shall shortly have another norm, it is no longer safe to omit the subscript W on "norm_W."), we have to calculate a different allowance for simple comparisons and multiply by a different norm.

If we restrict our attention to *contrasts*, that is to expressions

$$\Sigma c_i x_i \quad \text{with} \quad \Sigma c_i = 0\,,$$

which obviously includes all simple comparisons, then we calculate the

$$FSD = \text{fully significant difference,}$$

use $\pm FSD$ as an allowance for simple comparisons, and

$$\pm(FSD)\ \text{norm}_F \{c_i\}$$

where

$$\text{norm}_F \{c_i\} = (^1\!/_2\, \Sigma c_i^2)^{1/2}$$

as an allowance for the contrast $\Sigma c_i x_i$, with $\Sigma c_i = 0$.

If we wish to have allowances for all linear combinations, we calculate the

$$GSD = \text{generally significant differences}$$

as the *FSD* for $m + 1$ determinations, use ±GSD as an allowance for simple comparisons and

$$\pm(GSD)\ \text{norm}_F\{c_i\}$$

as an allowance for the contrast $\Sigma c_i x_i$.

Table 12

Values of [*FSD*/*LSD*] for a 5% Error Rate Familywise, and of Auxiliary Functions for Shifting 5% to any Rate Between 50% and 0.01% or Lower

$m-1$	m	[*FSD*/*LSD*]	A_F	B_F	C_F	D_F
1	2	1.00 + 0.0/*DDF*	.475 – .044*u*	1.09 + .68*u*	.202 + .0264*h*	.40
2	3	1.25 + 0.3/*DDF*	.362 – .033*u*	1.25 + .61*u*	.161 + .0157*h*	.53
3	4	1.42 + 0.8/*DDF*	.310 – .027*u*	1.36 + .57*u*	.136 + .0124*h*	.61
4	5	1.57 + 1.0/*DDF*	.276 – .024*u*	1.45 + .55*u*	.122 + .0104*h*	.68
5	6	1.70 + 1.3/*DDF*	.253 – .022*u*	1.53 + .53*u*	.110 + .0093*h*	.73
6	7	1.81 + 1.6/*DDF*	.234 – .020*u*	1.61 + .51*u*	.103 + .0083*h*	.78
7	8	1.91 + 1.9/*DDF*	.220 – .019*u*	1.66 + .50*u*	.096 + .0078*h*	.81
8	9	2.01 + 2.1/*DDF*	.208 – .018*u*	1.71 + .49*u*	.090 + .0074*h*	.83
9	10	2.10 + 2.4/*DDF*	.199 – .018*u*	1.76 + .48*u*	.086 + .0070*h*	.86
10	11	2.18 + 2.7/*DDF*	.190 – .017*u*	1.80 + .47*u*	.80 + .0067*h*	.90
11	12	2.26 + 3.0/*DDF*	.183 – .017*u*	1.84 + .47*u*	.077 + .0064*h*	.92
12	13	2.34 + 3.2/*DDF*	.176 – .016*u*	1.87 + .46*u*	.074 + .0061*h*	.94
13	14	2.41 + 3.4/*DDF*	.171 – .016*u*	1.90 + .46*u*	.072 + .0058*h*	.95
14	15	2.48 + 3.6/*DDF*	.166 – .016*u*	1.94 + .45*u*	.070 + .0056*h*	.97
15	16	2.55 + 3.8/*DDF*	.161 – .015*u*	1.96 + .45*u*	.068 + .0054*h*	.98
16	17	2.62 + 4.0/*DDF*	.156 – .015*u*	1.98 + .44*u*	.066 + .0052*h*	1.00
17	18	2.68 + 4.2/*DDF*	.152 – .015*u*	2.01 + .44*u*	.064 + .0050*h*	1.01
18	19	2.74 + 4.4/*DDF*	.149 – .014*u*	2.04 + .43*u*	.062 + .0048*h*	1.03
19	20	2.80 + 4.6/*DDF*	.146 – .014*u*	2.06 + .43*u*	.060 + .0046*h*	1.04
20	21	2.86 + 4.8/*DDF*	.142 – .014*u*	2.08 + .43*u*	.058 + .0044*h*	1.05
21	22	2.89 + 5.0/*DDF*	.139 – .013*u*	2.09 + .42*u*	.056 + .0044*h*	1.06
22	23	2.97 + 5.2/*DDF*	.137 – .013*u*	2.11 + .42*u*	.055 + .0044*h*	1.07
23	24	3.02 + 5.4/*DDF*	.134 – .013*u*	2.12 + .42*u*	.054 + .0043*h*	1.08
24	25	3.08 + 5.6/*DDF*	.132 – .013*u*	2.14 + .42*u*	.053 + .0043*u*	1.09
25	26	3.13 + 5.8/*DDF*	.129 – .013*u*	2.16 + .42*u*	.052 + .0043*h*	1.09
26	27	3.19 + 6.0/*DDF*	1.27 – .012*u*	2.17 + .41*u*	.051 + .0043*h*	1.10
27	28	3.24 + 6.2/*DDF*	.125 – .012*u*	2.18 + .41*u*	.049 + .0043*h*	1.11
28	29	3.28 + 6.4/*DDF*	.123 – .012*u*	2.20 + .41*u*	.048 + .0042*h*	1.12
29	30	3.33 + 6.6/*DDF*	.121 – .012*u*	2.21 + .41*u*	.047 + .0042*h*	1.13
30	31	3.38 + 6.7/*DDF*	.119 – .012*u*	2.23 + .41*u*	.046 + .0042*h*	1.14
35	36	3.59 + 7.4/*DDF*	.112 – .011*u*	2.26 + .40*u*	.043 + .0040*h*	1.17
40	41	3.82 + 8.2/*DDF*	.106 – .011*u*	2.32 + .39*u*	.040 + .0039*h*	1.20
45	46	4.01 + 8.8/*DDF*	.101 – .010*u*	2.36 + .39*u*	.037 + .0037*h*	1.22
50	51	4.20 + 9.5/*DDF*	.096 – .009*u*	2.40 + .39*u*	.035 + .0036*h*	1.24
60	61	4.54 + 10.8/*DDF*	.088 – .009*u*	2.44 + .38*u*	.030 + .0035*h*	1.28
80	81	5.15 + 13.2/*DDF*	.078 – .008*u*	2.53 + .37*u*	.023 + .0033*h*	1.35
100	101	5.70 + 15.3/*DDF*	.070 – .007*u*	2.58 + .37*u*	.018 + .0032*h*	1.41
200	201	7.80 + 24.8/*DDF*	.052 – .006*u*	2.74 + .36*u*	.009 + .0029*h*	1.49
500	501	12.0 + 40.0/*DDF*	.033 – .004*u*	2.87 + .36*u*	.002 + .0024*h*	1.59
∞	∞	∞+ (18*m* – 5)/*DDF*	.000 – .000*u*	3.27 + .26*u*	–.016 + .0020*h*	1.79

$m-1$ = degrees of freedom in numerator = number of determinations for *GSD*
m = number of determinations for *FSD*

The necessary values of [FSD/LSD], together with the factors required for error rate shifting are given in Table 12.
The latter are to be applied in connection with

$$u = \log_{10} r = -\log_{10} h$$

where the new error rate is $(5/r)\%$ or $5h\%$, and where for u positive we use

$$FSD_{(5/r)\%} = FSD_{5\%}\ [1 + u\ (A_F + \frac{B_F}{DDF})]\ , \qquad (r > 1)$$

while for u negative we use

$$\log_{10} FSD_{(5h\ \%)} = \log_{10} FSD_{(5\%)} + u\ (C_F + \frac{D_F}{DDF})\ , \qquad (h > 1)\ .$$

Examples of their use are as follows:

Beall [102, p. 246, Table 6] reports the results of an experiment by G. M. Stirrett and F. A. Stinson on the control of the tobacco hornworm *Phlegethontius quinquemaculata* Haw. by 6 insecticidal treatments in 6 replications. The counts (pooling plots treated alike in a block) were as follows:

	Treatment					
	1	**2**	**3**	**4**	**5**	**6**
Block 1	25	33	14	3	24	15
Block 2	22	15	18	3	12	16
Block 3	14	17	9	2	16	10
Block 4	11	5	11	2	4	13
Block 5	16	17	11	8	15	10
Block 6	14	14	11	8	19	17
Total	102	101	74	26	90	81

Treating this as a double classification we find

$TSSD = 505.7$

$DDF = 25$

$MSD = 20.2$

$$s^2 \text{ (totals)} = 121.2$$

$$s \text{ (totals)} = 11.0$$

$$[LSD/s\,] = 2.91$$

$$LSD = 32.0$$

$$[FSD/LSD\,] = 1.70 + \frac{1.3}{25} = 1.75$$

$$FSD = 56.0.$$

Thus we would be entitled to conclude, with a 5% error rate familywise, that, for example:

(1) treatment 1 exceeds treatment 3 in the number of insects surviving in a total over 6 blocks by

$$102 - 74 \pm 56 = 28 \pm 56 = \text{between } -28 \text{ and } +84;$$

(2) treatment 2 exceeds treatment 4 in the number of insects surviving in a total over 6 blocks by

$$101 - 26 \pm 56 = 75 \pm 56 = \text{between } 19 \text{ and } 131;$$

(3) the mean of treatments 1, 2, 5 and 6 exceeds treatment 4 in the number of insects surviving in a total over 6 blocks by

$$\frac{102 + 101 + 90 + 81}{4} - 26 \pm 56\sqrt{5/8} = 67.5 \pm 44.5 = \text{between } 23 \text{ and } 112\,.$$

If we wish an error rate of 1% rather than 5%, we have

$$r = 5$$

$$u = .70$$

$$A_F = .253 - .022\,(.70) = .238$$

$$B_F = 1.53 + .53\,(.70) = 1.90$$

$$A_F + B_F/DDF = .238 + \frac{1.90}{25} = .314$$

$$FSD_{1\%} = 1.22\, FSD_{5\%} = (1.22)\,(56) = 68.3.$$

Next, for comparison, we notice that if we had applied Section 16, we should have found

$$TSSD = 505.7$$

$$\sqrt{TSSD} = 22.5$$

$$\text{factor for totals} = 2.13$$

$$WSD\ \text{(totals)} = 47.9.$$

In this case ±48 would have replaced ±56 in (1) and (2), and ±46.6 in (3). With the amount of non-normality of distribution and inhomogeneity of variance which appears to be present in the original data, it seems reasonable to use the *WSD*. Thus the example illustrates the use of the *FSD*, but *not* the need for its use.

Finally, it should be noticed that it would be wise to apply a transformation to the insect counts before making either type of analysis. After transformation, the *WSD* would surely be safely applicable.

28. SIGNIFICANCE PROCEDURES

All the procedures so far discussed have been ***confidence*** procedures. The result has been a statement about, for example $x_2 - x_7$, — a statement that it lies between certain limits. Many investigators have been educated, far too many of them by statisticians, to use significance procedures, which answer "significant or nonsignificant?" instead of answering "how much?" as do the confidence procedures.

Part D is mainly devoted to a discussion of the pros and cons of significance vs. confidence. Suffice it here to say that the writer comes out strongly in favor of confidence in almost all cases. And let it be noted that many of his colleagues have drifted from significance toward confidence, but, to his knowledge, none have gone the other way.

It must be admitted, however, that there are exceptional situations in which significance procedures are necessary, and that there are a certain number of investigators who wish to use significance procedures where they had better not. In either case, the reader with such interests is referred to Part F.

29. ADVICE

We now attempt to produce concentrated and useful advice:

(1) avoid the short-cut procedures in the presence of substantial non-normality or inhomogeneity of variance;

(2) avoid the half-cut procedures in the presence of very extreme non-normality or inhomogeneity of variance.

In most individual fields of application, less than 20% of the analyses will involve "substantial" non-normality or inhomogeneity, and less than 10% will involve "extreme" non-normality or inhomogeneity, in the sense in which these terms are used here. Here the only cost of conservatism, as expressed by avoiding the short-cut or half-cut procedures, is a little more computation.

Thus the loss due to conservatism is small, *unless* the additional work means that some or all of the data is unanalyzed. In almost every situation the writer knows or imagines, it would be better to apply the short-cut procedure to *all* the experiments, rather than applying the full procedure to half of them. The doubtful reader is invited to analyze a few (or a few dozen) experiments both ways, compare them, and then make up his mind.

As between the *WSD* and the substitute *FSD* or *GSD* we conclude:

(3) if non-normality of distribution or inhomogeneity of variance is very extreme, the use of the substitute allowance may be wise;

(4) otherwise, the use of the substitute allowance is definitely wasteful.

In most individual fields of application, less than 5% of all analysis will involve "very extreme" behavior in the sense used here.

In this decision, conservatism is sometimes necessary, but often seriously wasteful.

The basis for these words of advice are to be found in the latter portion of Part E.

CHAPTER 6: OUTLINE AND SUMMARY

30. INTRODUCTION

Having finished our cookbook instructions for the simple cases, it is now time to complete Part A by giving a summary and outline of the contents and results of the entire paper.

We began with an introduction (§1), and account of the classical notions (§2), and continued with brief accounts of the new notions (§3), and of the distinction between batches and families (§4).

31. PART A

This part gives an account for the user, consisting of cookbook instructions for the use of allowance procedures and a summary of this and the remaining parts.

Chapter 1 gave a user's summary account of the calculation and use of allowances, beginning with an introduction (§5), and continuing with an account of the procedure for families (§6) and batches (§7) of determinations.

Chapter 2 discussed the applications of allowances to balanced single classifications (§8), treating the classical form (§9), the condensed form (§10), the short-cut form (§11), and the half-cut form (§12), and closing with a discussion of designing for prescribed accuracy (§13).

Chapter 3 discussed the applications of allowances to balanced double classifications (§14) treating the classical form (§15), the condensed form (§16), the short-cut form (§17), and the half-cut form (§18), and, again, closing with a discussion of designing for prescribed accuracy (§19).

Chapter 4 returned to the general applications of the procedure, and discussed extensions to graphical presentation (§20), to unequal variances in known ratios (§21), to error rates other than 5% for conventional analyses (§22) and short or half-cuts (§23).

Chapter 5 took up the cautions to be observed in the use of allowances, and the alternative procedures which may be followed (§24), discussing non-normality of distribution (§25) and inhomogeneity of variance (§26) as an introduction to the substitute system of allowances based on the *F* distribution (§27). The possible use of significance procedures is briefly mentioned, and the interested reader is (somewhat reluctantly) referred to Part F (§28). The chapter concludes with the best concentrated advice that the writer can supply (§29).

Chapter 6 summarizes the contents and results of the entire account (§30), treating Parts A (§31), B (§32), C (§33), D (§34), E (§35), F (§36) and G (§37).

32. PART B

Part B treats the principles and concepts which underlie the procedures set out in Part A.

Chapter 7 discusses the logical and philosophical consideration involved in parallel allowances. Beginning with a discussion of batched determinations and error rates per batch, the discussion continues with some illustrative and hypothetical examples (§40 and 41), and a demonstration of the irrelevance of independence, or lack of it, to error rates per batch (§42). In conclusion, error rates batchwise are introduced and studied, and it is suggested that "batchwise" be adopted as the standard basis (§43).

Chapter 8 carries forward the discussion for comparative allowances, beginning with the analogies with the batched case (§44), and going on to discuss the *LSD* and error rates per comparison (§45), intermediate bases (§46) and the graphical presentation of comparative error rates (§47). It next extends comparative allowances to contrast allowances, beginning with the definitions of a contrast and its norm (§48), and going on to point out that comparative allowances inevitably correspond to contrast allowances (§49). The application to coefficients in orthogonal expansions is touched on (§50).

Chapter 9 discusses discrepancies, including the one-way discrepancies which are constant multiples of ordinary deviations from the mean (§51), and the two-way and higher discrepancies which are constant multiples of interactions (§52). Numerical examples show how the use of discrepancies simplifies the treatment of interactions (§53) and may be used to avoid competition and overlapping between them (§54). This discussion provides the only seemingly reasonable technique for dealing with individual interactions known to the writer. The chapter closes with a mention of discrepancies related to fitted curves (§55).

Chapter 10 extends the analysis to universal allowances, beginning with a simple device of introducing a dummy determination not subject to fluctuation (§56) which settles the extension, and with a discussion of graphical presentation (§57), and continuing with a substantial number of examples from various fields.

Chapter 11 discusses the terminology used in this account, and its implications covering "allowances" (§64), "error rates" (§65), "per *X*" and "*X*-wise" (§66), and what attitude toward the inhomogeneous case is consistent (§67).

Chapter 12 discusses the historical background of allowances, pointing out the lack of novelty of the ideas and components used (§68), and then discussing confidence (§69), "per *X*" (§70), "*X*-wise" (§71), independence and non-independence (§72), range-based short-cuts (§73), approximate factors (§74), range in the numerator (§75), stragglers (§76), and multiple comparisons in general (§77).

33. PART C

This part presents the derivations and assesses the qualities of the approximations which underly the previous treatments.

Chapter 13 discusses the quantitative model which underlies the definition of allowances and develops the connection with % points of familiar or unfamiliar statistical distributions. It begins with a description of the quantitative model and a discussion of the role of distributional assumptions, such as normality in theoretical and applied statistics, concluding that security is only to be had when normality is one of many possible calibrating assumptions, all of which would lead to similar results (§78). All the "per *X*" allowances are then shown to be related to various % points of Student's $|t|$ (§79), while allowances batchwise lead to % points of the studentized maximum modulus (§80) and comparative allowances familywise lead to % points of the studentized range (§81). The automatic extension from comparisons to contrasts (§82) is followed by the examination of the related universal allowances and their connection with the studentized augmented range (§83). The chapter closes with the relation of the substitute allowances to the *F*-distribution (§84).

Chapter 14 deals with elementary relations and inequalities connecting the unfamiliar distributions with their more familiar analogs. The relation between maximum modulus, range and augmented range and the resulting inequalities are first established in general (§85) and then discussed numerically for the case of normality (§86), and the case of two determinations is then treated separately (§87). The studentization of these distributions is discussed in general (§88) and, numerically, for the special case of normality (§89). Finally the generality of the results is stressed(§90).

Chapter 15 deals with the approximate factors used to shift error rates, first to needed accuracy (§91), and then as to actual accuracy: for $|t|$ (§92), for $\sqrt{F}$ (§93), for the studentized range (§94) and for the studentized maximum modulus (§95), concluding with a further comment on needed and attained accuracy (§96).

Chapter 16 deals with the approximate factors used to shift multiplicity (to go from one determination or comparison to many), beginning with the reduction of "per *X*" factors to error rate shifting (§97), and then treating: the batchwise factor [*PA*/*IA*] (§98), the hindsight factor (§99), and the omnibus factor (§100).

Chapter 17 deals with the principles and numerical values underlying the short-cut methods. Beginning with an account of the use of a range multiplied by a factor as roughly equivalent to an *s* on an equivalent number of degrees of freedom (§101), it goes to discuss the factor (§102), the equivalent *DDF* (§103), error rate shifting (§104) and an example (§105).

Chapter 18 deals with the problem of designing experiments with known accuracy, beginning with a discussion of in what terms the accuracy is to be known, which concludes that, say 90% confidence that the allowance will be no wider than a given amount is the desired state (§106). The numerical procedure for simple designs is set forth (§107) and examplified (§108).

34. PART D

This part attempts to present the general considerations which affect the writer in choosing between significance and confidence or between different procedures. These considerations are of necessity of a philosophical nature, but of similar necessity are semiquantitative and rest on scientific experience. The guiding thread is the notion of definiteness of conclusion.

Chapter 19 discusses and classifies current statistical methodology and some of its sources (§110), beginning with an emphasis on science, and more specifically physical sciences, as a prime source of ideas and principles (§111), and emphasizing the distinction between administrative and scientific decisions — between decisions "for the moment" and those "for the record" — between wise gambles and established conclusions (§112). The central importance of definiteness of conclusion is stressed (§113), and its reflections in description (§114), significance (§115), confidence (§116), and the unstatisticated case (§117) are discussed. These considerations are contrasted with those underlying intelligent gambling (§118) and summarized (§119).

Chapter 20 deals with the separation of the *aims* of statistical procedures into action, indication and sanctification (§120), discussing the aims for which *p*% considerations are appropriate (§121), the reasons for a clear distinction between risk balancing and the use of $p\%$ (§122), the choice between sanctification and indication (§123), the role of sanctification in confusing scientific and administrative considerations (§124), and examples of combinations between the classifications of Chapters 19 and 20.

Chapter 21 deals with the considerations involved in the choice between significance and confidence procedures, beginning with the role of other variables (§126), and the causes of excessive pressure on the probability levels of significance procedures (§127). Next comes an attempt to argue the case for significance (§128), leading to a strong feeling in favor of confidence. The more specific applications to the multiple comparisons problems (§129) lead to the identification of straggler spotting as an example, possibly unique, where significance is to be preferred (§130). The lack of association of well-known statisticians with either side of the significance-confidence issue is emphasized (§131) and the conclusions of the chapter are summarized (§132).

Chapter 22 discusses the criteria for choice among various procedures of the same type and for measuring accuracy of conclusion, beginning with the power concept for significance procedures (§133), continuing with the tightness concept for confidence procedures (§134), and closing with an indication of the need for comparing the importance of various types of contrasts (§135).

Chapter 23 deals with this question of the importance of various contrasts (§136) emphasizing its diffuseness and difficulty, beginning with simple comparisons (§137), and continuing with stragglers (§138) and interactions (§139). The importance of subgroup means is discussed first in terms of apparent representativeness (§140) and then in other cases (§141). After a brief mention of regression coefficients (§142), the conclusion is reached that simple comparisons are outstandingly important, far more than any other kind of contrast, although stragglers, or perhaps interactions, are important in special cases (§143).

35. PART E

This part treats systems of allowances providing chosen error rates familywise, discussing various possibilities and their merits and demerits.

Chapter 24 discusses the general behavior and structure of universal and contrast allowances, taking the G-allowance as a leading example, and thus introducing the representation by vectors in an m-dimensional space, the *standard region* in this space corresponding to a given set of statements, and the corresponding *extent* which measures the extensiveness of any set of fluctuations.

The relations between a family of statements about linear combinations and a standard region are studied (§145) and each is found to determine the other. By using simple comparison as a reference standard, the notions of norm, extent, and size are freed of arbitrary size factors (§146). After introducing an arbitrary joint distribution of fluctuations, invariant under permutations of the determinations, studentization leads to a procedure in principle for setting allowances of any given shape through the upper $p\%$ points of a key statistic.

Chapter 25 discusses eigensystems of familywise allowances, initially covering the type of allowance proposed by Cornfield, Greenhouse and Halperin, which is particularly effective against straggling determinations, in terms of its basic geometry (§148), and of approximate % points of its key statistic for normality (§149). The chapter continues with eigensystems and eigenallowances in general, beginning with their definition as that symmetrical system of statements giving the tightest statement for some contrast (§150), continuing with an approximate calculation of the working % points of the corresponding key statistics (§151), and the properties and existence of a minimax regret allowance (§152).

Chapter 26 discusses the competition among various allowances in respect to tightness of statement, and the possibilities of compromise, beginning with a brief comparison of the relative tightness of the *C*-, *F*- and *W*-allowances for a few selected types of contrast (§153) followed by the introduction of compromises between any two allowances (§154), an inquiry into the cost of compromises (§155), the definition of compromise norms, and their values for a collection of reference contrasts (§156), and closes with a numerical comparison of the resulting allowances (§157).

Chapter 27 introduces the discussion of the effects of non-normality of fluctuation, almost entirely in the special framework of the single classification. An introduction (§158), is followed by a discussion of those cases depending, exactly or approximately, on the extreme % point of $|t|$ (§159). Comparative allowances in a precise sense are shown to require the *W*-numerator, the use of the *F*-numerator for 20 determinations being equivalent to spending an error rate of only 0.18% (for 10 *DDF*, but only 0.0001% for ∞ *DDF*) on simple comparisons (§160). The work of Pitman and Welch, in connecting *F*-procedures with randomization is reviewed, and found unlikely to have direct extension (§161). The average effects of one extreme fluctuation are then studied, first for a fixed set of fluctuations under randomization for *F* (§162) and *W* (§163) alone, and then combined (§164). Two extreme fluctuations are then considered (§165).

Chapter 28 considers the problem of non-normality further, shifting the background upon which one or two extreme fluctuations are superposed from *rc* arbitrary values to a normal sample (§166). The results are summarized for comparisons (§167) and a possible but difficult road to improvement identified (§168). Matters are summarized for contrast allowances (§169), and it is concluded (§170) (i) that *F* is far more insensitive than *C* which is slightly better than *W*, (ii) that the conventional denominator is safer than the half short-cut, which is safer than the full short-cut, and (iii) that compromises inherit most of the sensitivity of their most sensitive component.

Chapter 29 concludes the comparison of various allowances (§171) discussing inhomogeneity of variance (§172), simplicity of use (§173), case of interpretation (§174), flexibility of application (§175), and an unimportant property (§176). The summary (§177) goes a little against the compromises, and the balance between tightness where it matters and insensitivity to non-normality becomes all important (§178). The final decision is to use *W* (in some form) unless the dangers of non-normality and inhomogeneity enforce the use of *F* (§179).

36. PART F

This part attempts to give a coherent account of the many and various significance procedures. The guiding thread is the systematic use of the auxiliary concept of "apparent significance," whose use in a restricted form is due to Duncan [17,22].

Chapter 30 lays the general framework in which the procedures are to be placed (§180), a framework which implies a declaration of "significant" or "nonsignificant" for every group of determinations that can be made out of the family. The concepts of self-consistency of these declarations (§181) and of apparent significance (§182) lead to the definition and discussion (§183) of multilayer significance procedures. Next there are introduced the notions of apparent significance levels (§184), apparent and overall null hypothesis (§185), and overall and standard significance levels (§186). Next the case where only simple comparisons are of interest is discussed briefly (§187). This is followed by a discussion of the graphical representation for $m = 3$ (§188) and graphical examples of simple procedures (§189, §190), and the chapter closes with an acknowledgement of indebtedness to Duncan (§191).

Chapter 31 deals with the various systems of significance levels proposed to date, introducing $\gamma = \ell n\,(1-\alpha)$ as a convenient tool (§192), and treating constant apparent levels (§193) constant standard levels (§194), the independent comparisons basis (§195), the extreme case (§196), and their comparison (§197).

Chapter 32 discusses procedures involving ranges through studentized range tests of apparent significance (§198), beginning with Newman's procedure (§199) and Keuls' formulation of it (§200), and going on with a very attractive procedure (§201) and the quality of an interesting approximation in it (§202). Other possibilities are considered, including a range analog of Duncan's procedure (§203), Paulsen's top-down procedures and range-in-numerator versions of F (§204).

Chapter 33 discusses procedures involving F-tests of apparent significance, beginning with Duncan's procedure (§205), and discussing complexities of interpretation (§206), guiding considerations (§207), minimum weighted risk (§208), and how Duncan's procedure fits into the scheme of multilayer significance procedures (§209). This is followed by a discussion of the protected *LSD* (§210), other possibilities (§211), and the classical F-test (§212).

Chapter 34 deals with procedures involving gaps as indicators of apparent significance, beginning with a discussion of the procedure discussed in BIOMETRICS by the writer, covering some of its defects (§213), including those pointed out by John Mandel and Jerome C. R. Li, and showing how it can be slightly modified to become a multilayer procedure. The difficulties of gap procedures are discussed (§215) and a very attractive combined procedure is proposed, for which only rough approximations to critical levels are at present available (§216).

Chapter 35 deals with procedures involving discrepancies or deviations in tests of apparent significance and with other procedures for straggler spotting (§217). It begins with a discussion of the original Cornfield, Greenhouse and Halperin procedure (§218), and then points out how it may be converted into a multilayer procedure (§219). The relation between various distributions, and Moshman's procedure are then mentioned (§220).

Chapter 36 discusses briefly the various issues raised in previous chapters of this part, covering range vs F (§221), systems of significance levels (§222), the concomitant use of gaps (§223), recommendations when ordinary significance procedures are necessary (§224) and straggler spotting (§225).

37. PART G

Part G contains reference materials.

PART B
WHAT IS BEING DONE

CHAPTER 7: ALLOWANCES FOR BATCHES

38. INTRODUCTION

Having provided the cookbook account, it is now time to discuss the principles, concepts and mathematical results which underlie the techniques just discussed. In the present part we are going to concentrate on principles and concepts, on logical and philosophical considerations, leaving mathematical and probabilistic considerations for Part C. Here in Part B our treatment will be general enough to cover short-cut and half-cut procedures as well as more conventional ones, without further ado.

As noted earlier, we shall begin with batches of determinations, and proceed to families, since batches are somewhat simpler to treat. We do this recognizing that families are our main concern, but endeavoring to keep our discussion on as elementary level as we can, in order to keep the essential ideas as clear and simple as possible.

39. SINGLE AND BATCHED DETERMINATIONS

We begin by considering the case where (i) we have a number of determinations of different quantities, (ii) we recognize that, like all determinations, ours are subject to fluctuations, (iii) there is no thought of comparing the different determinations among themselves, and (iv) we have satisfactory information about the possible magnitude of the fluctuations. (Just what (iv) means will become clear shortly.) This case is simple, but the essentials of choice of basis stand out as well here as in the more complex cases.

It is the nature of fluctuations (or of errors, if you prefer that word) that they do not have a fixed upper limit — that arbitrarily large fluctuations occur with suitably small probabilities. This means that if x_i is the observed determination, then no matter how large we choose A, the statement "the result of leveling out all fluctuations lies between $x_i - A$ and

$x_i + A$" which we shall abbreviate to "the determinand lies in $x_i \pm A$" is not *certainly* true. By choosing A large enough, we can make the probability of its failure as small as we like, but we always must accept some risk of error. The most convenient and serviceable way of expressing this risk of error seems to be as an *error rate* — by stating that if we repeat such a statement thus-and-so many times we shall, in the long run, make errors at such-and-such a rate per repetition.

When the fluctuations are reasonably distributed, and knowledge of the magnitude of the fluctuations takes the form of an estimated (or known) variance, the problem of finding an A such that the determinand will be in $x_i \pm A$ a given percentage of the time is the classical problem solved by Student's "t." In Part C we shall return to Student's t and obtain specific solutions, limited to the case of reasonable distributions and estimated (or known) variance, to these problems which we are about to discuss in general. For the present, however, generality will gain us some simplicity, so that we will merely assume that (iv) at the start of this section means that, given $p\%$, there is a known way to calculate a number $A_{p\%}$ so that the true value will fall in $x_i \pm A_{p\%}$ except for $p\%$ of all cases (except for an event of probability $p\%$). We shall not need, for the present, to concern ourselves with how $A_{p\%}$ is calculated.

We have considered one type of error rate, defined by its numerator and denominator as follows:

$$\text{error rate per determination} = \frac{\text{number of statements untrue}}{\text{number of statements made}} .$$

We have not fixed $p\%$, once and for all, nor do we intend to. It is admitted, both by statisticians and by users of statistics that circumstances should alter $p\%$, although all realize that 5%, 1% and 0.1% are particularly popular values. Some would say that in defining an error rate of p per 100 determinations ($p\%$ per determination) we have used new and clumsy language to describe a $(100-p)\%$ confidence interval. If we were to stop with the one classical kind of error rate associated with a single, possibly complex, statement this position might have some justification, but, even in such a situation, emphasis on 5% rather than on 95% and on error rather than confidence has certain substantial advantages.

Let us next consider the man who always makes his determinations in batches of m. There are very many situations where he should be interested in numbers of wrong statements per batch rather than per determination — for it may be that one bad apple spoils the barrel. The error rate *per batch* is clearly to be defined as the long run value of the ratio

$$\frac{\text{number of statments about individual determinations in error}}{\text{number of batches } (= \frac{1}{m} \text{ times number of determinations})}.$$

Clearly error rate per batch = m times error rate per determination, so that, for $m = 6$, matters are as in the following examples:

Allowances	Error rates ($m = 6$)	
$A_{5\%}$	5 per 100 determinations 30 per 100 batches	5% per determination 30% per batch
$A_{1\%}$	1 per 100 determinations 6 per 100 batches	1% per determination 6% per batch
$A_{(5/6)}\%$	5 per 600 determinations 30 per 600 batches	(5/6)% per determination 5% per batch

In general, then, we require an allowance of $A_{(p/m)\%}$ in order to reach $p\%$ per batch This allowance is larger than $A_{p\%}$, which sufficed for $p\%$ per determination.

Should the user operate in terms of "error rate per batch" or "error rate per determination"? This question is intrinsically like the question about which p to use in $p\%$ in that its answer must depend on circumstances. But its ramifications are deeper and wider because it can and should be answered from two points of view, namely:

(1) Are two users with different values of m adopting procedures of "equal severity"?

(2) Does the statement of severity made by the user give the appropriate impression to his customers (who may include himself)?

and these answers should be given before attacking the question of what p to use.

40. ILLUSTRATIVE CASES

We shall try to clarify the issues involved by discussing a few artificial examples, beginning with the case of a medical laboratory determining blood sugar on samples of blood from persons more or less suspected of diabetes. We shall find conditions here under which either type of error rate seems appropriate.

First, as far as normal operation and individual patients go, it would seem entirely appropriate to use error rate per determination, since the doctor's action on John Jones would not depend on the other 19 determinations made at the same time by the same technician or on the other 47 determinations on samples from patients in Smithville. Each determination is an individual matter, and it is appropriate to set error rates accordingly.

Second, suppose that a medical investigator is going over the records with a certain fixed level, L, of blood sugar in mind, so that each case is classified as "low," "high" or "doubtful" according as $x_i + A < L$, $x_i - A > L$, or $x_i - A \leq L \leq x_i + A$. He is interested in the number of definite misclassifications between "low" and "high," and if he is studying cases in batches according to the technician involved, or according to the time between taking of the sample and analysis, for example, he will wish to allow for the fact that fluctuations of the known intensity will, on the average, produce a more extremely deviant value in a larger batch. Thus he will wish to set his A in terms of error rate per batch.

As a very different example consider a commercial testing laboratory and two situations it might regularly face. In the first, a batch of 10 determinations of the same characteristic are made, one on each of 10 samples. If the 10 samples come from 10 different customers, there is no reason at all for the testing laboratory not to report in terms of error rate per determination.

But, on the other hand, suppose that the testing laboratory is acting as an umpire between a suspicious buyer and a suspicious seller, each of whom will undoubtedly make check measurements in its own laboratories, with the intent of changing umpires whenever it finds too many "mistakes." Suppose further, that 10 characteristics of the material in question are to be determined, and that all 10 enter into the agreed-upon price. If the testing laboratory reports allowances based on, say, an error rate of 5% per determination, then nearly half the lots tested will be reported with at least one mistake. The discrepancy between 5% and 50% will clearly lead to trouble. In such a case, an error rate per batch would be more desirable than one per determination.

41. SIGNIFICANCE EXAMPLES

In dealing with batches, the distinction between significance and confidence is far less important than when dealing with families. Not because confidence has ceased, in the writer's opinion, to be more useful than significance, but because there is a perfect parallelism between significance procedures and confidence procedures. (For examples to the contrary for families, compare Part F, especially Chapter 32, Section 201). Thus, while significance examples might be misleading in distinguishing error rates per comparison from those per family, they may serve us well in distinguishing error rates per determination from those per batch.

Let us consider two sets of graduate students of education — each set is about to descend on a different Naval Air Station with psychological tests — each wishes to study the usefulness of their tests at predicting success as naval aviators, each will have only enough time to test and follow-up one class of about 100 students. One set is bringing a batch of 3 tests, the other set is bringing a batch of 250. Neither wishes to compare tests, they merely want evidence that some are useful. How should they set error rates? Let us examine the consequences of different choices when all the tests are useless and have zero predictive power.

Many investigators would feel that the natural and conventional thing to do would be to choose an error rate of 5% per determination. What would be the consequences of such a choice? The group with 3 useless tests would find an average of $3(0.05) = 0.15$ tests "significantly predictive of success in Naval aviation training." Only about once in seven such trips would any of their tests, all assumed useless as a basis for discussion, show false significance. The consequences for the group with 250 useless tests are more startling. They may expect to find an average of $250(0.05) = 12.5$ tests "significantly predictive of success in Naval aviation training"! On essentially every such trip, at least one test would appear falsely significant!

To the question "Are the two investigators using procedures of "equal severity" or "equal caution," the writer responds with a hearty "no." A procedure which is almost certain to make several errors each time that it is used is far from being equally cautious with one which makes occasional errors!

If, on the other hand, an error of 5% per batch were selected, then the average number of "significantly predictive" tests in either study would be 0.05. There would, however, be no guarantee that the 250-test study would find evidence that any tests were useful — perhaps, as a consequence, there would be no guarantee of a corresponding set of Ph.D's or Ed.D's for the graduate students with the 250 tests.

Which choice seems most likely to convey the situation to the consumers — the readers of the theses or of the paper or papers to which they lead? If the choice of error rate per determination were made, we could imagine the following statistically accurate footnote, which might be added to the report on the 250-test study — "We have found 18 (say) of our 250 tests significantly predictive of performance as naval aviators (at the 5% level). It must, of course, be kept in mind that we would expect an average of 12.5 tests to reach this level of significance by pure chance, and it is far from unlikely that as many as 18 would reach this level of significance if all 250 were totally devoid of predictive value." If this footnote would have no effect on the readers, then the choice of error rate per determination would have given a fair impression to the consumers. But I suspect that such a footnote would surprise and shock many readers and I regard such surprise and shock as strong evidence that the choice of "per determination" did not make a fair impression.

Now for the other side. There are two general arguments which can be identified with the labels "fruits of their labors" and "but think of the power." The first of these argues as follows: The 250-test study involves 83 times as many tests, 83 times as many statements of predictive value, and 83 times as much work in test administration. Therefore this group is entitled to make 83 times as many errors in its final statement as is the 3-test group. This doctrine of "equal errors for equal work" or "if you try enough different things you're entitled to a corresponding number of errors" seems fallacious and utterly repugnant to the writer. The aim of science, pure or applied, may vary between establishment of truth and picking a good gamble — but neither amount of unilluminated work nor number of varied attempts entitle the scientist to a corresponding number of errors. I reject this argument completely.

Then there are those who would talk of the loss in power — those who often, and sometimes wisely, point out in simple situations that a 10% or 20% error rate may be justified. In most cases, the ruling concepts here are "decision" and "null hypothesis" and when they are relevant, this argument has considerable weight. But in posing this problem, we deliberately avoided this question by stating that the studies sought evidence, not decision. The problem of decision vs. evidence is discussed further in Part D (Chapter 19).

There is another aspect which is often not brought up explicitly. It may be that the 250-test study is really the first phase in a program of *selection*, that the 25 tests which seemed to do best on the first class will be tried again on a second, and so on. In this case there would at first seem to be ground for assessing the evidence about each test without regard for any of the others, and thus for choosing error rate per determination (= per test). However, the problem of selecting has been analyzed as a whole [Cochran 1951, Yates 1950] and the present conclusion is that it is best done as an organized gambling scheme, picking the so-many apparently best at each stage without using any specific error rate. Selection is an important problem, but is not the place for allowances.

Having concluded that the testing example should lead to error rate per batch, we can summarize the writer's position as follows:

(A) The choice between error rates per determination and per batch depends on circumstances.

(B) Where the determinations are to be used really separately, error rates per determination are likely to be appropriate.

(C) Otherwise, and in all doubtful cases, error rates per batch (or batchwise) should be used.

(D) The best test of doubtfulness is to write the hypothetical footnote and consider its effect on the consumer.

42. INDEPENDENCE AND NONINDEPENDENCE

It has been some time since we wrote down

$$\text{error rate per batch} = m\,(\text{error rate per determination})$$

and some readers have no doubt worried about the absence of any statement about independence throughout the intervening section. In fact, no independence within batches is required! All we need to assume is independence between batches (or even only between superbatches, if there is batch-to-batch lack of independence). Take as an example $m = 7$ and an error rate of 5 per 100 determinations. Consider 100,000 batches (this is a large enough number so that "long-run" or "on the average" values will be reasonably accurate). In about 5,000 of these batches $x_1 \pm A_{5\%}$ will be in error (the true value will not be in $x_1 \pm A_{5\%}$). In about 5,000 of these batches $x_2 \pm A_{5\%}$ will be in error. No matter whether these 5,000 are exactly the same as, partly the same as, and partly different from, or all different from the first 5,000, there will be 10,000 errors among the first two determinations in each batch. Continuing in this way, 7(5000) = 35,000 errors (= false statements) will occur in the 100,000 batches and the error rate will be

$$\begin{aligned} \text{35,000 per 100,000 batches} &= \text{35 per 100 batches} \\ &= m(\text{5 per 100 batches}) \end{aligned}$$

since $m = 7$. This calculation is entirely independent of whether or not there is dependence *within* batches.

43. RATES BATCHWISE

There is ground for a slightly different sort of approach. There are users who might reasonably say: Once there is one false statement in a batch, I don't care how many more there are, what I want to control is the number of batches containing at least one false statement. The answer to this demand is error rate *batchwise*. We may contrast this with the other two bases so far considered by writing down the corresponding fractions:

$$\text{per determination} \qquad \frac{\text{false statements}}{\text{determinations}},$$

$$\text{per batch} \qquad \frac{\text{falsestatements}}{\text{batches}},$$

$$\text{batchwise} \qquad \frac{\text{batches with one or more false statements}}{\text{batches}},$$

In this last case, dependence does matter. If we have the greatest possible positive dependence, then all statements in a given batch are good or bad together, in this case

$$\text{batchwise} = \text{per determination} \qquad \text{(for complete positive dependence).}$$

If we have the greatest possible negative dependence, and the error rate is not too large (so that when one error is large all other errors are small), then all batches have zero false statements or one false statement, and

$$\text{batchwise} = \text{per batch} \qquad \text{(for strong negative dependence).}$$

These are the limits between which "batchwise" must lie.

Numerical computation will show us later that "batchwise" and "per batch" are very close indeed when "batchwise" is set in terms of what seems to be the natural standard — in terms of independent fluctuations in the various determinations of a batch. Thus there is no need for a further discussion about general pros and cons, since "batchwise" and "per batch" are equivalent for most practical purposes. The choice between them is a minor choice, perhaps even an esthetic one.

Besides the reaction of the user who adheres rigidly to the "one bad apple spoils a barrel" point of view, there is another reason to favor measuring error rates batchwise. If we think of the m determinations collectively, as components of one grand vector determination, then we should consider the m conclusions of the form $x_i \pm A_{5\%}$ as components of one grand vector conclusion $\{x_i \pm A_{5\%}, i = 1, 2, \ldots, m\}$. If we tally whether or not this composite vector conclusion is correct, we are tallying an error rate batchwise. (This point of view is developed further in Part E, Chapter 24). By taking error rates batchwise, we are really talking about a simultaneous confidence statement, if we wish to phrase things that way.

CHAPTER 8: COMPARATIVE AND CONTRAST ALLOWANCES

44. THE ANALOGOUS RATES

We come now to the case where we have m determinations which we wish to compare. Collectively they make up a family. We write $x_1, x_2, \ldots, x_m$ for the observed determinations, and observe that there are $m(m-1)/2$ observed differences of the form

$$x_i - x_j$$

so that there are $m(m-1)/2$ statements of the form

$$\mu_i - \mu_j \text{ is in } x_i - x_j \pm A$$

where μ_i and μ_j are the determinands corresponding to the determinations x_i and x_j. It is important to notice that $m(m-1)/2$ becomes rather large for moderate values of m. Thus, for example,

$m =$	2	3	4	5	10	15	20	25
$m(m-1)/2 =$	1	3	6	10	45	105	190	300

so that 50 to 300 comparisons are not uncommon.

By analogy with the rates set up for batches in the last chapter, it is not difficult to propose a number of kinds of error rates for families. Some of these are

$$\text{error rate per family} = \frac{\text{false comparisons}}{\text{families}}$$

$$\text{error rate familywise} = \frac{\text{false families}}{\text{families}}$$

$$\text{error rate per determination} = \frac{\text{false comparisons}}{\text{determinations}}$$

$$\text{error rate per degree of freedom} = \frac{\text{false comparisons}}{\text{degrees of freedom}}$$

$$\text{error rate per comparison} = \frac{\text{false comparisons}}{\text{comparisons}}$$

where "number of" may be prefixed to all numerators and all denominators, where a comparison is false if $\mu_i - \mu_j$ does not fall in $x_i - x_j \pm A$, where a family is false by definition if one or more of its comparisons are false, and where

$$\text{families} = \frac{\text{determinations}}{m} = \frac{\text{degrees of freedom}}{m-1} = \frac{\text{comparisons}}{m\,(m-1)/2}\,.$$

To which of these shall we give serious consideration?

By analogy with the conclusions of the previous chapter, we would expect to make *error rate familywise* the standard. This the writer does (and so has everyone who has used the F-test in an analysis of variance).

Those who believe in equal errors for equal work will naturally prefer error rate per determination. (As previously remarked, the writer rejects this position).

Those who are used to fixing their attention on $m-1$ independent contrasts in discussing the comparison of m determinations will fix their eyes on the $m-1$ degrees of freedom between the m determinations and, less specifically, on a set of $m-1$ independent contrasts. The "independence" of these contrasts is a mathematical concept, depending, for example, on the precise equality of fluctuation variances for the various determinations. Many, however, have come to feel that this corresponds to a logical independence, and that the experimenter should look at some such set of independent contrasts, and should look at each separately. (The present writer fails to understand such a position.) Given this doctrine, the experiment to compare m determinations becomes a source of $m-1$ contrasts for separate evaluation, and some support is given for an error rate per degree of freedom. Such a choice has been made by Duncan [17-22] and the issues are discussed further in Parts D and F (Chapters 21 and

31). Undoubtedly there are a few experiments where some one set of independent contrasts are natural and where it is appropriate to judge the separate contrasts individually, but the writer doubts that there are many.

In summary, the writer feels that error rates familywise should be standard, and that in the few exceptions something resembling error rates per degree of freedom or per determination would be appropriate.

45. PER COMPARISON AND THE *LSD*

We mentioned error rates per comparison, which may be tenuously related to error rates per (parallel) determination. The only logical connection seems to be by the philosophy that errors are allowed in proportion to the number of possible statements considered. (A horrible point of view.) If it were not for two other considerations, this would be all we would need to say about error rates per comparison.

The first consideration is the regrettable frequency with which error rate per comparison is used by those who have the *LSD* — the *least significant difference*. In some fields, e.g., see [1], the standard way to report on m comparative determinations is to give the m observed values and the *LSD*, which is the individual allowance for a single comparison of two particular determinations. Though it is not always stated, and even sometimes gently warned against, the elusive implication is always that it is proper to judge *any* comparison of two determinations by using the *LSD* as an allowance. This is no more and no less than using error rate per comparison.

The second consideration is that it seems good computation engineering to calculate the *LSD* on the way to many of the allowances which correspond to an error rate familywise.

There are occasional situations where error rate per comparison is appropriate. One such would be a situation where the true, secret purpose of the experiment was to compare 2 particular, preselected determinations, and the other $m-2$ have been added for camouflage — whether the camouflage is against a laboratory director, a competitor, or someone else. There is only 1 comparison of interest and that is identified in advance. The other $\frac{1}{2}m(m-1)-1$ are just dummies, and should be neglected. But such cases of appropriate use of the *LSD* are rare (other uses will be discussed in Chapter 34, in connection with gap analysis).

This example points up the distinction between the different kinds of error rates from another point of view. When one comparison of interest is selected after the values are available, it is quite likely to be that comparison which came out furthest from its true value. If the same identical comparison had been preselected, it would have been natural and appropriate to use the *LSD*. The larger allowance needed otherwise is an allowance for ***hindsight*** in the selection. With this in mind it is not unnatural to define the

$$\text{hindsight factor} = \frac{\text{allowance for error rate familywise}}{\text{allowance for error rate per comparison}} = [WSD/LSD].$$

On the rare occasions when we need a name for this ratio, we shall follow this suggestion.

46. INTERMEDIATE BASES

We have already mentioned error rates "per determination" and "per degree of freedom." Clearly various versions of "determinationwise" and "degree-of freedomwise" could be invented — to little purpose. All of these yield allowances of about the same size — larger than those "per comparison" and smaller than those "per family" or "familywise." Any one representative of this intermediate group of bases is about as good as any other.

We note that a very convenient intermediate basis can be implicitly defined by the use of the allowance *intermediatewise*, where

$$ISD = \frac{WSD + LSD}{2}$$

trusting that few will wish to use it.

47. GRAPHICAL PRESENTATION

Having prepared the ground for numerical calculation of allowances, we need also to consider presentation of results. Tabular presentation is easy, but graphical presentation requires a moment's thought. The scheme which seems best is to represent each determination by a bar of length equal to the allowance, as in Figure 1. As shown in that figure, comparing similar ends of two bars gives the apparent comparison (14 in the figure) while comparing dissimilar ends gives the limits obtained by adding and subtracting the allowances ($14 - 10 = 4$ and $14 + 10 = 24$ in the figure). Since only comparisons are being considered for the moment, no fixed vertical scale should be shown on such a figure.

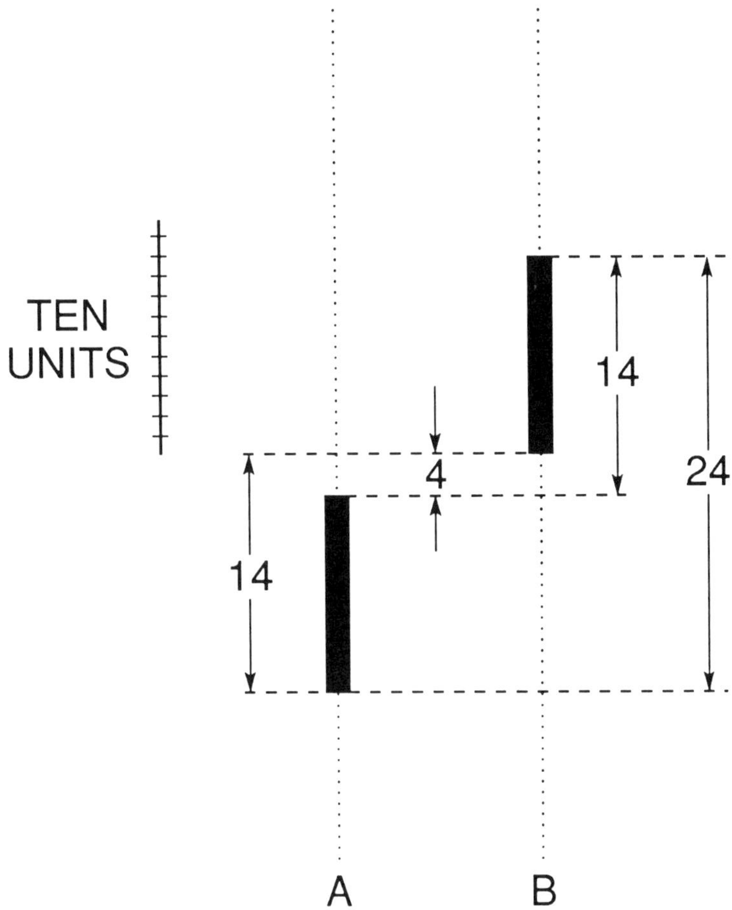

Figure 1. Graphical presentation of $B - A = 14 \pm 10$ (between 4 and 24). (Guide lines and dimensions added)

48. NORMS AND CONTRASTS

Once we are prepared to compare x_i with x_j, which means to consider $x_i - x_j$ (the observed difference) as telling us about $\mu_i - \mu_j$ (the difference in true values or determinands), we almost inevitably come to consider not only $x_i - x_j$, but also $x_1 - 2x_2 + x_3$, $-2x_1 - x_2 + x_4 + 2x_5$, $3x_1 + 2x_2 + x_3$, x_k, $\sqrt{2} - \pi x_2$, and so on, as telling us about the corresponding linear expressions in the true values. All of these are linear expressions in the given determinations (with coefficients which do not explicitly depend on the results of the particular experiment), and we would like to know how to extend our allowance procedure to cover them.

We consider, then, a general linear expression

$$c_1x_1 + c_2x_2 + \ldots + c_nx_n = \Sigma\, c_ix_i = \Sigma\, a_ix_i - \Sigma\, b_ix_i$$

where in the last formulation,

(i) all a_i and b_i are nonnegative; and

(ii) for each i, either a_i or b_i or both are zero.

(The latter condition can be expressed as $a_ib_i = 0$.) We are going to be particularly concerned with

A = sum of positive coefficients = sum of positive $c_i = \Sigma^+ c_i$

and

B = absolute value of sum of negative coefficients = – sum of negative c_i

$= -\Sigma^- c_i$.

If these two sums are equal, we call the linear expression a *contrast*. In any event, the larger of these two will be called the *norm* of the linear expression. A few examples may be helpful, and are given in Table 13.

It is perhaps worthwhile to note that the term "norm" is used by the mathematicians for any of a number of expressions meeting certain requirements. The "norm" we are discussing is one such, especially suited to our situation. In Chapter 3 we distinguished it as $\text{norm}_W\{c_i\}$ to separate it from

$$\text{norm}_F\{c_i\} = \tfrac{1}{2}(\Sigma\, c_i^2)^{1/2},$$

which was useful in connection with the substitute procedure for setting allowances. In Part E we shall have occasion to deal with still other norms.

Table 13

Examples of Contrasts and Norms

Examples	A	B	Contrast?	Norm
$x_i - x_j$	1	1	yes	1
$2x_i - 2x_j$	2	2	yes	2
$x_1 - 2x_2 + x_3$	2	2	yes	2
$-2x_1 - x_2 + x_4 + 2x_5$	3	3	yes	3
$3x_1 + 2x_2 + x_3$	6	0	no	6
x_k	1	0	no	1
$-2x_k$	0	2	no	2
$2x_1 - \pi x_2$	2	π	no	π

49. CONTRAST ALLOWANCES

If $y = \Sigma\, c_i x_i$ is a contrast, and $\eta = \Sigma\, c_i \mu_i$ is the corresponding true value, then it might seem natural to consider all statements of the form

$$(y - \eta) \leq (WSD)(\text{norm}\{y\}) \qquad (*)$$

which is equivalent to

$$y - (WSD)(\text{norm}\{y\}) \leq \eta \leq y + (WSD)(\text{norm}\{y\}) \qquad (**)$$

since these statements reduce to the familiar form when y is a simple contrast, since

$$\text{norm}\{x_j - x_k\} = 1 \;.$$

It will be shown in Part C that if (*) or (**) holds for all simple comparisons, it also holds for all contrasts. We know that (*) or (**) holds for all comparisons in all but $p\%$ of all families. Hence (*) or (**) holds for all contrasts in all but $p\%$ of all families.

Thus we may use

$$(WSD)\ (\text{norm}\{y\})$$

as an allowance for all contrasts with an error rate of $p\%$ familywise. As one listener once put it, in a doubtful and slightly injured tone — "Does that mean that you can mull the data around as much as you like"? To such questions the answer is a flat — "Yes, so long as you stick to contrasts and keep the norm in mind."

After the simple comparisons, the most significant classes of contrasts seem to be measures of discrepancy, deviation or interaction expressing the tendency of an individual determination to deviate from what would be expected from the results of other determinations. We shall discuss these cases in the next chapter.

There is one point, however, which deserves a brief word at this point. To some, the most interesting contrasts are comparisons between subgroup means. But the norm of

$$\frac{x_1+x_2+x_3+x_4}{4}-\frac{x_5+x_6+x_7}{3}$$

is 1, the same as the norm of $x_2 - x_5$. We are clearly giving no credit for the extra stability of means of several determinations in comparison with that of a single determination. At this point we intend only to say that we recognize this fact, and that pros, cons and implications will be discussed in Part D (Chapter 23) and again in Part E (Chapters 26 and 29). The writer is very content with this situation.

50. ORTHOGONAL COEFFICIENTS

Another source of contrasts is least-square fitting of linear expressions — that is expressions linear in the constants or parameters to be fitted. Such fitting can always be expressed in terms of orthogonal functions with considerably increased simplicity, both of computation and of interpretation. The most familiar case is, of course, that of orthogonal polynomials, but the situation is perfectly general.

If the expression contains a constant term, then we may, and will almost always choose to, take a constant as the first orthogonal function. If in addition, there are no weights (that is, all weights are equal, and may as well be put equal to unity), then the fitted coefficients of all other orthogonal functions are contrasts.

CHAPTER 9: DISCREPANCIES

51. DEVIATIONS AND DISCREPANCIES

If x in

$$x, 21, 27, 26, 22$$

represented a missing observation in a set of 5 taken under the "same" conditions, and if we felt compelled to supply a value for x, we would undoubtedly choose the mean of the other observed values, in this case 24. If instead of x we had the value 49, so that our observations ran

$$49, 21, 27, 26, 22$$

we might reasonably choose to measure the apparent unusualness of this value in two ways. We might calculate either the *deviation* of 49 from the mean, 29, of *all* 5 values, finding

$$49 - 29 = 20$$

or the *discrepancy* of 49 from the mean, 24, of the *other* 4 values, finding

$$49 - 24 = 25.$$

In general, if there are a total of k values, we shall have

$$\text{discrepancy} = \frac{k}{k-1}\ (\text{deviation}),$$

so that we can always change from one to the other easily. Both are obviously contrasts. We shall find it convenient to emphasize discrepancies, among other reasons because the discrepancy alters in a 1:1 rate with an alteration in the corresponding determination.

52. INTERACTIONS AND DISCREPANCIES

What we have just said about deviations and discrepancies will perhaps seem trite and tasteless to many. Our main purpose in discussing it was to lay a foundation for more complex cases. Consider the case of observations arranged by rows and columns, where something is in common in each row and something else is in common in each column. How do we naturally fill in a missing value? Table 13A illustrates the usual and reasonable procedure.

Table 13A

Filling in a Missing Value in a 2-Way Table

Row values			Means for boxes	
10	12	x	11	x
14	18	22	24	29
23	27	28	Suggested value for x	
30	32	37	$11 + 29 - 24 = 16$	

The philosophy is simple. In the same column with x the mean value was $(22+28+37)/3=29$, while in the other columns the row containing x had values which were, in mean, smaller than those in the comparison rows. The amount of this difference in mean is $24 - 11 = 13$. Thus it is reasonable to select $29 - 13 = 16$ as the value which might be filled in for x.

Thus the *discrepancy* caused by x is

$$x - 29 - 11 + 24 = x - 16$$

which is to be compared with the conventional *interaction*

$$\begin{aligned} x - \frac{x+22+28+37}{4} - \frac{x+10+12}{3} + \frac{x+10+\ldots+37}{12} &= \frac{1}{2}x - 8 \\ &= \frac{1}{2}(x-16). \end{aligned}$$

The mutual relation, if there are c columns and r rows, is easily seen to be

$$\text{interaction} = \left(1 - \frac{1}{c}\right)\left(1 - \frac{1}{r}\right)(\text{discrepancy})$$

and the norm of the discrepancy is easily shown to be 2, as may be seen in the example, since there

$$\text{discrepancy} = x - \frac{10+12}{2} - \frac{22+28+37}{3} + \frac{14+18+23+27+30+32}{6},$$

where the sum of the positive coefficients is

$$1 + {}^1/_6 + {}^1/_6 + {}^1/_6 + {}^1/_6 + {}^1/_6 + {}^1/_6 = 2$$

and that of the absolute values of the negative coefficients is also

$${}^1/_2 + {}^1/_2 + {}^1/_3 + {}^1/_3 + {}^1/_3 = 2.$$

In strict analogy with the one-way case, the sum of squares for interaction is the sum of the squares of the interactions, while the mean square for interaction is the mean of the squares of the (two-way) discrepancies.

We can continue the process for 3-way, 4-way, . . . , h-way tables, and we find that the norm of an h-way discrepancy is always 2^{h-1}, while its sensitivity (to changes in the determination) is unity.

In summary, then, an h-way discrepancy, apparent or "true,"

(1) is a fixed multiple of the corresponding interaction;

(2) contains the corresponding entry with coefficient +1; and

(3) requires an allowance 2^{h-1} times as great as the original entries.

53. NUMERICAL EXAMPLE

In order to illustrate *both* the simplicity of computation of multi-way discrepancies *and* the nature of certain other difficulties, which we shall shortly thereafter overcome, with interactions and multi-way discrepancies, we discuss next a numerical example. This artificial example is treated in Table 14, where Panel 1 gives original values and means, Panel 2 gives row-wise discrepancies, Panel 3 gives (as an alternative computational route) column-wise discrepancies, and Panel 4 gives 2-way discrepancies. Sample arithmetic runs as follows:

$^3/_2\,(10-16)=-9,\qquad {^3/_2}\,(12-16)=-6,\qquad {^3/_2}\,(26-16)=15.$

$^4/_3\,(10-19)=-12,\qquad {^4/_3}\,(12-22)=-13.33,\qquad {^4/_3}\,(26-28)=-2.67$

$$^4/_3\,(-9-(-6))=-4={^3/_2}\,(-12.00-(-9.33)),$$

$$^4/_3\,(-6-(-1.5))=-6={^3/_2}\,(-13.33-(-9.33)),$$

If the allowance for the original entries (Panel 1) were, say, 4, that for one-way discrepancies (Panels 2 and 3) would also be 4, while that for the two-way discrepancies would be 8. In that case, the "true" discrepancy for $i=3$, $j=1$ should be asserted to lie between $10-8=2$ and $10+8=18$.

Table 14

First Analysis of Interaction Example
(Panel 1 — original entries, Panel 2 — row-wise discrepancies,
Panel 3 — column-wise discrepancies, Panel 4 — two-way discrepancies.
All margins give corresponding means.)

(Panel 1)

10	12	26	16
14	18	22	18
23	27	28	26
29	31	36	32
19	22	28	23

(Panel 2)

–9	–6	15	0
–6	0	6	0
–4.5	1.5	3	0
–4.5	–1.5	6	0
–6	–1.5	7.5	0

(Panel 3)

–12.00	–13.33	–2.67	–9.33
–6.67	–5.35	–8.00	–6.67
5.33	6.67	0.00	4.00
13.33	12.00	10.67	12.00
0	0	0	0

(Panel 4)

–4	–6	10	0
0	2	–2	0
2	4	–6	0
2	0	–2	0
0	0	0	0

54. COMPETITION AMONG INTERACTIONS

Conventional analyses, which emphasize orthogonal contrasts, have always been rather unclear about interactions, and the nonorthogonality of the interactions seems to have been a major contributing cause. We have had no difficulty in setting up a procedure with known error rate, but we need to inquire further into the effects of nonorthogonality. Table 15 shows a very illustrative set of imagined data and the corresponding discrepancies. If the allowance were ±1 or ±2 as might seem reasonable in view of most of the determinations, then we would conclude that none of the discrepancies could be zero. This conclusion comes from the extreme deviance of just *one* determination — that one which has the value 120. This conclusion is likely to be legally correct — since by our definition of interactions and discrepancies, this one determination enters into each interaction and discrepancy — but it is not convenient or useful. Only one determination is discrepant, and we should like to come out with an expression of this fact.

Table 15

Example of Determinations and Discrepancies to Illustrate Mutual Relation of Interactions

Determinations			Discrepancies		
120	0	0	120	–50	–60
0	0	0	–40	20	20
0	0	0	–40	20	20
0	0	0	–40	20	20

Knowing from a large discrepancy that the corresponding determination is likely to be unusual, we should like to calculate modified discrepancies for the other cells which are freed from the influence of this determination.

Thus, in the numerical example previously discussed (Table 14), if we really believe that x_{31} and μ_{31} are behaving otherwise than as predicted from the other entries, we should like to exclude x_{31}. The simple way to do this is to drop either the first row *or* the third column and reanalyze. Table 16 shows the results. Panel 5 gives the apparent discrepancies dropping the 3rd column, Panel 6 those dropping the first row and Panel 7 the mean of these results. The differences between Panel 7 (Table 16) and Panel 4 (Table

14) show the effects of dropping x_{31} and μ_{31} from the prediction of other x_{ij} and μ_{ij}. The most extreme discrepancies on reanalysis are –3.0, 1.8 and –1.8 in comparison with –6.0, –6.0, –4.0 in the previous analysis. The reanalysis is more nearly fair, and suggests that the other discrepancies may be rather small, perhaps even quite near zero.

Table 16

Continuation of Interaction-Discrepancy Example-Analysis with Missing Value. (All entries are two-way discrepancies. Panel 5 — without 3rd column. Panel 6 — without first row. Panel 7 — mean of previous panels, rounded to 1 decimal. All margins give corresponding means.)

(Panel 5)				(Panel 6)				(Panel 7)			
1.33	–1.33	—	0	—	—	—	—	1.3	–1.3	—	0.0
–1.33	1.33	—	0	–1.50	0	1.50	0	–1.4	0.7	1.5	+0.2
–1.33	1.33	—	0	0.75	2.25	–3.00	0	–0.3	1.8	–3.0	–0.5
1.33	–1.33	—	0	0.75	–2.25	1.50	0	1.0	–1.8	1.5	+0.3
0	0	0	0	0	0	0	0	0.2	–0.2	0.0	0.0

It is easy to show that the norm of such modified and combined discrepancies remains 2^{h-1} (= 2 in this example) as before.

A comment on this procedure is in order. First, we chose to knock a particular entry out of consideration because we were, in some sense, suspicious of it. And then we reanalyzed. Ordinarily there would arise a question about the effects of repeated tests of significance. In our procedure that question does *not* arise. We are entitled to pull out ***any and all*** simple contrasts, and examine them separately and carefully. We have spent our error rate once and for all for the privilege of setting allowances on ***all*** contrasts and are entitled to mull over the x_{ij} as much as we wish.

55. DISCREPANCIES FROM CURVES

In addition to discrepancies from means, or from h-way additive structures, it is easy and natural to define and use discrepancies from fitted curves. Thus for example, if we have values of y for $x = 0, 1, 2, 3, 4$, we may fit a straight line to any four of these values and then compare the fifth value with that predicted from the corresponding line. Table 16a shows an example.

Table 16a

Example of Discrepancies from Lines

		Fitted lines omitting					
x	y	$x = 0$	$x = 1$	$x = 2$	$x = 3$	$x = 4$	Discrepancy
0	3	(2.5)	3.8	2.9	3.1	4.1	0.5
1	6	5.4	(6.1)	5.7	5.9	6.3	−0.1
2	8	8.3	8.4	(8.5)	8.7	8.5	−0.5
3	10	11.2	10.7	11.3	(11.5)	10.7	−1.5
4	15	14.1	13.0	14.0	14.3	(12.9)	2.1

In every case, the discrepancy is a contrast, thus

$$0.5 = 3 - 2.5 = 3 - (1\,(6) + {}^{1}/_{2}\,(8) - {}^{1}/_{2}\,(15))$$

$$-0.1 = 6 - 6.1 = 6 - (.57\,(3) + .29\,(8) + .14\,(10))$$

which we see as soon as we express the fitted value in terms of the other individual observations.

In this particular example, the norms of the contrasts are $^{3}/_{2}$, 1, 1, 1 and $^{3}/_{2}$ respectively. There does not seem to be any very simple general formula, and it will probably be advisable to work out the coefficients and norms separately in each case.

Similar computations may be made for discrepancies from parabolas and other more complex curves. We leave the details to the reader.

CHAPTER 10: UNIVERSAL ALLOWANCES AND COLLECTED EXAMPLES

56. THE SIMPLE DEVICE

So far we have only considered contrasts — linear expressions in the determinations with coefficients summing to zero — but there are various reasons why we might like to include other linear expressions among those to which the allowances apply. Some reasons are:

(1) The determinations themselves are not infrequently of interest as well as the contrasts between them.

(2) The results of polynomial or similar fitting and interpolation or extrapolation frequently yield linear expressions which are not contrasts.

(3) Other linear expressions which are not contrasts are sometimes of interest.

(4) A desire for mathematical symmetry.

All of these are of some mild value, but the gains they indicate would not warrant much in the way of compensating disadvantages.

We shall find that a simple device will allow us to make our allowances applicable to linear expressions, without great labor in derivation or appreciable loss in use.

We add a $(m+1)st$ determination x_o, to the m determinations, $x_1, x_2, \ldots, x_m$ which we already have, and define x_o to be identically zero. Any linear expression in $x_1, x_2, \ldots, x_m$ can easily be written as a *contrast* in the augmented set, thus

$$3x_1 + 2x_2 + x_3 = -6x_o + 3x_1 + 2x_2 + x_3,$$

$$\sqrt{2}\,x_1 - \pi x_2 = (\pi - \sqrt{2})x_o + \sqrt{2}\,x_1 - \pi x_2$$

and we see that we may always do this without changing the norm (the coefficient of x_o can be taken as $\Sigma\, b_i - \Sigma\, a_i$ in terms of our representation identifying positive and negative coefficients).

This new determination differs from the other m in one respect — it is always equal to its true value — which is obviously zero. Symbolically, we shall say that we are dealing with $m+$ determinations. There will be a little surprise when we learn for each specific case

$$(\text{allowance for } m) < (\text{allowance for } m+) < (\text{allowance for } m+1),$$

but there may be some surprise that the center and left-hand allowances are very nearly equal. We shall find for $m > 3$ that they are so nearly the same that separate tabulation of factors is not worthwhile. (For $m = 2$ we find that the 5% allowance for 2+ is about 2% longer than for 2. This difference is just barely worth tabulation.) What we have tabulated for $m > 3$ may also, then, be considered to be to practical accuracy

$$[USD/LSD] = \frac{\text{allowance for error rate familywise}}{\text{allowance for error rate per comparison}}$$

where the *USD* = usually significant difference is by definition the basis of a universal allowance of the form

$$\pm\,(USD)\ \text{norm}\ c_i$$

applicable to *all* linear expressions.

57. GRAPHICAL PRESENTATION

We wish a form of graphical presentation in which this common allowance is set forth simply, at least for both comparisons and determinations. We have already proposed a way to handle comparisons — we need to extend it to determinations — we must add a scale. If we add a single scale, we will be using only half as much allowance for determinations as for comparisons, and this is definitely not what we wish to do.

Figure 2 expounds a solution. It consists of vertical scales differing only by a vertical displacement equal to the allowance. The vertical bar corresponding to determination *A* has two ends, and each can be read on each of the two scales. Two of these readings yield 46, the apparent result, while the other yield the limits $46 - 10 = 36$ and $46 + 10 = 56$. Thus, just as in comparing two determinations, we can obtain the apparent value twice, and the two limits once each.

Figure 3 shows a more realistic example with 6 determinations and an allowance that is not simply related to 10. In this graphical procedure, it seems to the writer, we have the first simple procedure for graphically representing both uncertainties in comparisons and in values.

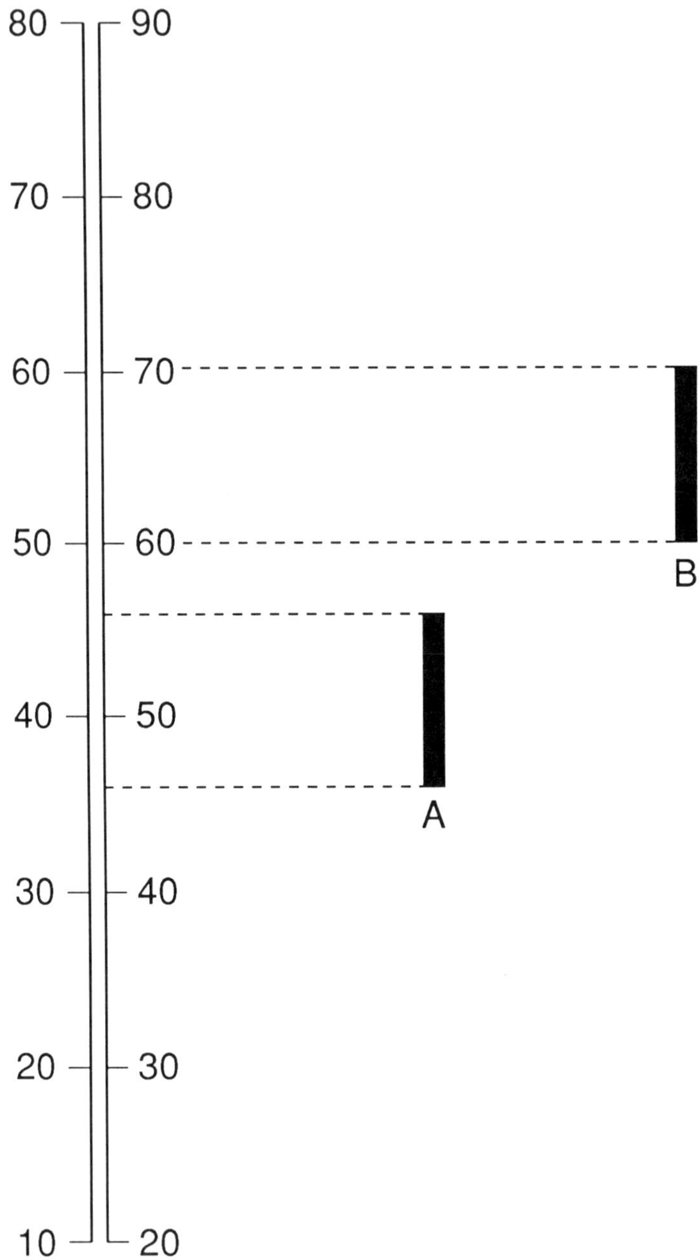

Figure 2. Graphical presentation of $A = 46 \pm 10$, $B = 60 \pm 10$ and $B - A = 14 \pm 10$. (Guide lines added)

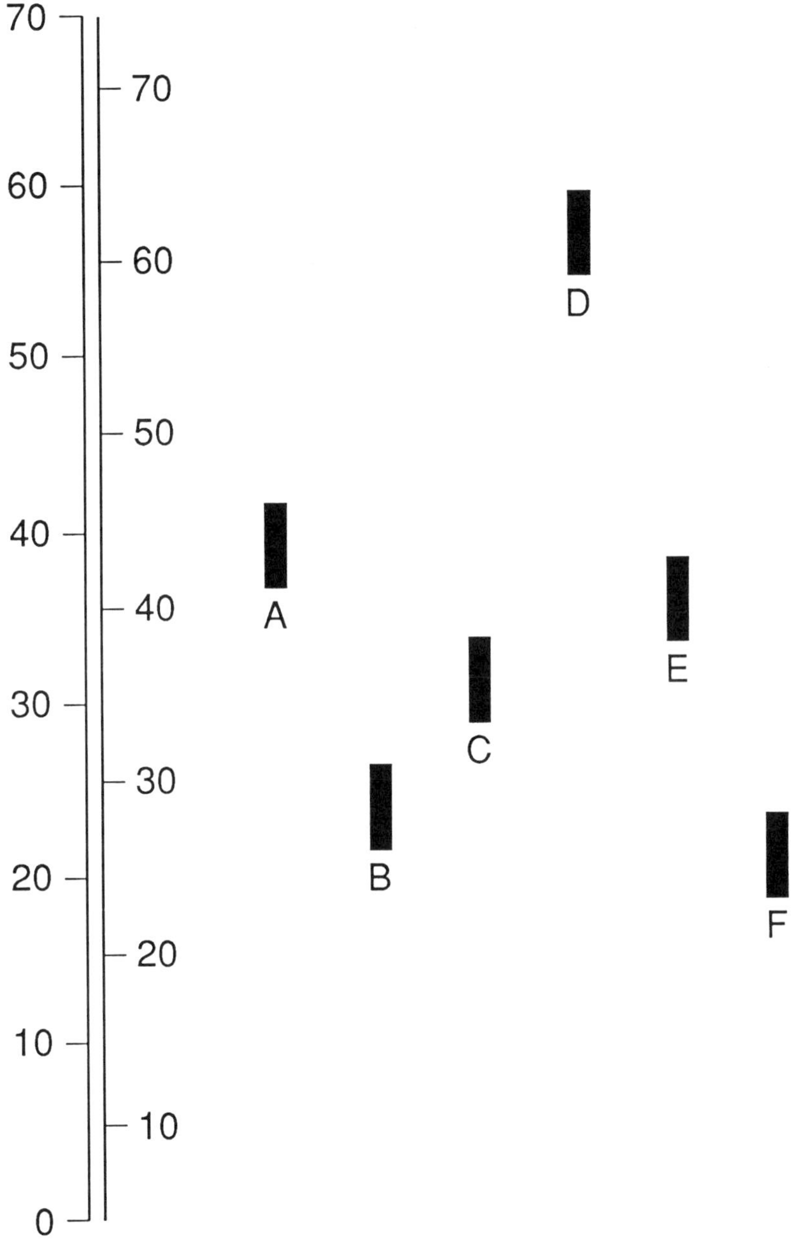

Figure 3. Graphical presentation of six determinations with allowance of 4.5 for comparison and position.

EXAMPLES

58. INTRODUCTION

In the remainder of this chapter we discuss a number of examples from a variety of fields which can be treated with the tools at our disposal. The emphasis is as much on the reasons for the selection of particular types of allowances and on the nature of the conclusions as on the computational aspects. The examples have been segregated by the type of analysis required, rather then by subject matter field.
The following examples seem never to have been prepared.

Section 59. Examples of use of parallel allowances.
60. Examples of use of comparative allowances.
61. Examples involving contrast examples.
62. Examples involving interactions.
63. Examples involving universal allowances.

CHAPTER 11: TERMINOLOGY AND IMPLICATIONS

64. ALLOWANCES

In this chapter we shall try to discuss briefly some of the reasons for our choice of terminology, and some of its implications. We have introduced a substantial number of new terms, and some explanatory discussion seems in order.

We have called our ±'s for the means, totals, comparisons or contrasts *allowances*. Why this choice? For one minor reason and one major one. The minor reason is that this is an intuitively reasonable word, which does not seem to have been preempted for some other use. The major reason is that "allowance" suggests "allowance for so-and-so" and in particular "allowance for the effects of such-and-such causes of variation." This seems to the writer to be a very desirable suggestion indeed. In any experiment or observational program, *internal* statistical analysis can only provide protection against those causes of fluctuation which were allowed a chance to manifest themselves in the experiment.

An experiment comparing 3 corn varieties in Iowa does not exhibit fluctuations due to changes in conditions from state to state. An analysis of its results will provide allowances which allow for the effects of many causes of fluctuation, but *not* for state-to-state changes.

An experiment comparing methods of chemical analysis in one laboratory does not exhibit fluctuations due to changes in practices or personnel from laboratory-to-laboratory. An analysis of its results will provide allowances which allow for the effects of many causes of fluctuation, but *not* for laboratory-to-laboratory changes.

An observational program comparing the weather consequences of certain (objectively-defined) weather patterns during a given winter does not exhibit fluctuations due to variations in general circulation from winter to winter. An analysis of its results will provide allowance which allow for the effects of many causes of fluctuation, but *not* winter-to-winter changes in general circulation.

Anything that might remind the user of the probable existence of such *systematic* errors or deviations seems worthwhile. Perhaps the term "allowance" may help here.

65. ERROR-RATES

In a statistical procedure which concludes with a single statement, as for example that a single interval contains a single five-mean difference, it is a matter of formal indifference whether we describe the chance of the statement being right or the chance of its being wrong. We can go back and forth among

> "we have 95% confidence that — lies in the interval,"
> "this is a 95% confidence interval,"
> "we should have 5% diffidence about — lying in the interval,"
> "this is a 5% diffidence interval,"
> "this interval has an error rate of 5%,"

with ease. The only reasons for choosing among them are psychological. Which produces the most useful reaction in the user?

In this simple case there seem to be two points of view among statisticians as to which produces in the user the reaction most useful to the statistician. There are those who feel that it is desirable to encourage the user to believe his apparent results by stressing the degree of "confidence" which may be placed in them. Certainly there have been such occasions, particularly where the user's background and thinking was qualitative rather than quantitative. But there are also those, and the writer is one, who feel that it is more often desirable to stress the chance of error:

(1) to discourage the user from overoptimism; or

(2) to make it clear to the user who has a strong quantitative "feel" that the statistical conclusions are really not more optimistic than his trained judgement; or

(3) to provide a basis to which additional error rates or allowances for systematic errors can be added.

Thus even in the case of a single statement, there *are* reasons to prefer the use of "error-rate" or some similar negative concept over "confidence" or any other positive concept.

Once we step away from single statements, there are great formal complications in making accurate and intuitively clear positive statements. (The reader can easily confirm this by trying to rephrase some of the early sections of this part in terms of numbers of correct statements rather than numbers of incorrect ones). We can avoid all these complications quite neatly and easily by using the idea of an "error rate," as we have seen. This would seem to be a conclusive argument.

66. PER *X* AND *X*-WISE

We have introduced several pairs of bases for error-rates which are very similar from the user's point of view, for example:

"per batch" and "batchwise"
"per family" and "familywise"

Was this duplication justified, since they are so similar? The writer's answer is a strong and vigorous "Yes, and particularly *since* they are so similar." The reasons for this view need a little elaboration, and exist at two levels of sophistication.

At the more elementary level, we notice that "per batch," "per family," or, in general, "per *X*" allowances can be obtained from simpler allowances merely by shifting error rates as we shall see in the next part. Not only in multiple comparisons, but in many other types of statistical problems, a "per *X*" solution can be had cheaply and rapidly. Such solutions are easily available. This is reason enough for the existence and use of "per *X*" bases. Next we notice that in many cases the neat and logically satisfying solution is an "*X*-wise" solution — it may be harder to reach, but is seems more closely what we want. Finally we notice that, for the case of independence in multiple comparison and for the natural cases in other problems, rates "per *X*" and "*X*-wise" *are* very nearly the same when either rate is small. This means that the two bases, one easy to jury-rig and the other more satisfying, lend support to each other.

At the more sophistical level we need to be concerned with the effects of non-normality of distribution which are discussed in more detail in Chapter 27. We shall summarize the relevant conclusions. Almost all statisticians are aware that extreme % points are greatly affected by non-normality. (Many realize that % points in the range from 5% to 1% are among the least sensitive.) Thus if we set up on a "per *X*" basis, it would seem that we are taking large risks with non-normality. In many cases, the technique best protected against non-normality is an "*X*-wise" technique (for example, the *FSD*). Again we have the easy-to-set-up and the desirable techniques, do they support each other? We know for normality (and independence) they come close to agreement, while for more general distributions we lack this assurance.

The situation can be sketched as in Figure 4. The *X*-wise error rate of the procedures calibrated to have a given per *X* error rate when the fluctuations are normally distributed varies, perhaps widely, as the fluctuations become less and less normally distributed. However, these error rates are always less than those of the corresponding procedures calibrated to have a given *X*-wise error rate when the fluctuations are normally distributed. These latter procedures, moreover, have *X*-wise error rates which vary only slightly as the fluctuations become less and less normally distributed. This property of *X*-wise procedures gives us confidence in the conservatism of the more easily available per *X* procedures.

In words, we may expect that a "p% per *X*" method, calibrated for normal distributions and used on non-normal ones may operate at "q% per *X*" where q is not close to p, but it will always be more conservative than a "p% *X*-wise" method, calibrated for normality. If this "p% *X*-wise" method is reasonably immune to non-normality, and operates at "r% *X*-wise" where r *is* near p, then we at least are sure that the "p% per *X*" method is conservative or almost conservative in comparison with a "p% *X*-wise" standard.

By this somewhat indirect route, the stability of "p% *X*-wise" against non-normality greatly supports the practical use of the "p% per *X*"

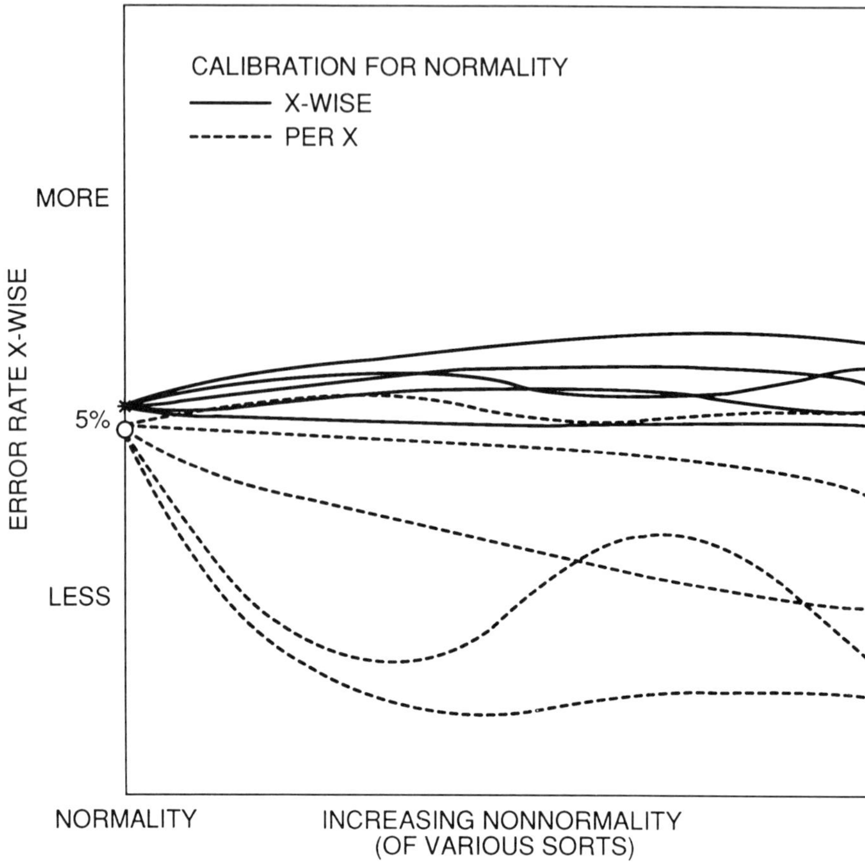

Figure 4. Non-normality and two bases for error rates.

method during the interim, which may be long, before tables are available for a "*p*% X-wise" method.

67. THE INHOMOGENEOUS CASE

We noted above (Section 26) that when there was unsuspected inhomogeneity of variance, our errors would tend to be concentrated in the determinations or comparisons of excessive variance. What attitude toward this apparent inequity is consistent with our general point of view?

Let us begin by examining a very simple case. One of the simplest is the case of 3 determinations of relative variance $1-u$, $1-u$ and $1+u$. If we use the same allowance on all three comparisons, and if u is large (say 0.5) then almost all the errors will be associated with the two comparisons of variance $2+u$ which will usually scatter far more than the comparison of variance $2-2u$. Is this unfair?

It seems to the writer that the answer must depend on the circumstances of the experiment or observational program. If, for example, the higher variance of the third determination is due to our deliberate planning, if with malice aforethought we had measured it less accurately than the other two, perhaps by repeating the individual measurement fewer times, then there is definite ground for a claim of unfairness. We have deliberately made the comparison of the first two determinations more accurate, so that this comparison has some claim to a shorter allowance. Some claim — but how strong a claim? Here the writer is still uncertain.

The opposite extreme arises when pure accident has diminished the accuracy of one determination, an accident which might equally well have happened to any of the other determinations. Here the writer denies the right of the accidentally less variable comparison to a shorter allowance. It may be wise under the circumstances to give it a differential allowance, but this is a matter of wisdom, not ethics or morality. If the difference in accuracy is small, say up to two-fold in variance, it may be wiser to maintain an equal allowance on the ground that the various comparisons are equally important. If the difference in accuracy is large, then different allowances will almost certainly be necessary.

Some would consider the "planned accident" a different case, although the writer does not. Suppose, as an extreme example, that we can "afford," because of pressure of time, money or materials, only 4 individual measurements. The best we can do with 3 determinations is to do one twice and the other two once each. If we select the one to be done twice by coin flipping or random numbers — by a "planned accident" — then the situation seems to the writer to be the same as for an unplanned accident, and if the differences in variance are not large he would be quite content to use an allowance of the same length for all three comparisons.

So long as we adopt an X-wise basis for our error rate, there would seem to be no injustice in a constant allowance for comparisons of different variance, for we are acting as if we had made one composite statement which we wanted to be true with the exception of an allowable rate of error. One error is much like another — any error makes the single composite statement wrong — why should we distinguish among errors?

The opposite point of view goes more naturally with the "per X" type of error rate. Here we are to count the number of individual statements in error. Each individual statement is an entity, and we should like to have nearly the same error rate for all the individual statements. In the case of inequality of variance we would hope for a corresponding inequality in allowance. Undoubtedly some statisticians and some users of statistical methods think this way — think in terms of individual statements, each of which should have its fair share, ***and no more***, of the overall error rate.

Once an area of investigation becomes sufficiently quantitative for an interest in contrasts other than comparisons to be important, this view is in difficulties. It is easy to extend "X-wise" from comparisons to contrasts, but no one has seen how to extend "per X." This is a real handicap, and perhaps the strongest argument against a "per X" — centered approach.

The problems we have discussed in this section seem likely to produce debate, heavy thought, perplexity and dissension among statisticians for some time.

CHAPTER 12: HISTORICAL BACKGROUND

68. INTRODUCTION

It is not the intention of the present account to claim or suggest that novel statistical ideas or components are here assembled into a coherent and workable approach to the multiple comparison problem and its relatives. The ideas and components are almost all old and well-known, the novel feature is their assembly. In this chapter we attempt to trace the history of some of these ideas and components and to list the other attempts which have been made to deal with multiple comparisons (further discussion of these other attempts is postponed to Parts E and F).

Our tracing is undoubtedly incomplete, and advice and suggestions as to missed references will be very welcome.

69. CONFIDENCE

The general notion of attaching more or less definite confidence to the conviction that the true value of some quantity lay in a certain neighborhood of its apparent value is of very respectable antiquity. The use of "probable errors" in surveying, geodesy, and astronomy goes back to Gauss and beyond. Only the more recent mileposts in its development are easy to recognize.

There seems to be two basic ideas involved in the modern idea of a confidence statement:

(1) the inversion of inequalities to convert them from statements about statistics to statements about parameters; and

(2) the recognition that the relevant standard error or sampling distribution is associated with the contemplated value of the parameter rather than with its apparent value.

In most cases the second of these is of central importance. In a few exceptional cases, the standard error or sampling distribution does not depend on the parameter in question, and (2) become irrelevant. Outstanding among these are the use of Student's t and the multiple comparison procedures discussed in the present account.

As Welch [82, 1939, pp. 58-59] has clearly pointed out, the first step had already been taken by Student in 1908 [71, p. 20], in connection with Student's t. In the present account we need rely only on the natural generalization of the procedure used by Student.

As Welch further points out, Student did not seem to recognize the importance of this procedure as a matter of principle, particularly since he attacked the related problem of estimating the correlation coefficient by way of inverse probability. Presumably, this was because of the lack of the second step, which seems to have been taken separately by E. B. Wilson in 1927 [84] (Wilson failed to combine the two steps).

Both a careful separation of confidence or fiducial argument from inverse probability and an appreciation of the importance of the principles involved were published by Fisher in 1930 [27] (the extent to which Ezekiel [cp. p. 534] had understood the situation is not clear from Fisher's paper), and followed-up in 1933 [28] with another example, wherein the words "Now this is a probability statement about the unknown parameter σ" appear on page 347.

The importance and usefulness of such procedures had apparently been recognized by Neyman by 1930 (see Clopper and Pearson [11, 1934, p. 404, footnote] for more complete references). However, Neyman's first publication on the subject dates from 1934 [52]. The divergence in approach of Fisher and Neyman began to become visible in the discussion of this paper, and grew steadily thereafter.

Their two divergent views can be characterized by two desires:

(1) to make an exhaustive statement about the parameter or parameters based on all the information in the observations; or

(2) to make some statement about the parameter or parameters within the framework of classical probability theory.

In the hands of Fisher, the first desire has led to considerations of fiducial probability. In the hands of Neyman, the second desire has led to the modern formulation of confidence procedures, intervals and statements. These two groups of ideas are now agreed upon as distinct.

It has been fashionable, at times, to allocate "priority" for the invention of confidence procedures to one statistician or another. More recently the fashion has been to allocate it to Neyman because of his definitive mathematical formulation of 1937 [54]. In earlier days it was the fashion to allocate it to Fisher (e.g., Neyman [52, 1934, p. 562 and footnote], Neyman [53, 1935, p. 111], Clopper and Pearson [11, 1934, p. 404], Pearson and Sukatme [61, 1935, p. 31], Wilks [83, 1938, p. 166], Bartlett [2, 1939, p. 129]). One purpose of the present brief account is to express the writer's view that "confidence," like so many other ideas in statistics, has grown through many stages, and that it would be a mistake to try to allocate priority to any one statistician.

70. ERROR-RATES "PER *X*"

The idea of using as a divisor the number of blocks, batches, etc. of events rather than the number of events themselves, must go back into the very early foundations of probability theory. By 1935, Fisher [29, Sec. 24] was indicating that a t-test at the $5/(^1/_2)m\,(m-1)\%$ level was reasonable for individual comparisons, so that the "per experiment" basis for simple comparisons was definitely in print. He did not explain why $^1/_2\,m\,(m-1)$, presumably thinking it clear, but several statisticians seem to have missed the point. Thus the application of "per *X*" rates to multiple comparisons is not new.

71. ERROR-RATES "*X*-WISE"

Walker's test for the significance of the largest periodogram ordinate [80A] was a definite use of the "X-wise" basis in 1914, as was Fisher's studentization of the problem in 1929 [26]. Fisher's distribution took the form

$$\text{Prob}\{\text{largest} > a\} = m\ \text{Prob}\{\text{single} > a\} - \binom{m}{2}\text{Prob}\{\text{pair} > a\}$$

$$+ \binom{m}{3}\text{Prob}\{\text{triple} > a\} - \ldots$$

in which the first term corresponds to a "per X" basis and the others to corrections back to an "X-wise" basis. While Fisher does not state this explicitly, it is certain that he understood this clearly at the time. He gave [26, p. 59] a table showing that the significance levels obtained by "using the first term only" were very close indeed to those obtained using the "exact formula." Thus, in our terminology, he showed explicitly for this case that "X-wise" and "per X" were practically identical.

The same pattern of in-and-out corrections have occurred in many distributions of the largest this or the largest that (cp. Cochran [12], Nair [47-48]). (Among probabilists it is known as the method of inclusion and exclusion, cp. Feller [23, Chapter 4]).

A table of batchwise limits for 4 error rates was included in Part II of "Tables for Statisticians and Biometricians" [62, 1931]. Thus it seems fair to say that the basic idea of an "X-wise" basis for an error rate has been available for more than twenty years, as have examples showing that "X-wise" and "per X" give practically equivalent results at and beyond the 5% point.

The use of "per X" as a substitute for "X-wise" in the outlier problem was suggested by Thompson [72] in 1935.

The novelties of the present account seem to lie in (1) introducing a convenient language in which to talk about these kinds of error rates, (2) taking the view that they are practically almost identical, and (3) applying them to the multiple comparison problem.

72. INDEPENDENCE AND NONINDEPENDENCE

Fisher's 1929 discussion [26] of the studentized largest periodogram ordinate was perhaps the first statistical treatment of the largest, and thus by implication all, of a set of nonindependent quantities. As mentioned above, it has been followed by a number of other treatments of distributions of "the largest . . . ." In most of these cases the dependence has been relatively simple, so that the emphasis has fallen on the literal mathematics involved in building up a distribution, rather than on the principles involved.

The novelty in the development of the *WSD* in this account, of the generalized *FSD* by Scheffé [68-70], of the *CSD* by Cornfield, Halperin and Greenhouse [14], and the suggestions of Nandi [49], is apparently minor, but the 20-year delay suggests that it is nontrivial. The novelty lies in facing a problem with interdependencies sufficiently complex to make the principle of inclusion and exclusion messy enough to be unattractive, and observing that a seemingly different statistical problem has already been solved or nearly solved which solves the comparison problem. This seemingly different problem is that of the distribution of the studentized range (*WSD*), variance ratio (*FSD*) or studentized extreme deviate (*CSD*).

A further novelty lies in the extension from the $m(m-1)/2$ comparisons to the continuous infinity of all contrasts. This again is in nowise difficult, but it seems to the writer as though there must have been a substantial psychological barrier. Part of this barrier must be traced to the conventional insistence on independent comparisons and the spendthrift policy of alloting 5% error rate to each.

73. RANGE-BASED SHORT-CUTS

Here the original steps are due to Lord [42-43], Hartley [34], and Patnaik [56]. Lord developed [42] the range-in-the-denominator version of the t-test, which had been suggested by many others [42], to a high degree of usefulness and then showed [43] that its power compared very satisfactorily with the conventional t-test. Patnaik [56] dealt with the principles and practice of using mean ranges in denominators. Hartley [34] extended the procedure to analysis of variance considered as a significance testing procedure. He used ranges in both numerator and denominator, and appeared to consider both range uses as short-cuts.

At about this time, the present author was invited to discuss the "Uses of the range in industrial statistics" before a one day conference of the ASQC in Baltimore. He prepared an account [73] which included some short-cut analyses for single classification designs, basing the treatment on the work of Lord and Patnaik and on some of his own work on the comparison of individual means [73].

This technique developed through a number of stages [74-79]. By May 1951 [75, p. 190], the fundamental idea of allowances for simple comparisons was clearly stated. At about this time the writer recognized that this was a fundamental idea, and could be applied to a mean-square-in-the-denominator analysis as well as to a range-in-the-denominator analysis.

The details of this development seem likely to be of interest, since this is a case where the conventional form arose from the short-cut form rather than vice versa. This may encourage others to pursue short-cuts.

74. APPROXIMATE FACTORS

A noticeable feature of the present account is the use of simple approximations, especially of the form

$$A + \frac{B}{DDF},$$

for quantities which might otherwise require extensive tabulation. The use of the reciprocal of the number of degrees of freedom is not novel, it has always been a cardinal principle of Fisher's basic tables. The only novelty has been to choose B so as to give a small error of changing sign over the working range. This has been a familiar idea in other fields (e.g., Herget and Clemence [36]) but has not been much used in statistics.

While these approximations are in nowise essential for the approach taken here, they make it very much handier, since with two well-chosen pages of tables we can carry out almost any analysis.

75. RANGE IN THE NUMERATOR

The introduction of range into the numerator, and hence of the studentized range, is due to Newman [51, 1939], whose significance testing procedure has recently been clarified and expounded by Keuls [39, 1952]. In the meantime the studentized range has also been used by Patnaik [56], Hartley [34], and H. A. David [15]. The use of the studentized range for several simultaneous comparisons is due to Paulsen [57], who showed that the apparently highest determination could be compared with all others. The extension to all comparisons is due to the writer [76, 77, 78].

76. STRAGGLERS

We distinguish between outliers and stragglers. An outlier is so labelled because of its unusualness in comparison with the spread between the other determinations. A straggler is so labelled because of its unusualness in comparison with an "independent" (internal) estimate of variability. The outlier problem is not being discussed in the present account.

The basic work on stragglers is that of K. R. Nair [48], who found the relevant distribution. This was approximately incorporated in his first general treatment [73] by the writer, and has been used in a short-cut procedure by Moshman [46].

77. MULTIPLE COMPARISONS IN GENERAL

The classical references are of course to the work of R. A. Fisher on the analysis of variance, whose early papers are referred to in his book [25].

More recently, besides those already mentioned above in connection with more specific topics, there are the somewhat distinctive approaches of Duncan [17,18,19,20,21,22] and of the writer [73]. (The writer feels that, this, his original approach, is obsolete and should be entirely replaced by the present one.)

The fact that the *F* procedure could be used to make a joint confidence statement about all *m* determinands was observed by Hotelling in 1931 [37, p. 378] but was not followed-up.

PART C
DERIVATIONS AND APPROXIMATIONS

CHAPTER 13: RATIONALE FOR ALLOWANCES

78. THE MODEL

It is time to restate the model under which we develop the comparative allowances for contrasts and universal allowances for all linear expressions. We write

$$\text{observed determination} = \text{determinand} + \text{fluctuation},$$

or in literal notation

$$x_i = \mu_i + e_i\,,$$

where the fluctuations $\{e_i\}$ form a sample from some distribution of variance σ^2. We further assume that we have available an estimate s^2 of σ^2 with suitable properties. Our problem in this chapter is to make a qualitative connection between the numerical values appropriate for allowances and the fluctuation distribution and the stability properties of the estimate s^2. The next two chapters will be concerned with successively more quantitative aspects of this connection.

The phrase used above reads "form a sample from some distribution" — what do we and shall we mean by this? The problem of remaining honest and ethical *both mathematically and applicationally* about distribution assumptions is one of the hardest in present-day theoretical statistics. Three rather different positions might run as follows:

(A) Let us assume everything is exactly normal, then the mathematics will be precise (and, since no application corresponding to exact normality is known, no application is well founded).

(B) Let us assume that the distribution is "reasonable" and calculate as if it were normal, then the dangers of the application are before our eyes (and the mathematical statisticians are unhappy).

(C) In special circumstances, it has been proved that the distributions or quantities in which we are concerned are relatively independent of the fluctuation distribution, let us rely on this result and choose some convenient distribution to calibrate our procedures by fixing particular numbers, remembering that any other choice would yield very similar and practically equivalent values. (In practice, of course, we shall almost always choose the normal distribution for calibration.)

In all 3 of these positions we calculate our operating numbers from results based on the normal distributions. But our points of view differ greatly. The major defects of the three positions seem to be these:

(A) leaves conscientious applied statisticians very uncomfortable;

(B) leaves conscientious mathematical statisticians very uncomfortable;

(C) is patchy — theorems are available for some cases, and are known to be impossible for others, while the remaining cases must be dealt with by plausibility arguments of varying degrees of weakness.

For my own part, I find it hard to see how a theoretical statistician can avoid being eventually driven to a position near (C). In statistics, as in other sciences and in contrast to mathematics, we must work with unestablished assumptions, but it is better that they be assumptions which have some hope of being establishable and that we are careful to describe their plausibility.

The special cases in which some of the quantities which we consider are known (and others are hoped) to be relatively insensitive to the underlying distribution, is the case where each determination is the mean of individual measurements, and thus each fluctuation is the mean of individual fluctuations. The original investigations were by Pitman [64-66] and Welch [81] and are discussed in the recent book of Kempthorne [38]. We shall delay further discussion of this problem until we deal more generally with such situations in Part E (Chapter 27), hoping to have made it clear in the present section:

(i) that we hold a position very close to (C);

(ii) that we want to carry our discussions as far as convenient in terms of a general distribution;

(iii) that we shall calculate our working numbers from distributions based on exact normality.

Our failure to be more specific about the properties of s^2 is deliberate. These properties are, in the special cases of most interest, closely related to the normality of the fluctuations, and it is simplest to reserve their discussion to Part E also. In the standard case of exact normality, which we shall use for calibration, we naturally assume that s^2 follows a mean-square (constant times chi-square over degrees-of-freedom) distribution with mean σ^2 and *DDF* degrees of freedom.

79. THE SIMPLE ALLOWANCES

In the following four cases we can set allowances in terms of Student's t distribution:

Parallel = p % per determination,
p % per batch,

Comparative = p % per comparison,
p % per family.

The first case in each pair is very classical, and the relation of the second to the first has already been discussed. The resulting values for the allowance are the product of s by a suitable percentage point of $|t|$ for the parallel cases and by $\sqrt{2}$ times a suitable percentage point of $|t|$ in the comparative cases. (Here $|t|$ denotes the absolute value of Student's t.)

These percentage points are most easily described as upper % points. (Thus the upper 5% of $|t|$ is a number c such that prob$\{|t| > c\} = 5\%$.) The number of degrees of freedom is always *DDF* and the % levels are:

Parallel; per determination: $-p\%$

Parallel; per batch: $-(p/m)\%$

Comparative; per comparison: $-p\%$

Comparative; per family: $-(2p/m\,(m-1))\%$.

The only comments we need to add relate to the ways in which the assumptions may be further relaxed. In dealing with the two parallel allowances, we need make no assumption of independence of the e_i, as was pointed out in Chapter 1. In dealing with the two comparative allowances, we may obviously allow the addition of an additional contribution to the model, one common to all the determinations of the family in any one experiment.

80. BATCHWISE ALLOWANCES

In this case we wish to choose a factor k so

$$\text{Prob}\{|x_i - \mu_i| < ks, \text{ all } i\} = (100-p)\%.$$

The probability can immediately be rewritten

$$\text{Prob}\{|e_i| < ks, \text{ all } i\} = \text{Prob}\{\max\{|e_i|\} < ks\}$$

$$= \text{Prob}\left\{\frac{\max\{|e_i|\}}{s} < k\right\}.$$

Thus we see that we should choose k as the upper $p\%$ point of the quantity

$$\max\{|e_i|\},$$

which is the result of studentizing the maximum absolute value or maximum modulus

$$\max\{\ |\ e_i\ |\ \}$$

of a set of m fluctuations drawn from a distribution with "true value" zero. Thus we shall be concerned with the distributions of the (normal) maximum modulus and the (normal) studentized maximum modulus.

81. FAMILY-WISE COMPARATIVE

Here we seek to choose a factor k so that

$$\text{Prob}\{\ |\ x_i - x_j - (\mu_i - \mu_j)\ |\ < ks, \ \text{all } i \text{ and } j\} = (100-p)\% \ .$$

This probability can be easily rewritten

$$\text{Prob}\{\ |\ (x_i - \mu_i) - (x_j - \mu_j)\ |\ < ks, \ \text{all } i \text{ and } j\}$$

$$= \text{Prob}\{\ |\ e_i - e_j\ |\ < ks, \ \text{all } i \text{ and } j\}$$

$$= \text{Prob}\{\max\{e_i\} - \min\{e_j\} < ks\}$$

$$= \text{Prob}\{\text{range}\ \{e_i\} < ks\}$$

$$= \text{Prob}\left\{\frac{\text{range}\ \{e_i\}}{s} < k\right\}$$

and we see that we should choose k as the upper $p\%$ point of the *studentized range*

$$\frac{\text{range}\{e_i\}}{s} .$$

82. EXTENSION TO CONTRASTS

In Section 46 we asserted that if WSD was a comparative allowance with $p\%$ error rate familywise, then (WSD) norm $\{c_i\}$ was a contrast allowance with $p\%$ error rate familywise for all contrasts. We now prove this by showing that if

$$|(x_i - x_j) - (\mu_i - \mu_j)| = |e_i - e_j| < WSD, \text{ all } i \text{ and j}$$

for all i and j, then

$$|y - \mu| = |\Sigma c_i x_i - \Sigma c_i \mu_i| = |\Sigma c_i e_i| \quad (WSD) \text{ norm} \{c_i\}$$

for all contrasts, y. Since the first condition holds for 95% of all families so does the second, and the result follows.

If y is a contrast of norm 1, then

$$y = \Sigma c_i x_i = \Sigma a_i x_i - \Sigma b_j x_j$$

where $a_i \geq 0$, $b_j \geq 0$ and $\Sigma a_i = \Sigma b_j = 1$. Now

$$\Sigma c_i e_i = \Sigma a_i e_i - \Sigma b_j e_j = \Sigma \Sigma a_i b_j e_i - \Sigma \Sigma a_i b_j e_j = \Sigma \Sigma a_i b_j (e_i - e_j),$$

where $\Sigma \Sigma a_i b_j \geq 0$. Hence

$$|\Sigma c_i e_i| = |\Sigma \Sigma a_i b_j (e_i - e_j)| \leq \Sigma \Sigma a_i b_j |e_i - e_j| \leq \Sigma \Sigma a_i b_j (WSD) \leq (WSD).$$

Thus the result is proved for contrasts of norm 1, and since, when y is a contrast,

$$y' \equiv \frac{y}{\text{norm}\{y\}}$$

is a contrast of norm 1, the result follows in general.

83. UNIVERSAL ALLOWANCES

In Section 56 we showed how the general linear expression reduced to a contrast if we adjoined a $(m+1)st$ (dummy) determination with $x_o = \mu_o = e_o = 0$. A familywise allowance for the new situation requires the use of the upper $p\%$ point of

$$\frac{\text{range}\{o, e_1, e_2, \ldots, e_m\}}{s}$$

which we shall refer to as the *studentized augmented range of m*. The augmented range of m being the range of m independent fluctuations augmented by an $(m+1)st$ value always equal to zero.

84. SUBSTITUTE ALLOWANCES

One of the most powerful elementary inequalities is the Schwartz inequality

$$|\ \Sigma\, a_i b_i\ |\ \leq (\Sigma\, a_i^2)^{1/2}\ (\Sigma\, b_i^2)^{1/2},$$

where equality holds if and only if the a_i are proportional to the b_i. If we are concerned with the substitute universal allowance, we put $a_i = \frac{1}{\sqrt{2}}\, c_i$, $b_i = \sqrt{2}\, e_i$ to obtain

$$|\ \Sigma\, c_i e_i\ |\ \leq (1/2\ \Sigma\, c_i^2)^{1/2} \cdot (\Sigma\, 2e_i^2)^{1/2}$$

and since $\text{norm}_F\,\{c_i\} = (1/2\ \Sigma\, c_i^2)^{1/2}$, this yields $y - \mu = \Sigma\, c_i x_i - \Sigma\, c_i \mu_i = \Sigma\, c_i e_i$, and

$$\frac{|\ y - \mu\ |\ =\ |\ \Sigma\, c_i e_i\ |}{\text{norm}_F\,\{c_i\}} \leq (2\ \Sigma\, e_1^2)^{1/2}$$

with equality when the c_i are proportional to the e_i. Hence

$$A\ \text{norm}_F\,\{c_i\}$$

is a universal allowance with $p\%$ error rate if and only if familywise

$$\text{Prob}\{(2\ \Sigma\, e_i^2)^{1/2} < A\} = (100 - p)\%.$$

and we are thus led to consider the distribution of

$$\frac{(2\ \Sigma\, e_i^2)^{1/2}}{s} = \sqrt{\frac{2\ \Sigma\, e_i^2}{s^2}}$$

which is the distribution of $\sqrt{2mF}$, where F follows the F (or variance ratio) distribution with m and DDF degrees of freedom.

Hence

$$GSD/LSD = \frac{p\ \%\ \text{point of } \sqrt{2mF}}{p\ \%\ \text{point of } t\sqrt{2}}$$

where F has m and DDF degrees of freedom and t has DDF degrees of freedom.

If we wish a substitute contrast allowance, then because $\Sigma\, e_i = 0$, we have

$$y - \mu = \Sigma\, c_i e_i = \Sigma\, c_i (e_i - e_.)$$

where $me_. = \Sigma\, e_i$, and hence

$$|\ y - u\ | = |\ \Sigma\, c_i (e_i - e_.)\ | \leq (^1/_2\ \Sigma\, c_i^2)^{1/2}\ (2\ \Sigma\, (e_i - e_.)^2)^{1/2}$$

so that

$$FSD = p\ \%\ \text{point of } (2\ \Sigma\, (e_i - e_.)^2)^{1/2}$$

$$= s\ (p\ \%\ \text{point of } \frac{(2\ \Sigma\, (e_i - e_.)^2)^{1/2}}{s}).$$

Now

$$\frac{(2\ \Sigma\, (e_i - e_.)^2)^{1/2}}{s} = \sqrt{\frac{2\ \Sigma\, (e_i - e_.)^2}{s^2}}$$

and follows the distribution of $\sqrt{2(m-1)F}$ with $m-1$ and DDF degrees of freedom.

Hence

$$FSD/LSD = \frac{p\ \%\ \text{point of } \sqrt{2(m-1)F}}{p\ \%\ \text{point of } t\sqrt{2}}$$

where F has $m-1$ and DDF degrees of freedom and t has DDF degrees of freedom.

Thus tables of GSD/LSD and FSD/LSD are easily computed.

CHAPTER 14: DISTRIBUTIONS

85. MAXIMUM MODULUS AND AUGMENTED RANGE

In this chapter we shall discuss various distributions associated with a sample from a distribution of fluctuations. Part of this discussion will apply to general distributions, part to distributions symmetric around zero, and part will be specialized to the normal distribution.

Given a sample of m fluctuations $e_1, e_2, \ldots, e_m$, the *range*

$$w_m = \max\{e_i\} - \min\{e_i\}$$

of the sample is a familiar quantity. We have just seen that there are two related quantities that we wish to consider, in connection with applications to allowances, namely, (i) the *augmented range*.

$$w_{m+} = \max\{0, e_1, \ldots, e_m\} - \min\{0, e_1, \ldots, e_m\}$$

that is, the range of the $m+1$ fluctuations obtained by augmenting the original m by an $(m+1)st$ equal to zero, and (ii) the *maximum modulus*

$$v_m = \max\{\mid e_i \mid\},$$

the greatest absolute value of any of the fluctuations.

For reasons which are clear from simple algebra, or from the cases in Figure 5, we have

Augmented range = larger of range and maximum modulus

that is,

$$w_{m+} = \max\{w_m, v_m\}.$$

This establishes a relatively strong relation between these three quantities.

The distribution of the range is moderately complicated, being expressed by

$$1 - W_m(x) = \text{Prob}\{w_m \geq x\} = \int\int m\,(m-1)\,\{F(z) - F(y)\}^{m-2} f(z)\,dydz$$

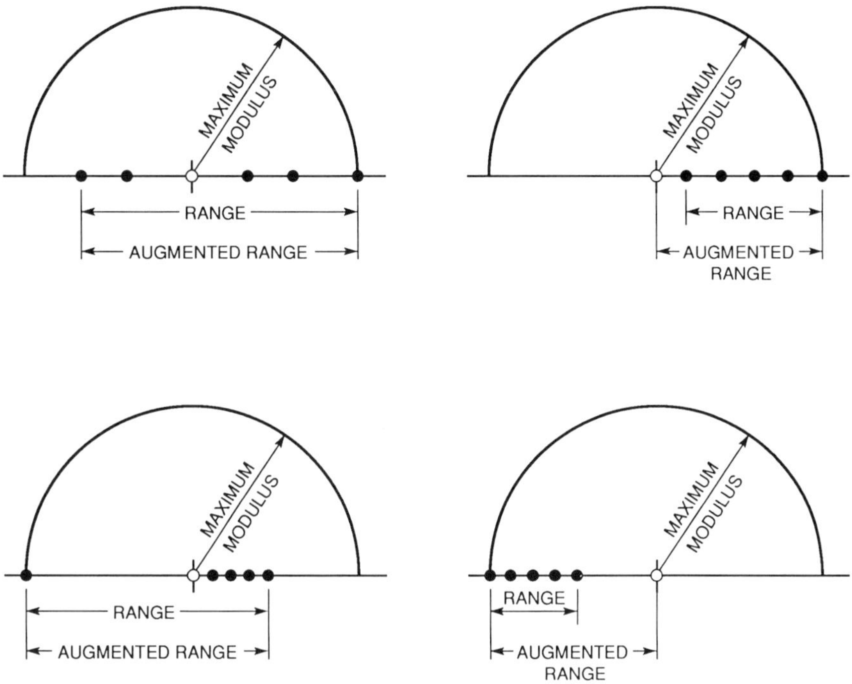

Figure 5. Examples showing relation between range, augmented range, and maximum modulus (• = observed point, ⊕ = origin).

where $F(x)$ is the cumulative and $f(x)$ the density of the underlying distribution of fluctuations, and the integral is taken over y and z satisfying $z \geq y + x$.

It can be obtained in closed form for a few simple distributions, thus for a rectangular distribution of length 1 it becomes

$$1 - W_m(x) = \int\int m\,(m-1)\,(z-y)\,dydz = 1 - (mx^{m-1} - (m-1)x^m),$$

while it has been rather carefully tabulated for the normal distribution [32, 59, 60].

The distribution of the maximum modulus is always easy to obtain numerically. If the cumulative distribution of the absolute values of the fluctuations is

$$G(x) = F(x) - F(-x), \text{ then}$$

$$V_m(x) = G^m(x),$$

and if the distribution of fluctuations around zero is symmetric, $G(x) = 2F(x) - 1$ so that

$$1 - V_m(x) = 1 - (2F(x) - 1)^m.$$

The distribution of the augmented range is only a little more complicated than that of the range, being given by the same integral, where the limits of integration are now $z > y - x$ when $y < 0$ and $z > x$ when $y > 0$. For the unit rectangular distribution centered at zero this yields the same distribution for $x > 1/2$ and an altered one for $x < 1/2$, as may be seen directly. (Adding zero cannot increase the range beyond $1/2$ in this case!) The distribution for the normal case has not been tabulated to the writer's knowledge.

In order to get useful inequalities for the distribution of augmented range, we return to the relation

$$w_{m+} = \max\{w_m, v_m\}$$

whence

$$\text{Prob}\{w_m > x\}, \text{Prob}\{v_m > x\} < \text{Prob}\{w_{m+} > x\} < \text{Prob}\{w_m > x\} + \text{Prob}\{v_m > x\}$$

that is

$$1 - W_m(x), 1 - V_m(x) < 1 - W_{m+}(x) < 2 - W_m(x) - V_m(x).$$

Now if the fluctuations are distributed symmetrically about zero, so that the 2^m samples $(\pm e_1, \pm e_2, \ldots, \pm e_m)$, obtained by all choices of signs, are equally likely, then we can obtain a closer inequality. For in 2 of the 2^m cases (all 2^m of which have the same value of v_m) all m of $\pm e_1$, $\pm e_2$, $\pm e_3, \ldots, \pm e_m$ are on the same side of zero, that $w_{m+} = v_m > w_m$, while in the other $2^m - 2$ cases $w_{m+} = w_m > v_m$. Now

$$\begin{aligned} \text{Prob}\{w_{m+} > x\} &= \text{Prob}\{w_m > x\} + \text{Prob}\{w_m < x, v_m > x\} \\ &\le \text{Prob}\{w_m > x\} + \text{Prob}\{w_m < v_m > x\} \\ &= \text{Prob}\{w_m > x\} + 2^{1-m}\,\text{Prob}\{v_m > x\} \end{aligned}$$

where we have used $2/2^m = 2^{1-m}$ as the fraction of those cases with any given v_m which have $w_m < v_m$. In terms of cumulatives this yields

$$1 - W_{m+}(x) < 1 - W_m(x) + 2^{1-m}\,(1 - V_m(x))$$

and so finally, for any distributions of fluctuations symmetric about zero, we have

$$W_m(x) - 2^{1-m}\,(1 - V_m(x)) < W_{m+}(x) < W_m(x),\ V_m(x),$$

a very useful and surprisingly tight double inequality.

86. THE CASE OF NORMALITY

In the case of normality, we may use easily available tables to obtain various cumulative distributions and bounds. Since we are interested in cumulatives, we find it convenient to make plots on arithmetic probability paper, on which every normal distribution gives a straight line. Figures 6, 7, and 8 show the results for $m = 2$, 3, and 5. For $W_{2+}(x)$ the bounds are only moderately close together. For $W_{3+}(x)$ the bounds are much closer together for the higher values of W_{3+} while for $W_{5+}(x)$ the bounds coincide closely for $W_{5+} > 1/2$. The situation at the percentage points of W_m is roughly as given in Table 17 where y is the solution of $W_{m+}(y) = W_m(x)$. We see that except for $m = 2$ or $W_{m+} < .95$ we can reasonably neglect the difference between $W_m(x)$ and $W_{m+}(x)$. (The most doubtful corner of the remainder is the 95% point for $m = 3$, where the difference between y and x amounts to less than 0.3%. This is not of practical importance.

Table 17

Bounds for % Points of Augmented Range

W_m	m	x	Lower bound for $W_{m+}(x)$	Upper bound for $y - x$	y to 2 decimals
.80	2	1.81	.732	.23	1.81 to 2.04
	3	2.42	.788	.05	2.42 to 2.47
	5	3.04	.7993	.003	3.04
.95	2	2.77	.944	.07	2.77 to 2.84
	3	3.331	.9493	.007	3.33 to 3.34
	5	3.86	.94996	.0004	3.86
.99	2	3.64	.9897	.008	3.64 or 3.65
	3	4.12	.98996	.002	4.12
	5	4.60	.9899987	.00005	4.60

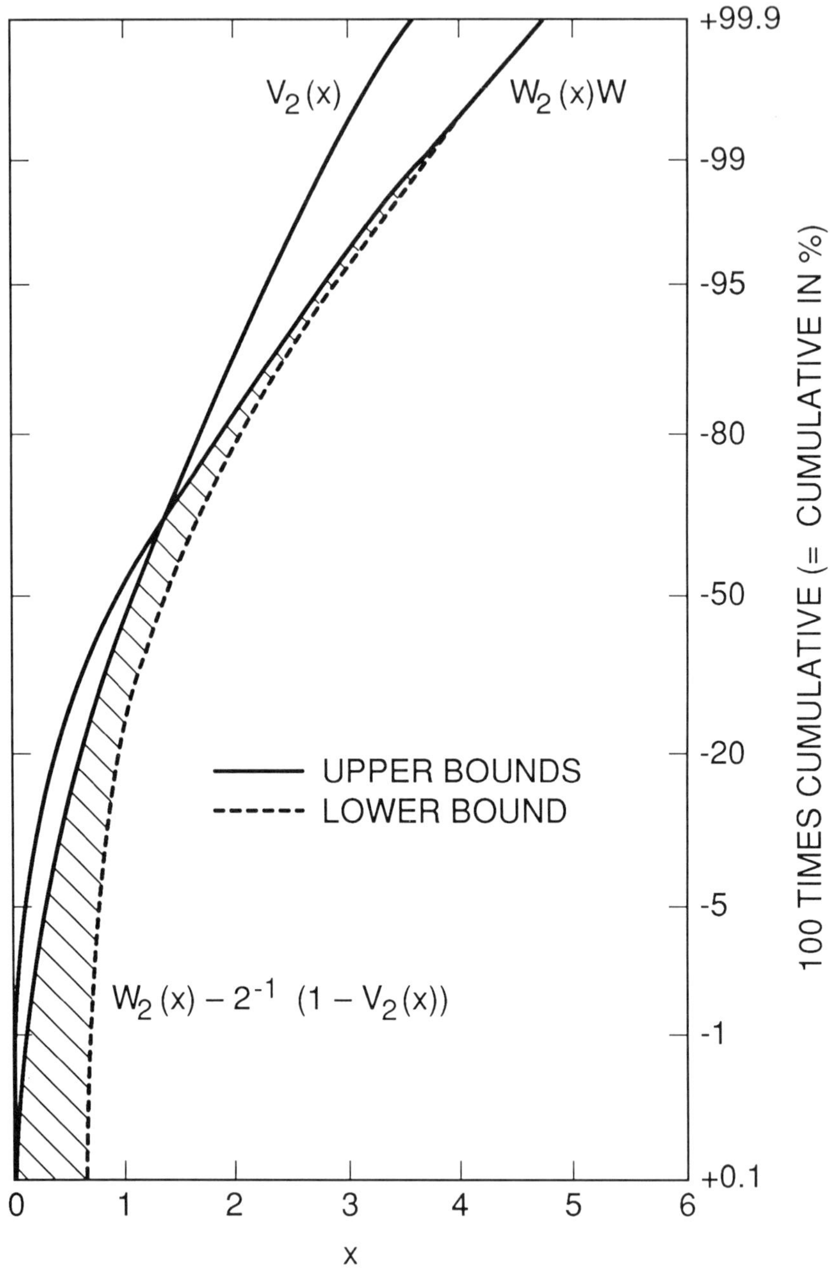

Figure 6. Bounds for $W_{2+}(x)$ for an underlying unit normal.

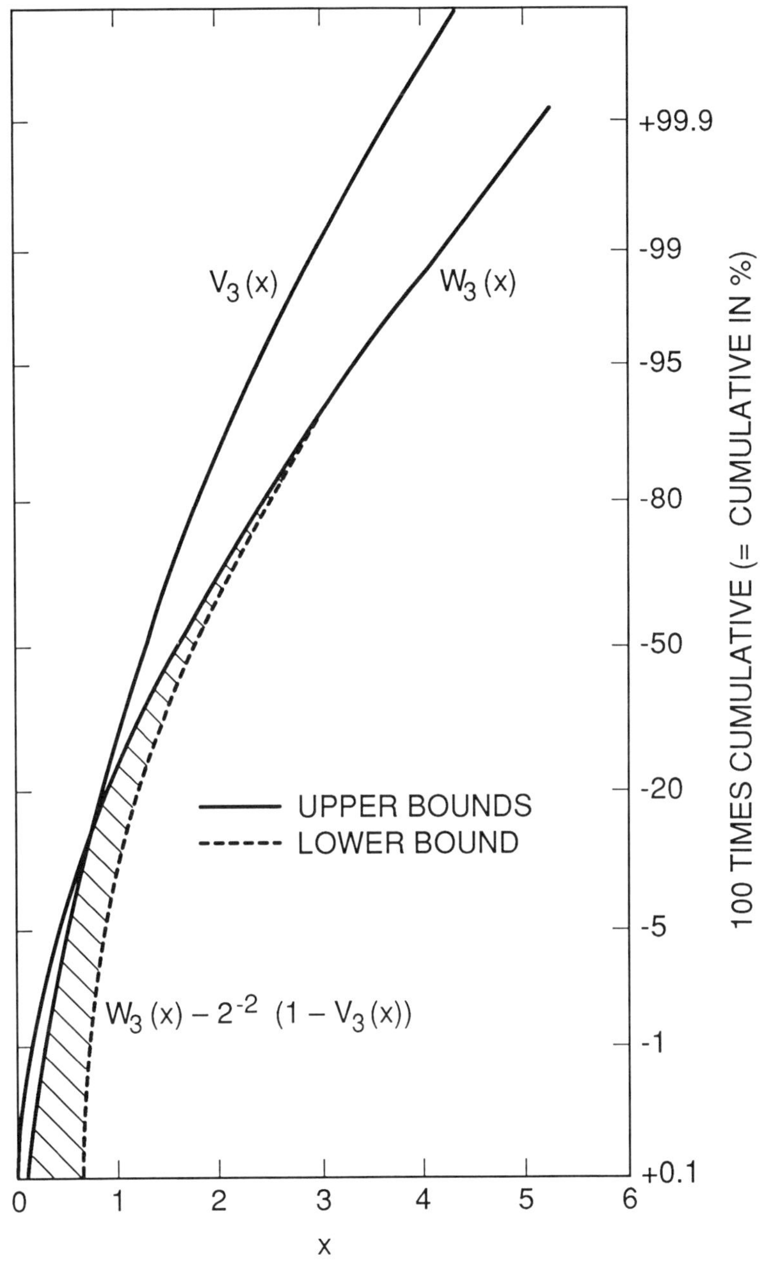

+99.9
-99
-95
-80
-50
-20
-5
-1
+0.1
100 TIMES CUMULATIVE (= CUMULATIVE IN %)
V3 (x)
W3 (x)
UPPER BOUNDS
LOWER BOUND
W3 (x) – 2-2 (1 – V3 (x))
0
1
2
3
4
5
6
x

Figure 7

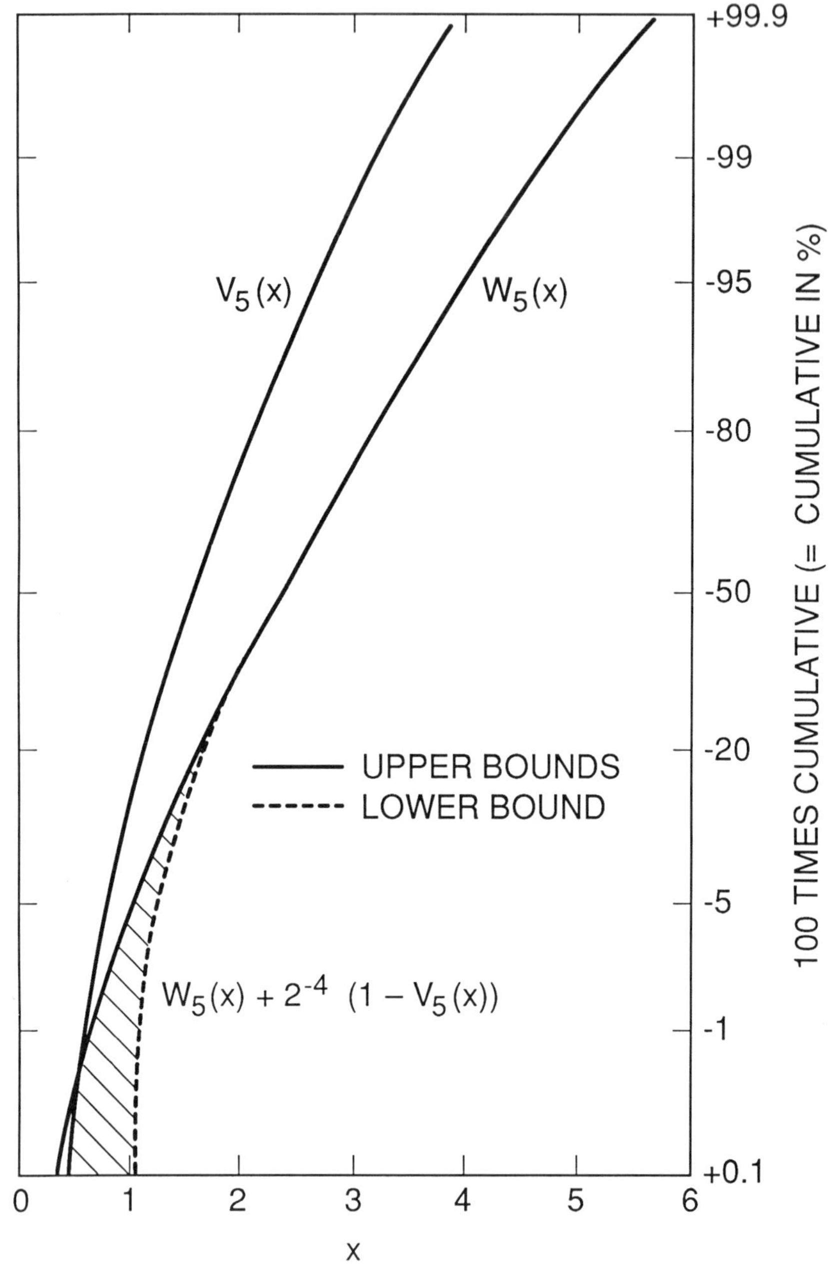

Figure 8. Bounds for $W_{5+}(x)$ for an underlying unit normal.

Since we shall rarely be concerned with the range below 95% we may conclude that, for $m > 3$ and practical accuracy, we may use the percentage points of the range in place of those of the augmented range. The case $m = 2$ requires further examination.

87. THE CASE $M = 2$

In this case the range reduces to the absolute value of the difference, whose cumulative is $K(x) - K(-x)$, where $K(x)$ is the cumulative for the difference of 2 fluctuations. In the symmetrical case this reduces to $2K(x) - 1$ and in the special case where the fluctuations have a unit normal distribution, $K(x)$ is of course half of a normal distribution with zero mean and variance two.

In view of this simplification, we might seek for a comparable simplification for the distribution of the augmented range of two. We shall find that we shall not do as well as for the range itself, but can make real progress. Let use write x and y, instead of θ_1 and θ_2, for the two fluctuations, and, instead of considering

$$\text{Prob}\{\text{range } 0, x, y \leq A\}$$

directly, let us express this in the equivalent form

$$\text{Prob}\{\,|x| \leq A,\ |y| \leq A, \text{ and } |x-y| \leq A\}, \tag{$*$}$$

which reminds us that we are basically concerned with allowances, although the connection has not yet been made explicitly.

By symmetry between (x, y) and $(-x, y)$, this last probability is the same as

$$\text{Prob}\{\,|x| \leq A,\ |y| \leq A, \text{ and } |x+y| \leq A\}. \tag{$**$}$$

If x, y satisfy both $|x-y| \leq A$ and $|x+y| \leq A$ then they satisfy all the conditions in both ($*$) and ($**$). If they do not satisfy both $|x-y| \leq A$ and $|x+y| \leq A$, but do satisfy both $|x| \leq A$ and $|y| \leq A$, then they satisfy just one of ($*$) and ($**$), but not both.

Twice the common probability in (*) and (**) is given by

$$\text{Prob}\{\,|x| \le A \text{ and } |y| \le A\,\} + \text{Prob}\{\,|x-y| \le A \text{ and } |x+y| \le A\}.$$

$$u = \{\,|x| \le A,\ |y| \le A\,,\ |x-y| \le A\}$$

$$v = \{\,|x| \le A,\ |y| \le A,\ |x+y| \le A\}$$

$$P_u = P_v = P \to \text{Prob}\{\text{range } 0, x, y \le A\}.$$

Remember:

$$|x-y| \le A,\ |x+y| \le A \to |x| \le A,\ |y| \le A,\ 2P = P_u + P_v =$$

$$P_r\{\,|x| \le A \text{ and } |y| \le A\} + P_r\{\,|x-y| \le A \text{ and } |x+y| \le A\}.$$

In this last of these, we may set $x - y = u\sqrt{2}$, $x + y = v\sqrt{2}$, where u, v are again a sample of 2 from a unit normal. Hence

$$\text{Prob}\{\,|x-y| \le A \text{ and } |x+y| \le A\}$$

$$= \text{Prob}\{\,|u|\sqrt{2} \le A \text{ and } |v|\sqrt{2} \le A\}$$

$$= \text{Prob}^2\{\,|x| \le \frac{A}{\sqrt{2}}\}$$

and since

$$\text{Prob}\{\,|x| \le A \text{ and } |y| \le A\} = \text{Prob}^2\{\,|x| \le A\},$$

We conclude that the probability we seek is given by

$$\tfrac{1}{2}\,\text{Prob}^2\{\,|x| \le A\} + \tfrac{1}{2}\,\text{Prob}^2\{\,|x| \le \frac{A}{\sqrt{2}}\}.$$

This is the cumulative at A for the augmented range of 2 from a unit normal.

If we choose $A = 1.96\sqrt{2}$ then for the range of 2 we have the well-known result

$$\text{Prob}\{\,|\,x - y\,| < 1.96\sqrt{2}\,\} = \text{Prob}\{\,|\,x\,| < 1.96\} = 0.950.$$

The corresponding value for the augmented range may be found from $\text{Prob}\{\,|\,x\,| < 1.96\sqrt{2}\,\} = 0.99445$ to be approximately

$$\tfrac{1}{2}(.950)^2 + \tfrac{1}{2}(.99445)^2 = \tfrac{1}{2}(.9025) + \tfrac{1}{2}(.98892) = .9457$$

which may be compared with the lower bound of .944 obtained in the last section.

Further computation shows that we need to use not $1.96\sqrt{2}$ but $2.0005\sqrt{2}$ in order to reach the 95% point of the augmented range of 2. The ratio

$$\frac{2.0005}{1.960} = 1.020$$

is thus the value of *WSD*/*LSD* corresponding to 5%, $m = 2+$ and known variance ($DDF = \infty$). This is to be compared with the upper bound of

$$\frac{2.77 + 0.07}{2.77} = 1.025$$

obtained in the last section.

88. STUDENTIZATION

All the distributions we have so far discussed have been appropriate to the case where the variance is known. In practice we must deal with the case where the variance is estimated. It will usually be possible to indicate the stability of the estimate by stating an equivalent number of degrees of freedom. We denote this number by f or by *DDF* for denominator degrees of freedom. It may help to have some idea of how much stability corresponds to different values of *DDF*. Table 18 gives 10%, 30%, 50% 70% and 90% points for s^2/σ^2 when s^2 has 5, 7, 10, 20, 70 and ∞ *DDF*. In practice we shall wish to work with various studentized distributions, that is with the ratios of various quantities to the square root of an independently distributed s^2. Some of these are:

$$(1) \quad \frac{\text{unit normal}}{\text{independent } s^2} = \text{Student's } t;$$

which is tabled almost everywhere

$$(2) \quad \frac{\text{range}}{\text{independent } s^2} = \text{Studentized range};$$

for which a revised table has recently been calculated and published by May [44]

$$(3) \quad \frac{\text{maximum modulus}}{\text{independent } s^2} = \text{Studentized maximum modulus};$$

some of whose percentage points have recently been computed by Nemenyi [50], and for which certain reasonable approximations will be discussed shortly,

$$(4) \quad \frac{\text{augmented range}}{\text{independent } s^2} = \text{Studentized augmented range};$$

which has not been tabulated to the writer's knowledge, but which we can manage in the same general way that we managed the augmented range itself.

Table 18

Illustrative % Points of s^2/σ^2

DDF	10%	30%	50%	70%	90%
5	.32	.60	.87	1.21	1.85
7	.40	.67	.91	1.20	1.71
10	.49	.73	.93	1.18	1.60
20	.62	.82	.96	1.14	1.42
70	.79	.89	.99	1.08	1.22
∞	1.00	1.00	1.00	1.00	1.00

We shall use the following notation:

Name	Symbol	Cumulative
Student's t	t_f	$T_f(x)$
Studentized range	$w_{m,f}$	$W_{m,f}(x)$
Studentized maximum modulus	$v_{m,f}$	$V_{m,f}(x)$
Studentized augmented range	$w_{m+,f}$	$W_{m+,f}(x)$

here m is the number of fluctuations and $f = DDF$ is the number of degrees of freedom for s^2.

We begin by discussing the approximate behavior of the Studentized maximum modulus, $v_{m,f}$. We have

$$v_{m,f} = \frac{\max\{\,|\,e_i\,|\,\}}{s} = \max\left\{\frac{|\,e_i\,|}{s}\right\}$$

which is nearly the same as the maximum modulus of t

$$v^*_{m,f} = \max\{\,|\,t_i\,|\,\} = \max\left\{\frac{|\,e_i\,|}{s_i}\right\},$$

whose cumulative is $(2T_f(x) - 1)^m$. It will be illuminating to compare $v^*_{m,f}$ with $v_{m,f}$.

We can always consider the s in $v_{m,f}$ to be a random sample of one from the m s_i's which appear in $v^*_{m,f}$ Let us fix m s's and m e's. Then there are $m!$ ways of pairing the m s's with the m e's, which fall into m equally probable classes according as to which s is associated with e_h, which we take to be the e_i with the largest absolute value. There are m equally likely ways of selecting one s out of the m s's to define $v_{m,f}$. We shall match m classes with the m ways in the natural manner. Then given a $v^*_{m,f}$ expressed in terms of e_i's and s_i's, we associate with it the $v_{m,f}$ defined by the same e_i and $s = s_h$ where $|\,e_h\,| = \max\{\,|\,e_i\,|\,\}$. This association preserves probability, and since

$$\max\left\{\left|\frac{e_i}{s_i}\right|\right\} \geq \left|\frac{e_h}{s_h}\right| = \left|\frac{e_h}{s}\right| = \max\left\{\left|\frac{e_h}{s}\right|\right\},$$

we always have

$$v^*_{m,f} > v_{m,f}\,.$$

Since probability has been preserved we must have

$$V_{m,f}(x) > V^*_{m,f}(x) = (2T_f(x)-1)^m$$

which establishes the convenient and close inequality

$$1 - V_{m,f}(x) < 1-(2T_f(x)-1)^m$$

for all symmetrical fluctuation distributions.

We next observe that

$$w_{m+} = \max\{w_m, v_m\}$$

is immediately translatable into

$$w_{m+,f} = \max\{w_{m,f}, v_{m,f}\}$$

and that the argument that led to

$$W_{m+}(x) > W_m(x) - 2^{1-m}(1-V_m(x))$$

is immediately translatable to give

$$W_{m+,f}(x) > W_{m,f} - 2^{1-m}(1-V_{m,f}(x))$$

since $w_{m+,f}$, $w_{m,f}$ and $v_{m,f}$ always are obtained from w_{m+}, w_m and v_m by division by the same s.

Hence for symmetrical fluctuation distributions, we have

$$W_{m,f}(x) > W_{m+,f}(x) > W_{m,f}(x) - 2^{1-m}(1-(2T_f(x)-1)^m)$$

which enables us to relate the distribution of the studentized augmented range quite closely to that of the studentized range itself.

89. THE CASE OF NORMALITY

We can again give a few numbers for the case of normality, and choose to do so for $DDF = 10$ and $m = 2, 3 \ldots$. Since detailed distributions of the studentized range are not available, we shall have to confine ourselves to working at the upper 5% (= lower 95%) and upper 1% (= lower 99%) points of $w_{m,f}$, since these are what is available.

Interpolation in the t-table leads us to the results in Table 19. We see that for the upper 5% points and $m = 3$ the ratio y/x, where $W_{m,f}(x) = .95 = W_{m+,f}(y)$ amounts to no more than

$$\frac{3.88 + .05}{3.88} = 1.013.$$

This is a tolerable approximation to 1.000. Our conclusion is that for $m > 3$, we can, for practical purposes, neglect the difference between $W_{m+,f}(x)$ and $W_{m,f}(x)$.

There remains again the case $m = 2$. In Section 87 we obtained certain relations between probabilities for fluctuations x and y distributed normally around zero. The main part of that development made no use of independence, and can be easily translated to give

$$\text{Prob}\left\{\left|\frac{x}{s}\right| \leq A, \left|\frac{y}{s}\right| \leq A, \left|\frac{x-y}{s}\right| \leq A\right\}$$

$$= {}^1\!/_2 \,\text{Prob}^2\left\{\left|\frac{x}{s}\right| \leq A, \left|\frac{y}{s}\right| \leq A\right\}$$

$$+ {}^1\!/_2 \,\text{Prob}^2\left\{\left|\frac{x}{s}\right| \leq \frac{A}{\sqrt{2}}, \left|\frac{y}{s}\right| \leq \frac{A}{\sqrt{2}}\right\}.$$

The left-hand side is $W_{2,f}(A)$ and the right-hand side clearly refers to studentized maximum moduli, being ${}^1\!/_2\, V_2(A) + {}^1\!/_2\, V_2\,(A/\sqrt{2})$. Thus we deduce, writing x for A,

$$W_{2,f}(x) = {}^1\!/_2\, V_2(x) + {}^1\!/_2\, V_2(x\,/\,\sqrt{2}).$$

Thus the calculation of the studentized augmented range of 2 is reduced so that of the studentized maximum modulus, and we may use Nemenyi's results [50]. (It is to be noted that at the same session in which the present

Table 19

Bounds for % Points of Studentized Augmented Range

Upper %	m	x	Lower bound for $W_{m+,f}(x)$	Approximate bound for $y-x$
5%	2	3.15	.942	.13
$(DDF=10)$	3	3.88	.9478	.05
	5	4.66	.9491	.02
10%	2	4.48	.9892	.06
$(DDF=10)$	3	5.26	.9897	.02
	5	6.14	.9899	.01

material was presented [79], Bose and Roy reported the same application [8a]. Tabulation of the distribution of the studentized maximum modulus is believed to also have been undertaken at Chapel Hill.)

90. COMMENT

One interesting thing about the results developed in this chapter is the great extent to which they hold for any symmetrical distribution of fluctuations. Only the numerical results, and the special formulas for $m=2$ require normality. For any symmetrical distribution for which we can deal with the studentized maximum modulus and studentized range numerically, we can apply rather similar calculations to dispose of the need of treating the studentized augmented range separately, except for very small values of m.

CHAPTER 15: ERROR-RATE SHIFTING FACTORS

91. INTRODUCTION

A feature of the present account of allowances is its free use of empirical approximations to factors of one kind or another. This feature should be a good feature in the eyes of many but will undoubtedly be a bad feature in the eyes of a few. For some will feel that it is more moral, or clever, or in some other way superior to leave answers in terms of inaccessible or even unknown percentage points of unfamiliar distributions, rather than to introduce approximations of peculiar form and, to such readers, doubtful accuracy. The writer disagrees, and for at least three reasons.

In the first place, the present account is directed to the actual analysis of observations, and as such is only successful when its procedures can actually be carried out. Procedures can only be carried out in terms of approximations, for even a 15 decimal place table yields only approximations, even if they are to many decimals. A working procedure is an approximate procedure, and at the working level the question is only as to the quality of the approximation, and not to its existence.

In the second place, our position about distributional assumptions was set forth in Section 78 — namely that we choose the normal distribution only as a matter of calibration, that we try to work with quantities which depend only a little on the distribution, and that we would gladly calibrate in terms of some other distribution if it were easy to use. Since we regard the normal distribution assumption as itself an approximation, how can we object to further approximations?

Third, we are concerned with allowances, we wish to come to statements with $\pm A$ at their closes. It is the nature of these statements, and of those who use them that small % fractional changes in the lengths of the allowances make small differences in the conclusions reached. Thus, from the direction of eventual use we have further encouragement for the use of approximations.

So long as each factor is within ±1% or ±2% of its normal theory value, and so long as no more factors are combined than we do here, the writer feels that the accuracy obtained is entirely sufficient. Some may think otherwise, but these are asked to consider a few real examples.

We have two general types of factors in our analyses. One type is concerned with shifting error rates away from 5%; factors of this type are discussed in the present chapter. Those of the second type are concerned with shifting m from its minimum value of 1 or 2 up to its actual value. These multiplicity shifting values are discussed in the next chapter.

While the multiplicity shifting factors are reasonably immune to the detailed nature of the distribution, and error rate shifting from 5% to 1% may sometimes be so, it must be admitted that extreme error rate shifting is very sensitive to the exact distribution involved. This means that accuracies of even a few percent in approximating the normal theory values of extreme error rate shifting are highly illusory, since no factor can work that well for the varying fluctuation distributions corresponding to practical situations. Two conclusions are to be drawn from this:

(1) Do not take the accuracy of extreme error rate shifting too seriously; and

(2) Do not take extreme error rates too seriously either.

With this foundation we are ready to go ahead.

92. ERROR-RATE SHIFTING FOR $|t|$

The simplest case, and one which deserves somewhat more detailed study than most, is the case of error rate shifting for the absolute value of Student's $|t|$. We have here to approximate the ratio

$$\frac{(5/r)\% \text{ point of } |t|}{5\% \text{ point of } |t|}$$

where both t's are on DDF degrees of freedom. A little experimentation shows us that the form

$$A + \frac{B}{DDF}, \qquad (*)$$

where A and B depend on r does a quite satisfactory job. If this is the case, then we should be quite satisfied. For we need at least two constants, when r is fixed, in order to represent the dependence on DDF at all. Furthermore, the factor $1/DDF$ is both likely to be small, and is easy to compute (arising naturally on occasion).

In order to examine this type of approximation in a little detail we observe that the values $r = 2.5$, 5, 25 and 50 correspond to $(5/r)\% = 2\%$, 1%, 0.2%, 0.1% and the corresponding percentage points of t_f are widely tabled (e.g., Fisher and Yates [26], Hald [32]). The results of computing the corresponding batching factors are shown in Table 20. Lines labelled (a) contain values computed from the tables just referred to, those headed (b) contain values obtained from (*) where A_m and B_m are chosen to reproduce the values at $DDF = \infty$ and $DDF = 10$, while those headed (c) contain values obtained from the two-decimal formula given at the foot. Lines (d) contain the result of an alternate formula also given at the foot of the table. Lines (e) give the result of substitution in the formula from Table 21.

For $DDF >$ the greater of 4 and m the working formula gives results within ±2% and usually within ±1%. This accuracy is sufficient for most practical purposes.

Table 20

Values and Approximations to the Batching Factor

(a) actual values computed from 3-decimal % points
(b) values from linear interpolation
(c) values from two-decimal formula
(d) values from alternate formula
(e) values from wide range formula

DDF	Case	$r = 2.5$	$r = 5$	$r = 25$	$r = 50$
∞	(a) = (b)	1.186	1.314	1.566	1.679
	(c) = (d)	1.19	1.32	1.57	1.68
	(e)	1.183	1.303	1.568	1.681
20	(a)	1.211	1.363	1.702	1.846
	(b)	1.213	1.368	1.713	1.869
	(c)	1.22	1.375	1.72	1.87
	(d)	1.214	1.368	1.709	1.852
	(e)	1.210	1.357	1.710	1.871
10	(a) = (b)	1.240	1.422	1.860	2.059
	(c)	1.24	1.43	1.87	2.06
	(d)	1.240	1.420	1.869	2.061
	(e)	1.237	1.412	1.853	2.057
7	(a)	1.267	1.480	2.023	2.285
	(b)	1.263	1.468	1.986)	(2.223)
	(c)	1.261	1.477	1.999)	(2.224)
	(d)	1.261	1.468	2.038	2.282
	(e)	1.260	1.459)	(1.975)	(2.225)
5	(a)	1.309	1.569	2.292	2.667
	(b)	1.294)	(1.530)	(2.154)	(2.439)
	(c)	1.29	1.54)	(2.17)	(2.44)
	(d)	1.293)	(1.536)	2.308	2.667
	(e)	1.191)	(1.521)	(2.138)	(2.441)
4	(a)	1.350	1.659	2.585	3.103
	(b)	(1.321)	(1.584)	(2.301)	(2.629)
	(c)	(1.32)	(1.595)	(2.32)	(2.63)
	(d)	(1.322)	(1.600)	2.620	3.124
	(e)	(1.318)	(1.575)	(2.281)	(2.631)
Two Decimal Formula		$1.19 + \frac{0.5}{DDF}$	$1.32 + \frac{1.1}{DDF}$	$1.57 + \frac{3.0}{DDF}$	$1.68 + \frac{3.8}{DDF}$
Alternate Formula		$1.19 \frac{DDF}{DDF-0.4}$	$1.32 \frac{DDF}{DDF-0.7}$	$1.57 \frac{DDF}{DDF-1.6}$	$1.68 \frac{DDF}{DDF-1.85}$
Wide-range Formula		$1.183 + \frac{.54}{DDF}$	$1.303 + \frac{1.09}{DDF}$	$1.568 + \frac{2.85}{DDF}$	$1.681 + \frac{3.80}{DDF}$

The alternate formula yields even closer results, but is a little more complex to handle. We shall not discuss it further here. Failures in approximation by more than 1% are marked ")" and those of more than 2% are enclose in parentheses.

If we examine the results in detail we find the following accuracies:

Approximation	Range	Tolerance
Two-decimal	$DDF \geq 3 + 3\log_{10} r$	±1%
Alternate	$DDF \geq 6$ $DDF \geq 4,\ r \geq 10$	±1%
Wide-range	$DDF \geq 3 + 3\log_{10} r$	±2%

The first two cases illustrate the accuracy which can be obtained by adjusting two constants separately for each value of r. However, this is both tedious and impractical, since the necessary t-tables are not available. The wide-range formulas are the result of substituting

$$u = \log_{10} r$$

in the ratio $FSD_{(5/r)\%} / FSD_{5\%}$ of Table 12 for $m = 2$, that is, substituting u in

$$1 + u \left\{ .475 - .044u + (1.09 + 0.68u)\frac{1}{DDF} \right\}.$$

This relatively simple formula works over a very wide-range to ±2%, so that we are quite content to use it.

We do not know that this is the best simple approximation, or even the best in this form. Here, as in other cases we are content to know that the approximation is reasonably satisfactory.

We may also wish to adjust the error rate upward, and Table 21 gives the expression for this as

$$\log_{10} \frac{(5h)\% \text{ point of } |t|}{5\% \text{ point of } |t|} = u\{.202 + .0264h + \frac{.40}{DDF}\}$$

where $u = -\log_{10} h$ has essentially the same significance as before. As an indication of the effectiveness of this formula, consider the comparison in Table 21 between the Fisher and Yates values and those furnished by this

Table 21

Accuracy of Wide-Range Formula for Increasing Error Rates with $|t|$.
(Entries (a) from Fisher and Yates table.
Entries (b) calculated from lines of Table 12 for $m = 2$.)

DDF	Kind	$h = 2(10\%)$	$h = 4(20\%)$	$h = 6(30\%)$	$h = 8(40\%)$	$h = 10(50\%)$
∞	(a)	1.645	1.282	1.036	.842	.674
	(b)	1.642	1.278	1.028	.831)	.670
20	(a)	1.725	1.325	1.064	.860	.687
	(b)	1.728	1.324	1.058	.849)	.683
10	(a)	1.812	1.372	1.093	.879	.700
	(b)	1.818	1.378	1.088	.868	.695
7	(a)	1.895	1.415	1.119	.896	.711
	(b)	1.903	1.428	1.122	.890	.709
5	(a)	2.015	1.476	1.156	.920	.727
	(b)	2.038)	1.500)	1.169)	.922	.730

Values in error by more than ±1% are marked with a half parenthesis.

formula for $DDF = \infty$, 20, 10, 7, 5 and $h = 2, 4, 6, 8, 10$ (for $(5h)\% = 10\%, 20\%$, 30%, 40%, and 50%). Again the errors are less than ±2%. This is again a good working approximation.

93. ERROR-RATE SHIFTING FOR $\sqrt{F}$

We now come to dealing with the body of Table 22. We recall from Section 84 that

$$\frac{FSD}{s} = p\ \%\ \text{point of}\ \sqrt{2(m-1)F}$$

$$= (\sqrt{2(m-1)})(p\ \%\ \text{point of}\ \sqrt{F})$$

so that

$$\frac{FSD_{(5/r)\%}}{FSD_{5\%}} = \frac{(5/r)\%\ \text{point of}\ \sqrt{F}}{5\%\ \text{point of}\ \sqrt{F}}$$

Table 22

Comparison of Tabular Values of $F_{(5/r)\%}$ with Those Obtained From the Formula of Table for $\sqrt{F_{(5/r)\%}/F_{5\%}}$. (All F's on $m-1$ and F degrees of freedom. Entries (a) based on Hald's values. Entries (b) based on formula in Table 21.)

			r=2	*r*=5	*r*=10	*r*=50	*r*=100
DDF	*m*		2.5%	1%	0.5%	0.1%	0.05%
20	3	(a)	4.46	5.85	6.90	9.95	11.4
		(b)	4.44	5.91)	7.09	10.20)	11.7)
	8	(a)	3.01	3.70	4.26	5.69	6.38
		(b)	3.00	3.74	4.35)	(5.98)	(6.80)
	13	(a)	2.68	3.23	3.68	5.08*	5.66*
		(b)	2.67	3.23	3.79	5.25)*	(5.90)*
	18	(a)	2.52	3.02	3.42	4.29**	4.75**
		(b)	2.50	3.00	3.41	4.40**	(4.94)**
	51	(a)	2.25	2.64	2.96	3.77	4.15
		(b)	2.24	2.63	2.96	3.82)	4.24)
	201	(a)	2.13	2.48	2.76	3.48	3.82
		(b)	2.11	2.43)	2.72)	(3.40)	3.82
7	3	(a)	6.54	9.55	12.4	21.7	27.2
		(b)	6.52	9.49	12.2	(20.4)	(24.6)
	8	(a)	4.99	6.99	8.89	15.0	18.7
		(b)	4.95	6.85)	8.70)	(14.1)	(17.2)
	13	(a)	4.67	6.47	8.18	14.1*	17.5*
		(b)	4.00	6.30)	7.89)	(13.2)*	(16.1)*
	18	(a)	4.52	6.24	7.87	12.9**	16.0**
		(b)	4.48	6.10)	7.69)	(12.2)**	(14.8)**
	51	(a)	4.28	5.86	7.35	12.2	15.1
		(b)	4.29	5.84	7.29	(11.5)	(13.9
	201	(a)	4.18	5.70	7.15	11.8	14.6
		(b)	4.20	5.76)	7.20	(11.5)	(14.0)

Errors of 1% to 2% in $\sqrt{F}$ are marked with a half parenthesis, those or more than 2% with a full parenthesis.

* indicates $m = 11$ not 13.

** indicates $m = 21$ not 18.

where all F's are on $m-1$ and DDF degrees of freedom. In case $m=2$, this reduces to the

$$\frac{(5/r)\% \text{ point of } |t|}{5\% \text{ point of } |t|},$$

a fact which we used in applying the line for $m=2$ of Table 21 in the last section.

We have now to verify the corresponding approximation of Table 21 for $m>2$. Like most of our approximations in the general form $A+B/DDF$, this was based on values for $DDF=\infty$ and $DDF=10$. It seems conservative to verify it for $DDF=20$ and $DDF=7$. (Here A_F and B_F are linear in u.) Similarly, it is natural to choose $m=3$, 8, 13, 18, 51, 201. The results of the comparison, for all values of r provided directly by Hald's tables, are shown in Table 22. The results are not as good as for $m=2$, since a few of the comparisons given have errors of 1% or more in $\sqrt{F}$. This is substantial but probably not serious.

Similar results for increases in error rate are set forth in Table 23. (Here C_F is linear in h rather than in u.) Here only one error reaches ±2% for $m<51$ and the approximation is quite satisfactory.

94. ERROR-RATE SHIFTING FOR THE STUDENTIZED RANGE

Here we must go out into the unknown, and the only reasonable policy is boldness. Tabular values are available for $DDF=\infty$ and m between 1 and 20 for a number of values of $(5/r)\%$ and $5h\ \%$. These can be approximated quite well by expressions of the form

$$\frac{(5/h)\% \text{ point of the studentized range}}{5\% \text{ point of the studentized range}} \sim 1+u\,(A+B/DDF)$$

$$\log_{10}\frac{5h\ \% \text{ of the studentized range}}{5\% \text{ point of the studentized range}} \sim u\,(C+D/DDF)$$

where A and B are linear in u while C is linear in h.

Table 23

Comparison of Tabular Values of $F_{5h\%}$ with these Obtained From the Formulas of Table for $\log_{10}\sqrt{F_{5h\%}}/\sqrt{F_{5\%}}$. (All F's on $m-1$ and DDF degrees of freedom. Entries (a) are tabular. Entries (b) are based on $F_{5\%}$ from Hald and Table 21.)

DDF	*m*		$h=2$ 10%	$h=6$ 30%	$h=10$ 50%
20	3	(a)	2.59	1.28	.718
		(b)	2.58	1.28	.720
	8	(a)	2.04	1.30	.938
		(b)	2.04	1.31	.942
	11	(a)	1.94	1.29	.966
		(b)	1.94	1.30	(.988)
	21	(a)	1.79	1.27	1.00
		(b)	1.81	(1.31)	(1.04)
	51	(a)	1.69	1.25	1.02
		(b)	1.70	(1.29)	(1.07)
	201	(a)	1.63	1.24	1.03
		(b)	1.64	(1.30)	(1.11)
7	3	(a)	3.26	1.44	.767
		(b)	(3.19)	1.45	.778
	8	(a)	2.78	1.51	1.00
		(b)	2.76	1.50	1.00
	11	(a)	2.70	1.51	1.03
		(b)	2.69	1.49	(1.07)
	21	(a)	2.59	1.51	1.07
		(b)	2.55	1.49	1.08
	51	(a)	1.69	1.25	1.02
		(b)	1.70	(1.20)	(1.07)
	201	(a)	1.63	1.24	1.03
		(b)	1.64	(1.30)	(1.11)
	51	(a)	2.52	1.50	1.09
		(b)	(2.45)	(1.43)	(1.06)
	201	(a)	1.48	1.50	1.10
		(b)	(2.36)	(1.38)	(1.05)

When we compare A, B, C, D, with A_F, B_F, C_F, D_F where

$$\frac{(5/r)\% \text{ point of } \sqrt{F}}{5\% \text{ point of } \sqrt{F}} \sim 1 + u\,(A_F + B_F/DDF)$$

$$\log_{10}\left\{\frac{5r\ \% \text{ point of } \sqrt{F}}{5\% \text{ point of } \sqrt{F}}\right\} \sim u\,(C_F + D_F/DDF).$$

We find considerable similarity. We are thus fairly comfortable about this part of the approximation.

For finite degrees of freedom, the only trustworthy tables are those of May [44], and these give only 5% and 1% points of $[WSD/s]$. Thus we can only check for $r = 5$. To do this, we may calculate the apparent value of B for $u = 0.699$ in the form

$$J = \frac{10}{\log_{10} 5}\ \frac{(1\% \text{ point for } DDF = 10}{(5\% \text{ point for } DDF = 10} - \frac{1\% \text{ point for } DDF = \infty)}{5\% \text{ point for } DDF = \infty)}\,.$$

We may calculate a comparable J_F from the behavior of $[FSD/s]$, which is proportioned to $\sqrt{F}$. We expect J for m determinations, perhaps for $m' < m$, where m' is defined as a function of m as in the next chapter. Table 24 compares the actual values of J for m with those of J_F for m' and m'', denoted respectively by J'_F and J''_F. Here m is the same function of m' as m' is of m.

Clearly J'_F is consistently larger than J. The relation between J''_F and J is less clear cut, and we have taken B for m equal to B_F for m'', not only for $1 \le m \le 20$, where we are, at least in part, interpolating, but also for $m > 20$ which is pure extrapolation.

By analogy, we have taken D for m equal to D_F for m''.

The values of A and C can be based on the % points of the range, which are available for $1 \le m \le 20$ in Hald [32, based on Pearson [60]], and for $20 \le m \le 100$ in Pearson [59]. The values for $200 \le m$ involve extrapolation.

The resulting formulas are of course doubtful extrapolations, but the writer is quite content with them for the present. One concrete piece of evidence helps to reduce the extent of the doubt. Similar considerations about the relation of studentized range and $\sqrt{F}$ were the cause of the writer's belief that the Pearson and Hartley table was in error for the 1% level and thus indirectly led to May's recomputation, which confirmed this belief. At one time and at 1%, the approximation was better than the current tables. Now and beyond 1% it is surely much better than the nonexistent tables.

Table 24

Values of J, J'_F, J''_F and Their Differences

m	J	m'	J'_F	$J'_F - J$	m''	J''_F	$J''_F - J$
2	1.55	2.0	1.55	.00	2.0	1.55	0.00
3	1.66	2.9	1.64	.05	2.8	1.63	.04
4	1.73	3.8	1.74	.01	3.6	1.73	.00
5	1.79	4.6	1.81	.02	4.2	1.78	–.01
6	1.85	5.3	1.87	.02	4.8	1.83	–.02
7	1.90	6.1	1.92	.02	5.4	1.87	–.03
8	1.93	6.8	1.96	.03	5.9	1.90	–.03
9	1.95	7.4	1.99	.04	6.3	1.93	–.02
10	1.93	8.0	2.02	.09	6.8	1.96	.03
11	1.98	8.6	2.04	.06	7.2	1.98	.00
12	2.00	9.2	2.07	.07	7.5	1.99	–.01
13	2.02	9.8	2.09	.07	7.9	2.01	–.01
14	2.00	10.4	2.11	.11	8.2	2.03	.03
15	2.02	10.9	2.12	.10	8.6	2.04	.02
16	2.04	11.5	2.15	.11	9.0	2.06	.02
17	2.02	12.0	2.17	.15	9.3	2.07	.05
18	2.08	12.7	2.19	.09	9.6	2.08	.00
19	2.04	13.4	2.81	.17	10.0	2.10	.06
20	2.06	14.3	2.23	.17	10.5	2.12	.06

95. ERROR-RATE SHIFTING FOR *PA*

There remains one more possible care of error rate shifting which needs to be mentioned. This arises when we wish to use a parallel allowance at some % other than 5%. $PA_{5\%}$ is essentially dependent on a % point of $|t|$, but on a more extreme % than 5%. Actually $PA_{5\%}$ depends on the 5% point of the studentized maximum modulus, but as we saw in Section 88, and as Nemenyi has shown numerically, this is very close to the $q\%$ point of $|t|$, where $(100 - q\,\%^m) = 95\%$. This justifies the use of "essentially."

A brief numerical investigation of

$$\frac{(5/mr)\% \text{ point of } |t|}{(5/m)\% \text{ point of } |t|}$$

shows that we can put, with reasonable accuracy

$$\log_{10} \frac{(5/mr)\% \text{ point of } |t|}{(5/m)\% \text{ point of } |t|} \sim \frac{(1.25 + \frac{24.6}{DDF})\,(3 + \frac{20}{DDF})}{\left\{\log_{10} m + 3 + \frac{20}{DDF}\right\}\left\{\log_{10} m + 3 + \frac{20}{DDF} + \delta\right\}}$$

where

$$\delta = \log_{10} r + 0.11 \left(1 - \frac{1}{r}\right) = u + 0.11 \left(1 - \frac{1}{r}\right)$$

when $0 \le u \le 2.7$.

We have the following chain of approximations. First,

$$\frac{(5/m)\% \text{ of studentized maximum modulus}}{5\% \text{ of studentized maximum modulus}}$$

is approximately equal to

$$\frac{q^* \text{ point of } |t|}{q\,\% \text{ point of } |t|},$$

where $((100 - q^*)\%)^m = (100 - \frac{5}{rm})\%$. But $q\,\% \sim (5/m)\%$ and $q^*\% \sim (5/rm)\%$, so that the last ratio of % point is approximately the same as

$$\frac{(5/rm)\% \text{ point of } |t|}{(5/m)\% \text{ point of } |t|}$$

and thus we are led to the formula resulting from our brief numerical investigation.

We leave the details of the overall accuracy of this chain of approximations to the reader.

96. CONCLUDING REMARKS

We close this chapter by reminding the reader that, while the present approximate error rate shifting factors are not too close approximations to normal theory, and while they will undoubtedly be refined, they are still much better approximations to normal theory than normal theory is to practice. Thus they are already adequate for their intended use.

CHAPTER 16: THE MULTIPLICITY-SHIFTING FACTORS

97. INTRODUCTION

We come now to the factors which make allowance for the number of determinations in a batch or family — the factors which convert an *LSD* or an *IA* into allowances on other bases. Again we use simple approximations, as was already clear in Chapter 1.

Many of the cases we have mentioned:

per determination → per batch

per comparison → per degree of freedom

per comparison → per family

are just matters of changing the denominator of the error rate in a fixed way, and thus are just matters of error-rate shifting. The factors of the last chapter will serve here. These factors are worth attention, not because they are moderately close, oversize approximations to the factors which correspond to something other than mere error rate shifts. This is true not only in the confines of the present paper, but in a much wider area, where we find that error rate shifting gives us a simple first approximation to the factors or critical levels we require.

98. THE BATCHWISE FACTOR *PA*/*IA*

We call the ratio

$$\frac{PA}{IA} = \frac{p\,\%\text{ point of the studentized maximum modulus}}{p\,\%\text{ point of } |\,t\,|}$$

Table 25

Values of [PA/IA] Obtained Directly from Nemenyi's 5% Points of the Studentized Maximum Modulus ($DDF = 5, 10, 20, \infty$)

m	[PA/IA] based on Nemenyi				Fitted
	$DDF = 5$	$DDF = 10$	$DDF = 20$	$DDF = \infty$	Approx.
1	1.00	1.00	1.00	1.00	$1.00 + 0.0/DDF$
2	1.20	1.17	1.16	1.14	$1.14 + 0.3/DDF$
3	1.32	1.27	1.25	1.22	$1.22 + 0.5/DDF$
4	1.39	1.33	1.30	1.27	$1.27 + 0.6/DDF$
5	1.47	1.39	1.35	1.31	$1.31 + 0.8/DDF$
6	1.53*	1.43	1.39	1.34	$1.34 + 0.9/DDF$
7	1.57	1.45**	1.41*	1.37	$1.37 + 1.0/DDF$
8	1.62*	1.51*	1.45	1.39	$1.39 + 1.1/DDF$
9	1.65	1.53	1.47	1.41	$1.41 + 1.2/DDF$
10	1.68*	1.56	1.49	1.43	$1.43 + 1.3/DDF$
15	1.82	1.66	1.58	1.50	$1.50 + 1.6/DDF$
20	1.87***	1.72	1.63	1.54	$1.54 + 1.8/DDF$

Number of *'s indicates deviation from fitted approximation in units of 0.01.

the batchwise factor. Here we can use the recent results of Nemenyi [50] $1 \le m \le 20$. We find the results set forth in Table 25, with the exception of two values which may involve copying errors, the simple approximate formula fits within 1% down to $DDF = 5$. A very satisfactory fit.

The values of [PA/IA] for $DDF = \infty$ can be calculated as the upper q % points of the normal. There

$$((100 - q)\%)^m = (100 - 5\%)$$

and if $q\ \% = (5(\mu)\%$, $u = \log_{10} r$, then

$$u = \log_{10} m - .025(1 - \frac{1}{m})$$

with good accuracy.

For $DDF = 10$, we observe that for the known values we have, approximately,

$$[PA/IA] = (\text{upper } (5/r')\% \text{ of } |t|)$$

where $u' = \log_{10} r'$ is given by

$$u' = \log_{10} m \; (1 - .05 - .03 \log m).$$

This formula, which checks well for the values of m in Table 25, was used for other values of m. Then an $A + B/DDF$ formula was fitted to the values for $DDF = \infty$ and $DDF = 10$.

The results are rather extrapolated, but are probably of better than working accuracy.

99. THE HINDSIGHT FACTOR *WSD/LSD*

The allowance for error rates familywise, as we saw in Section 81, is s^* times the upper $p\%$ point of the Studentized range.

Thus allowance for error rate familywise = (upper 5% point of $w_{m,f}$) (s). Now the variance of $e_i - e_j$ is 2^{σ^2}, so that allowance for error rate per comparison = 2 (upper 5% point of t_f) (s), and hence,

$$[WSD/LSD] = \text{hindsight factor} = \frac{\text{upper 5\% point of } w_{m,f}}{\text{upper 5\% of } \mid t_f \mid \text{ times } \sqrt{2}}$$

$$= \frac{\text{upper 5\% point of } w_{m,f}}{\text{upper 5\% point of } w_{2,f}}.$$

We can now resort to May's tables and calculate hindsight factors easily for all $m \leq 20$. The results for $m = 3, 4, 5, 10, 15$ and 20 and $f = DDF = \infty, 20, 10, 7, 5$ are shown in Table 26. The meat of the matter lies in the % errors of the working formula which are given in Table 27. We recall that

(1) we really intend to deal with the augmented hindsight factor;

(2) the basic values from May are to 2 decimal places, and so values in Table 27 are subject to effects of 0.1 to 0.2% due to rounding.

In connection with (1), we recall that the hindsight factor for $m = 3+$ was about 0.3% greater than that for $m = 3$ when $f = \infty$ and more nearby 1.0% greater for $f = 10$. Thus the larger percentage errors for $m = 3$ and $m = 4$ are of almost exactly the right amount and sign to allow for the fact that we

Table 26

Values of an Approximations to [WDS/LSD]

(a) = value from tables (not accurate to 3D)

(b) = linear interplate

(c) = working formula

	$DDF = \infty$		$DDF = 20$			$DDF = 10$		$DDF = 7$			$DDF = 5$			Working
m	(a) = (b)	(c)	(a)	(b)	(c)	(a) = (b)	(c)	(a)	(b)	(c)	(a)	(b)	(c)	formula
3	1.198	1.20	1.214	1.211	1.22	1.243	1.24	1.243	1.234	1.257	1.249	1.256	1.28	$1.20 + 0.4/DDF$
4	1.311	1.32	1.343	1.342	1.35	1.374	1.38	1.399	1.401	1.406	1.425	1.437	1.44	$1.32 + 0.6/DDF$
5	1.393	1.40	1.437	1.437	1.44	1.480	1.48	1.513	1.517	1.514	1.553	1.567	1.56	$1.40 + 0.8/DDF$
10	1.614	1.62	1.691	1.696	1.70	1.778	1.78	1.839	1.848	1.848	1.905	1.942	1.94	$1.62 + 1.6/DDF$
15	1.732	1.74	1.841	1.837	1.845	1.942	1.95	2.018	2.032	2.040	2.101	2.152	2.16	$1.74 + 2.1/DDF$
20	1.808	1.81	1.935	1.930	1.945	2.053	2.06	2.141	2.158	2.167	2.236	2.298	2.31	$1.81 + 2.5/DDF$

Table 27

Percentage Errors of Working Formulas for Hindsight Factor

			DDF		
m	∞	20	10	7	5
3	0.2	0.5	0.2	1.1	2.5
4	0.7	0.5	0.4	0.5	1.1
5	0.5	0.2	0.0	0.1	0.5
10	0.4	0.5	0.1	0.5	1.8
15	0.5	0.2	0.4	1.1	2.8
20	0.1	0.5	0.3	1.2	3.3

really want to work with the augmented hindsight factor. In general, it seems that for

$$DDF > \text{the smaller of 10 and } m$$

the working formula fits with an error of less than ±0.5% (almost always in the conservative, positive direction), which is entirely satisfactory accuracy for practical purposes.

For $20 < m \leq 100$ and $DDF = \infty$ we can use the % points of the range as given by Pearson [59] to determine the A's in the $A + B/DDF$ formulas of Table 1. Before making use of this table, approximate A's were obtained by observing that

$$[WSD/LSD] \text{ for } m \leq [PA/IA] \text{ for } {}^{1}/{}_{2}\, m(m-1)$$

and that the ratio of the left-hand to the right-hand side of this inequality increases from .91 for $m = 3$ to about .938 for $10 \leq m \leq 20$. A curve of this ratio was extrapolated, and the values of A obtained differed from those based on Pearson's table by at most 0.01. Thus the extrapolated values for $m = 200$ and $m = 500$, which were obtained with the aid of the extrapolated ratio, are believed to be quite reliable.

Since no studentized range tables are available for $m > 20$, some other method had to be used for the B values. It is reasonable to suppose that the ratio B/A for $[WSD/LSD]$ should be related to the corresponding ratio B_F/A_F for $[FSD/LSD]$, and that the relation would be of the form

$$(B/A \text{ for } m) = (B_F/A_F \text{ for } m')$$

where m' would be expected to be a smooth function of m, with $m' < m$.

Investigation of m' as a function of m showed that m'/m decreased steadily from 1.00 for $m = 2$ to 0.71 for $m = 20$. It is almost certain that $m'/m \to o$ as $m \to \infty$, but rather slowly, so that $m' \to \infty$. However in order to be both definite and conservative, it was decided to calculate the B values for $m > 20$ in Table 1 on the arbitrary assumption that $m' = .707m$ for $m > 20$. The true m' values will be somewhat smaller, so that the B values in Table 1 are somewhat too large for $m > 20$. This will ordinarily be unimportant since for $DDF \geq m$ these oversize B values contribute no more than 6.5% to the size of the WSD for $m = 20$. 3.9% for $m = 50$, and 1.5% for $m = 200$.

100. THE OMNIBUS FACTOR *FSD/LSD*

In order to make the substitute contrast (and universal) allowances available, we must tabulate either

$$FSD/s = p\ \%\ \text{points of}\ \sqrt{2(m-1)F}$$

or

$$FSD/LSD = \frac{p\ \%\ \text{point of}\ \sqrt{2(m-1)F}}{p\ \%\ \text{point of}\ |\,t\,|\ \sqrt{2}}\ .$$

As in the case of the hindsight factor, it proves simpler to tabulate the latter. Approximations in our standard form $A + B/DDF$ have already appeared in Chapter 5. Our task here is merely to examine the quality of the approximation. Table 28 sets forth comparing $(m-1)$ times the 5% of F on $m-1$ and DDF degrees of freedom with

$$\text{(approximate omnibus factor)}^2\ (5\%\ \text{point of}\ |\,t\,|\,)^2$$
$$= \text{(approximate omnibus factor)}^2\ (5\%\ \text{point of}\ F\ \text{on 1 and}\ DDF)$$

which is the corresponding approximation. If

$$DDF \geq 6 \log_{10} (0.4m),$$

which is really a very weak condition (note that $\geq DDF\ m-1$ already implies this), the error in the length of the omnibus factor is less than ±2%.

Table 28

Tabular Values of the Upper 5% Point of $(m-1)F$, Where F has $m-1$ and DDF Degrees of Freedom, Compared With Approximate Values of the Form $(A+B/DDF)^2$. (Entries (a) are Tabular. Entries (b) are Approximate.)

m	$m-1$		$DDF=\infty$	$DDF=20$	$DDF=10$	$DDF=7$	$DDF=5$	$DDF=3$	$DDF=2$
2	1	(a)	3.84	4.35	4.96	5.59	6.61	10.1	18.5
		(b)	3.84	4.35	4.96	5.59	6.61	10.1	18.5
3	2	(a)	6.00	6.98	8.20	9.48	11.8	19.1	38.0
		(b)	6.00	6.95	8.10	9.32	11.4)	18.4	36.1
5	4	(a)	9.48	11.5	13.9	16.4	20.8	36.5	76.8
		(b)	9.48	11.4	13.8	16.4	20.7	36.5	79.0
7	6	(a)	12.6	15.6	19.3	23.2	29.7	53.6	116
		(b)	12.6	15.5	19.2	23.2	30.0	55.7)	(126)
10	9	(a)	16.9	21.6	27.2	33.1	42.9	79.3	175
		(b)	16.9	21.5	27.0	33.4	44.0)	(85.0)	(189)
15	14	(a)	23.7	31.1	40.0	49.4	65.0	122	
		(b)	23.5	30.7	40.0	50.0	(67.6)	(136)	
20	19	(a)	30.2	40.7	52.8	65.7	86.8		
		(b)	30.0	40.0	52.6	66.5	(91.1)		
51	50	(a)	67.5	98.5	132	166			
		(b)	67.9	95.0	132	172)			
201	200	(a)	234	376	512				
		(b)	234	(355)	522	(720)			

Approximate values where the error in $\sqrt{(m-1)F}$ is between 1 and 2% are marked with a half parenthesis. Those where it exceeds 2% are placed in full parenthesis.

CHAPTER 17: SHORT-CUTS

101. INTRODUCTION

In this chapter we outline very briefly the derivation of the tables and formulas for short-cuts and semishort-cuts. This account is based in large part on the work of Kurtz, Link, Tukey, and Wallace [40]. Details will, in general, be found in that paper. A few details are given here in order to allow the reader to construct at least rough short-cuts for more complex situations than single and double classification.

In general, our purpose is to use a sum of ranges to replace the square-root of a mean square. The ranges may be of groups of observations (as in single classifications) or of groups of differences of observations (as in double classifications) or of groups of more complicated expressions.

In a complex design we may decide to consider only some of the groups which make up a symmetrically related family of groups. Thus the particular double classification method described in Chapter 3, uses only the groups of differences between adjacent pairs of columns, although there are $\frac{1}{2}m(m-1)$ pairs of columns, and the differences of any pair stand on an equal footing with those of any other pair. Such partial use is likely to be most pronounced for designs with larger numbers of observations.

We are going to substitute a sum of ranges for the square-root of a mean square of deviations. In fact we are going to multiply the sum of ranges by a factor and treat it as if it were the square-root of mean square deviations with an equivalent number of degrees of freedom. What we need are two values, the factor and the equivalent degrees of freedom.

102. THE FACTOR

When we say that we shall treat the sum of ranges w_+, as if it were the square-root of a mean square deviation, *MSD*, we have a number of choices. We could choose a factor k_1, so that

$$\text{ave}\{k_1 w_+\} = \text{ave}\{(MSD)^{1/2}\},$$

or we could choose a factor k such that

$$\text{ave}\{(kw_+)^2\} = \text{ave}\{MSD\} = \sigma^2$$

or we could try to match in some other way. While others have chosen to use k_1, Kurtz, Link, Wallace and the writer [40] chose k for reasons which they discuss, and we shall follow this practice.

If

$$w_+ = \Sigma\, w_i\,, \qquad i = 1, 2, \ldots,$$

where for each individual range

$$\text{ave}\{w_i\} = y\sigma,$$

$$\text{ave}\{w_i\} = \beta^2\sigma^2,$$

and where ρ is the correlation between any two ranges, then

$$\text{ave}\{w_+^2\} = \text{ave}\{\Sigma\,\Sigma\, w_i w_j\} = (m^2 a^2 + m\,(1 + (m-1)\rho)\beta^2)\sigma^2$$

as may be verified by a simple computation. If we wish to have

$$\text{ave}\{(kw_+)^2\} = k^2 \ \text{ave}\{w_+^2\} = \sigma^2,$$

then we must take

$$k = \frac{1}{\sqrt{m^2 a^2 + m\,(1 + (m-1)\rho)\beta^2}}\,.$$

We can write

$$\begin{aligned} k &= \frac{1}{m \cdot a}\left\{1 + \left\{\frac{1}{m} + \left\{1 - \frac{1}{m}\right\}\rho\right\}\frac{\beta^2}{a^2}\right\}^{-1/2} \\ &\approx \frac{1}{m \cdot a}\left\{1 - \left\{\frac{1}{2m} + \frac{\rho}{2}\right\}\frac{\beta^2}{a^2}\right\}, \end{aligned}$$

and a glance at a table of a and β (as for example, Hald [32], Table VIII) shows us that β^2/a^2 is reasonably small, being ≤0.1 for ranges of 7 or more. Thus we need not know ρ very accurately, since it appears with a small coefficient. Hartley [34] has obtained the approximate ρ for ranges of correlated values, and the sort of accuracy considered sufficient in the present paper is usually obtained by

(1) assuming independence;

(2) using Hartley's values; or

(3) making a judgement interpolation between (1) and (2).

For a table of the k's underlying Tables 10 and 11 see [40].

103. EQUIVALENT DEGREES OF FREEDOM

Those who choose k_1 as a normalization for their sum of ranges, naturally establish the equivalent number of degrees of freedom by matching variances (or, what is equivalent, by matching second moments). Those who choose k, as we have, determine the equivalent number of degrees of freedom by matching variances of w_+^2 (or, what is equivalent, by matching $(w_+^2)^2$, that is by matching 4th moments).

For a mean square with DDF degrees of freedom, normal theory gives

$$\text{ave}\{MSD\} = \sigma^2$$

$$\text{var}\{MSD\} = \frac{2}{DDF}\sigma^4$$

so that we naturally combine

$$\text{ave}\{(kw_+)^2\} = \sigma^2$$

with

$$\text{var}\{(kw_+)^2\} = \frac{2}{DDF^*}\sigma^4,$$

thus defining DDF^*, the equivalent number of degrees of freedom.

Tables of DDF^* for various standard situations are given by Kurtz, Link, Tukey, and Wallace [40], for the convenience of those who may wish to extend short-cuts, we give condensed tables of

$$\frac{DDF^*}{DDF}$$

where DDF is the number of degrees of freedom that would be extracted from the design by a corresponding sum of squares of deviations or by the corresponding mean square. Table 10 gives this fractional recovery for independent ranges. Table 11 gives this fractional recovery for a chain of ranges where the correlation of nonadjacent ranges is zero and that of adjacent ranges derives from a correlation of ± 0.5 across the groups whose ranges are involved. Table 29 gives this fractional recovery for a group of equally correlated ranges whose correlation arises from equal correlation of $\pm 1/(m-1)$ across the groups.

In the half short-cut situation, where we sum the squares of ranges, the analysis is even simpler, since we have

$$\text{ave}\{\Sigma\, w_i^2\} = m\ \text{ave}\{w_i^2\}$$

$$\text{var}\{\Sigma\, w_i^2\} = m\ \text{var}\{w_i^2\} + m\,(m-1)\ \text{cov}\{w_i^2,\, w_j^2\}$$

Table 29

Values of [(equivalent *DDF*)/($m-1$) ($r-1$)] for m Ranges of r, Based on Values Correlated $-1/(m-1)$ Across the Ranges

r = number of measurements per range

m	2	3	4	5	6	8	10	12	14	16	18	20
2	.96	.98	.98	.95	.94	.92	.88	.85	.82	.79	.77	.74
3	.95	.98	.97	.94	.93	.91	.87	.84	.81	.78	.76	.73
5	.92	.95	.95	.94	.92	.89	.86	.83	.80	.77	.75	.72
10	.91	.93	.93	.92	.91	.89	.84	.82	.79	.76	.74	.72
20	.90	.93	.93	.92	.91	.88	.84	.82	.79	.76	.74	.71
∞	.90	.93	.93	.92	.91	.88	.84	.81	.78	.75	.73	.70

(m = number of ranges) [approx. $1-.15r+.08/m$, for $m>4$; Author's note]

if the groups are symmetrically related. Thus

$$k_+^2 = k_i^2/m,$$

$$\text{var}\{k_+^2 \,\Sigma\, w_i^2\} = \frac{1}{m}\,\text{var}\{k_i^2 w_i^2\}\,(1+(m-1)\rho),$$

$$DDF_+^* = DDF_i^* \,\frac{m}{1+(m-1)\rho_2}$$

where

$$\rho_1 = \frac{\text{cov}\{w_i^2, w_j^2\}}{\text{var}\{w_i^2\}}$$

is the correlation between squared ranges, which may be estimated as rather roughly equal to

$$\rho_1 = \frac{\text{cov}\; w_i,\, w_j}{\text{var}\; w_i}\,.$$

Thus DDF_+^* is roughly the same as for the short-cut, although more careful analysis shows that the value for the half-cut is slightly larger. The approximation of using the same number of DDF^* for both forms seems, however, to be satisfactory.

104. ERROR-RATE SHIFTING

Since we are consistently treating an adjusted combination of ranges as an s or s^2 with an equivalent number of degrees of freedom, we clearly use the error-rate shifting factors for *WSD*, replacing *DDF* by *DDF**, when we wish to shift error-rates in the short-cut procedures.

105. EXAMPLE

An example of how the methods and tables of this chapter may be used, we consider an example. Let us suppose that we have a double classification with 20 columns and 10 rows, which we wish to analyze on the basis of the following incomplete set of ranges

$$w_1 = \text{range}\{x_{11} - x_{21},\ x_{12} - x_{22},\ x_{13} - x_{23},\ x_{14} - x_{24},\ x_{15} - x_{25}\}$$

$$w_2 = \text{range}\{x_{26} - x_{36},\ x_{27} - x_{37},\ x_{28} - x_{38},\ x_{29} - x_{39},\ x_{2,10} - x_{3,10}\}$$

$$w_3 = \text{range}\{x_{31} - x_{41},\ x_{32} - x_{42},\ x_{33} - x_{43},\ x_{34} - x_{44},\ x_{35} - x_{45}\}$$

$$w_4 = \text{range}\{x_{46} - x_{56},\ x_{47} - x_{57},\ x_{48} - x_{58},\ x_{49} - x_{59},\ x_{4,10} - x_{5,10}\}$$

where the ranges of odd-minus-even column differences are over rows 1 to 5 and the ranges of even-minus-odd column differences are over rows 6 to 10. The result is 20 ranges of 5 differences.

Each difference has variance

$$2\,(\text{variance of an individual fluctuation}) = 0.2\,(\text{variance of column total})$$

$$= 0.2\sigma^2$$

and all difference are uncorrelated, hence the ranges are uncorrelated, so that

$$\text{ave}\{(\Sigma\, w_i)^2\} = (m^2\mu_{\bar{5}}^2 + m\beta_{\bar{5}}^2)2\sigma^2$$

$$= \{361(2.326)^2 + 19(.864)^2\}\, 0.2^2$$

and

$$[s/\Sigma\, w_i] = k = \frac{1}{19(2.326)\sqrt{0.2}} \left\{ 1 + \frac{1}{19} \left\{ \frac{.864}{2.326} \right\}^2 \right\}^{-1/2} = .0504.$$

From Table 10 we find

$$\text{equivalent } DDF = 20(4)(0.92) = 74.$$

Hence

$$[LSD/s\,] = 2.82$$

$$[WSD/LSD\,] = 1.84$$

and so, finally

$$[WSD/\Sigma\, w_i] = [WSD/LSD\,]\,[LSD/s\,]\,[s/\Sigma\, w_i]$$

$$= 1.84(2.82)\,.0504$$

$$= .262.$$

Here we have the short-cut factor from the sum of ranges of such a set of differences to the allowance for column totals.

CHAPTER 18: DESIGNING FOR ACCURACY

106. DESIGNING FOR KNOWN ACCURACY

The bulk of the present paper is devoted to analysis — to the analysis of data by means of allowances. This chapter is the part devoted to design. One aspect of the designed experiment is planning for predetermined accuracy. This is surprisingly simple with allowances, if we interpret "predetermined accuracy" in the proper way. This choice not only leads to numerical simplicity, but also seems highly appropriate from the point of view of the user.

Those who take the decision approach to analysis will predetermine accuracy in terms of "power." More precisely they will ask to have at least a certain probability of asserting "significance" when some value or difference deviates from the null hypothesis by a given amount. Those who, like ourselves, take the estimation approach will find it natural to predetermine accuracy differently, probably in terms of a "confidently bounded allowance." More precisely, we shall seek for a chosen degree of confidence (that is, a chosen probability) that the allowance (individual, parallel or comparative) that we seek will be no larger than a chosen bound. We suggest, tentatively, the selection of 90% for this chosen level of confidence and the abbreviation *CBA* for the confident bound for the allowance.

We note in passing that a confidently bounded allowance implies a minimum corresponding power. Consider the case of a comparative allowance for k determinations. Let us suppose that the experiment has been planned to give 90% confidence that the allowance is less than B, which is thus the *CBA*. Then in 90% of all experiments we shall assert that

$$| \; x_i - \mu_i - (x_j - \mu_j) \; | \; \leq A$$

where $A \leq B$, which implies

$$| \; x_i - \mu_i - (x_j - \mu_j) \; | \; = \; | \; (x_i - x_j) - (\mu_i - \mu_j) \; | \; \leq B \; .$$

In all but 5% of all families, if A is based on an error rate of 5% familywise, these assertions will all be true. Thus if we except 5% of all families, if

$$\mu_i > \mu_j + 2B$$

which is equivalent to

$$\mu_i - \mu_j \geq 2B$$

then

$$x_i - x_j \geq B \; ,$$

but

$$| \; (x_i - x_j) - (\mu_i - \mu_j) \; | \; \leq B$$

which implies the assertion

$$\mu_i - \mu_j > 0$$

which is equivalent to the assertion

$$\mu_i > \mu_j .$$

Thus in at least 85% of all families $\mu_i > \mu_j + 2B$ implies that among our particular conclusions will be $\mu_i > \mu_j$. The "power" of the procedure is at least 85% for a true difference of $2B$.

The last long paragraph was inserted for deciders and not for estimators. As among those who estimate, we prefer the precise interpretation of the *CBA* in the form — "In 90% of all experiments we will select intervals of length at most $2B$ within which we assert the corresponding true difference to be, in 95% of all families all our assertions will be correct." This we find more tasty and more useful than a statement phrased in terms of power.

107. GENERAL PROCEDURE

We have set the problem, how do we solve it? The general form for a comparative allowance is the product of two factors, namely:

$$\text{percentage point of studentized range} = w\,(r)$$

and

$$(\text{estimated variance of determination})^{1/2} = s = f\,(r)s_w\,.$$

Here r is an indicator of the affective amount of replication in the design, which enters into $w\,(r)$ because $f = DDF$ depends on this amount of replication, and which enters into $f\,(r)$ because the ratio of σ^2, the variance of a determination, to σ_w^2, the variance of an individual fluctuation, also depends on this amount of replication.

Thus we had a single classification, with each of m determinations of mean of r replicate measurements, we would have:

$$DDF = m\,(r-1),$$

$$\sigma^2 = \sigma_w^2 / r,$$

$$s = s_w / \sqrt{r},$$

$$f\,(r) = 1/\sqrt{r},$$

a case to which we shall return in the next section.

If we fix r, then the allowance is proportional to s_w. If we wish a confident bound for the allowance, then we must find a confident bound for s_w. If our knowledge of σ_w^2 can be represented by

(1) an estimate or guess S_w^2 of s_w^2; and

(2) an equivalent number of degrees of freedom, DDF^*, representing the accuracy of this guess or estimate;

then we are in a position to act as if

$$s_w^2 / S_w^2$$

follows the F-distribution with DDF and DDF^* degrees of freedom. This would correspond to 90% confidence that

$$s_w^2 / S_w^2 \le F_r$$

or

$$s_w^2 \le S_w^2\, F_r$$

where F_r is the 90% point of F with DDF (which depends on r) and DDF^* degrees of freedom. This corresponds to 90% confidence that

$$\text{allowance} = w\,(r) f\,(r) s_w \le w\,(r) f\,(r) \sqrt{F_r}\; S_w\,.$$

If we can accept this type of summarization, the problem of predetermined accuracy is easy in principle.

Should we accept this sort of a summary? Perhaps, and perhaps not. The ratio s_w^2 / S_w^2 is, at best, the quotient of two mean square from *different* samples. This means that its distribution is rather sensitive to lack of normality in the underlying distribution of fluctuations. If we took the 90% confidence that

$$s_w \leq \sqrt{F_r}\; S_w$$

too seriously, we should have to be seriously concerned about this sensitivity. In many applications, however, we may be perfectly content to neglect this sensitivity (which is equally prevalent in *any* procedure for obtaining predetermined power for any nonsequential method of analysis).

In the remaining situations, we must either:

(1) guess a 90% confident bound for s_w directly; or

(2) have on hand a large enough collection of observed values of s_w to estimate a 90% point directly; or, possibly,

(3) plan to calculate our allowance from a long term estimate of σ_w^2 rather than from the estimate, s_w^2, of σ_w^2 to be obtained in the experiment being designed.

These are not the usual cases. Usually we shall use the $\sqrt{F_r}$, and hope it suffices.

108. SIMPLE DESIGNS

The two simplest designs to consider are the single and double classifications, for which

$$DDF = \begin{cases} m\,(r-1), & \text{(single classif.)} \\ (m-1)\,(r-1), & \text{(double classif.)} \end{cases} \qquad (*)$$

$$f\,(r) = \frac{1}{\sqrt{r}}\,.$$

If we desire B as a CBA (90% confident bound for the allowance) we must choose r so that

$$w\,(r)\,\frac{1}{\sqrt{r}}\,\sqrt{F_r}\;S_w \leq B$$

which may be written

$$r \geq w^2(r)F_r \frac{S_w^2}{B^2} .$$

The advantage to this form lies in the relatively slow dependence of the product $w^2(r)F_r$ on r for fixed m, and DDF^*.

We recall that

(1) $w(r)$ is the 95% point of the studentized range of m with DDF degrees of freedom;

(2) the figure 95% arises from our standard choice of a 5% error-rate;

(3) F_r is the 90% point of tabular F with DDF and DDF^* degrees of freedom;

(4) the figure 90% is directly related to the desired confidence;

(5) if an even chance, rather than 90% confidence of an allowance as short as B is desired, F_r is to be replaced by the corresponding 50% point.

In Tables 5 and 8 of Chapters 2 and 3, respectively, we presented approximate values of $w^2(r)F_r$ for balanced single and double classifications, with directions for use and examples. Similar tables can be prepared for other classes of designs, or complete trial and error procedures may be used.

PART D
BASIC ISSUES

CHAPTER 19: STATISTICAL METHODOLOGY

110. INTRODUCTION

In this part we shall try to discuss some broad underlying issues of statistical methodology and philosophy. In the present connection they bear on the choice between systems of allowances on the one hand (op. Part E) and multilayer significance procedures on the other (op. Part F). In general they bear on the choice between confidence and significance procedures and on other broad decisions.

Some readers may feel that we are adopting an anti-mathematical point of view or that we are raising issues better left ignored. For we shall not be concerned in this part with a mathematical formulation or, directly, with a choice between alternative mathematical formulations. Rather we will be mainly concerned with the aims of our statistical procedures and with the extent to which certain broad types of procedures meet the various aims. These considerations come before either mathematical formulation or mathematical analysis so far as the logic of the situation goes. Historically, it was to be expected that they would first be actively discussed after two or more alternative broad types of procedures had been mathematically formulated and put to good use.

To date, the statement of statisticians' insight into such basic issues has most usually taken the forms:

"Ah, at last I see it, *take it from me*."

or the form

"My, isn't this a neat mathematical formulation, *it will be so nice to teach*."

The writer feels that appeals to authority and appeals to mathematical elegance are both inadequate. Both provide useful evidence, but neither is adequate.

The words which follow are an attempt to explain *why* the writer "sees it at last," and are provided in the hope that they will aid the readers to insights of their own (which may or may not agree with the writer's).

111. SOURCES

The development of statistical methodology and philosophy has rested on two sorts of insight-insight into how people *should* behave and insight into how people *actually* behave. The latter sort has perhaps been a more significant contributor to the progress of statistics than the former. Formal statistics has always profited from careful examination of the ways of the best practitioners of informal statistics.

Where have these best practitioners been found? Undoubtedly in some branch of science, biological, physical or social. It is not accidental that Student was a chemist, that Fisher studied under Eddington, that Neyman was attached to an agricultural college for 11 years, that Wald had an extensive background in econometrics. For it is in the various branches of science that most attention has been paid to methodology — it is there that methodology has had its most extensive empirical testing. It will pay us, as statisticians, to look to the practice and philosophy of able scientists for some time to come.

While modern statistics took its impulse and its initial support from the needs and difficulties of biological scientists, pure or applied, this need not imply that biological scientists have a monopoly on, or even a plurality of, skill in informal statistics. The writer believes, indeed, that it is the physical scientists who have gone furthest in this, and he believes that we, as statisticians, still have much to learn from them. (Their long-continued lead in principles has, until recently, served to keep them unnecessarily unaware of modern statistical techniques.)

Thus the writer feels that, after those examples which are clear to almost all readers, the best forms of evidence for or against a statistical procedure or item of philosophy are the actual actions of physical scientists and a consensus of their opinion. (The physical scientist may or may not use efficient estimates, and he may or may not judge his probabilities correctly, but he is very likely to *try* to do the right thing.) This is undoubtedly a biased view, for the two strongest influences which have affected the writer's statistical philosophy are undoubtedly (1) his early training in physical science, and (2) his association with Charles P. Winsor, himself originally trained in physical science. Biased — but perhaps usefully so.

The frequent reference to science and the working philosophy of its practitioners which follow are not intended to imply that the statistical techniques which are the ultimate product are designed for science alone. By no means. They are an appeal to the informal specialists, in the hope of constructing a better formal tool for all.

112. SCIENCE VS. ADMINISTRATION

Before we go on to discuss science as a source of guidance for statistical methodology, we need to make a clear distinction between science and administration, between scientific decisions "for the record" and administrative decisions "for the moment." We do not imply an antithesis here between science and administration. A good administrator uses all the relevant scientific facts, and science grows fastest when research is wisely administered. But there is a sharp distinction between the types of decisions that are made.

Science, historically (and the writer believes essentially) has built its structure a stone at a time, realizing that greater or smaller parts of the pattern must be reassembled into a better pattern from time to time, and realizing that a small fraction of all the stones will have to be discarded as defective. In general, however, it expects the results of an experiment or observational program to be of continuing validity. A small percentage will prove to be wrong, either from experimentalist's errors in method or from the random accumulation of fluctuations. Most, however, will stay in the record as useful information, although the interpretation will change from time to time.

Administrative decisions, on the other hand, are made *for the moment* on the basis of the best available data. The decision to spend time and money gathering more data before taking definite action is parallel to the decisions to take various definite actions. The decision between them is to be made on the basis of estimated gains and losses. The decisions of military, naval and air strategy and tactics, of diplomacy, government and politics, of day-to-day administration and the acceptance sampling of lots of raw or manufactured material are all of this character.

The rise of decision theory, and the discovery that the conventional statistical analyses appropriate to "for the record" decisions can be discussed in the framework of statistical decision functions has tended to blur the distinction between analyses "for the record" and analyses "for the moment." The writer feels that this distinction is important, and worth maintaining.

It is entirely appropriate to base both sorts of conclusions on the same evidence. Thus, when the first work on yellow fever was completed, an appropriate conclusion for the record would have been:

> The *Anopheles* mosquito is a carrier of yellow fever. Its population can be drastically reduced by removing or spraying pools of water. There may be other carriers.

while an appropriate administrative conclusion would have been:

> If we are to go forward successfully with the Panama Canal, we must remove or spray all the pools of water we can reach.

The administrative decision was made successfully, but years later it was learned that it would not have been successful in all parts of the world, because there *are* other carriers of yellow fever.

The fact that scientific judgement, experience and knowledge can be essential in making a decision does not make that decision a scientific one (at least not in the sense in which we use the term here). The yellow fever example shows this clearly. Conclusions "for the moment" are not necessarily conclusions "for the record."

Another way to formulate the difference between procedures intended for administrative and scientific uses is to say that the procedures differ in the amount of focusing among alternatives. Procedures intended "for the moment" emphasize *choice* or *selection*, with a single remaining alternative out of few or many, while procedures intended "for the record" emphasize *concentration*, the elimination of *as many* alternatives as may reasonably be eliminated.

113. DEFINITENESS

An important principle of science is that a more definite statement, if true and proved, is more useful than a less definite one. One of the great advances in the development of scientific method was the recognition that "A is different from B" is a more definite statement than "A is the same as B." More definite, because susceptible of finite proof! (Attempts to prove the second with a finite body of evidence always seemed to come down to "A seems to be the same as B, as far as I can tell after x year's work," a statement which is intuitively and clearly less definite than "A is different from B."

To the original statement of least definiteness "the difference between A and B *seems* to be in this direction," there was added the statement of greater definiteness "A is different from B" or perhaps "A is more — than B."

But the recognition of this fact of relative definiteness does not stand alone. With the development of quantitative measurement, not only "A possesses between 3.24 and 3.25 units more of this property than B" but also "A and B differ in the amount of this property they possess by less than 0.01 unit" can be recognized as very definite statements. The answer to "By how much?" was found to be at least as definite as the answer to "Could they be the same?" (The writer finds "how much" the source of much the more definite statement!) For those who doubt the interest of statements about nearness, we recall the Michaelson-Morley-Miller "ether drift" experiment, whose importance in ushering in the theory of relativity lay in its proof of near equality of the apparent velocity of light in different directions.

The physical scientist made a fourth discovery of a class of definite, and thereby useful, statements. He introduced the general law. The statement "All bodies attract one another inversely proportional to the square of their mutual distance." is very definite indeed, and its usefulness stems in large measure from its definiteness.

114. DESCRIPTION

The first sort of definiteness was reflected in statistical methodology in a wide variety of descriptive procedures, which sought to extract from the observations and summarize what "seemed" to be the case. In some fields of application, notably economic indices, statistics is still mainly concerned with problems of pure description.

Science began with description, but early in its development it became realized that, without some indication of how wide variations are likely, "seems" alone was of very little use. Thus there was a steady, tacit shift from description toward some informal sort of confidence procedure.

Description is still a scientific technique in many fields, but in others the bare "seems" has become an administrative statement, good enough for the moment, but no longer good enough for the record. More and more fields are entering this latter state all the time. Description, *accompanied* by even a vague estimate of accuracy is concentration, and is appropriate "for the record," although of weak definiteness. Description alone is selection of one alternative and is appropriate "for the moment."

115. SIGNIFICANCE

The second sort of definiteness was reflected in statistical methodology both early and often. Its final flower was the formal test of significance, whose logical structure runs as follows:

(1) If A and B are the same in these respects, we can calculate the probabilities of the various outcomes.

(2) The outcome observed was very extreme — an outcome more or equally extreme would occur only rarely if A and B are the same.

(3) Therefore A and B should be considered different.

Such a test of significance is a direct and useful formalization of the second sort of definiteness.

116. CONFIDENCE

The second sort of definiteness appears almost as early in the history of statistics as the first sort. For a long time, however, it appeared less frequently, although today it is very important. It has flowered into the confidence interval, which although usually described differently, has the following logical pattern:

(1) If we *contemplate* a particular numerical difference between A and B, we can calculate the probability of various outcomes.

(2) We can assess the probability, for each particular contemplated numerical difference, of an outcome more or equally extreme than the observed outcome.

(3) We can discard those contemplated differences for which the observed result is unduly extreme in the probability sense.

(4) We can assert that the "true" difference lies among the remaining numerical differences, confident that a regular user of the whole procedure will be in error with no more than a chosen probability.

Here we have a clearly formulated way to pick out, and in some sense validate, statements of the third sort of definiteness — answers to "By how much"?

117. THE FOURTH SORT

By analogy, we might expect some statistical procedures related to the fourth sort of definiteness — to the development of general principles. But when we study the actual actions of scientists, and the reasons why they accept general principles, we find a serious difficulty not yet overcome (and one which seems quite likely not to be overcome!). For the evidence for or against a general principle in science seems always to be two-fold, consisting of both

(1) the observation of a number of isolated cases; and

(2) the extent to which the principle fits into the scientific scheme of things.

Either of these *can* dominate. Enough isolated cases can and do change the scientific, or at least the scientist's scheme of things. But an eminent theoretical physicist has been quoted as saying that the dying off of older physicists helps more to bring new theories into vogue than does the arrival of new empirical data. We lack today, and may well continue to lack for a long time, any reasonably objective way of including (2) in a formal procedure for weighing the strength of evidence concerning a possible general law.

Until this lack is met, statistical procedures must relate themselves to other sorts of definiteness. Until new sorts of definiteness or new sorts of procedures are developed, statistics arising from science will live mainly by significance or confidence. Which is preferable?

118. INTELLIGENT GAMBLING

Administration (and military command) long ago discovered another kind of definite statement — "It is time to stop measuring (or scouting or gathering intelligence) and take the action which seems best on the data so far available." This statement or its negative answers a question of the form "which is the better gamble?," namely the question "Is it a better gamble to act or to measure (scout, etc.) further"?

The formalization of such situations began with sampling inspection of industrial production and continued through sequential analysis to the theory of decision functions. Its most highly developed state turned out to be so general and flexible that the mathematical theories of significance and confidence could be and have been incorporated as special portions. At a purely formal level they have been incorporated into the theory of decision functions. At a logical level, however, the writer believes them to be still quite distinct. For the insights into definiteness on which they were built came from science rather than administration.

119. SUMMARY

The main conclusions of this chapter are these:

(1) The actions and philosophy of scientists, especially physical scientists, are a very good source of suggestions as to statistical methodology.

(2) Three broad sorts of statistical methodology have stemmed from the recognition in science of three types of definite sentences:

descriptive procedures — "A seems to be . . ."

significance procedures — "A is different from B"

confidence procedures — "The difference between A and B lies between c and d."

The latter two are most important.

(3) Another sort of statistical methodology, decision procedures, stems from administration rather than science.

(4) While all three important sorts, significance, confidence, and decision can be embedded in the mathematical theory of decision functions, their logical difference is large enough to make separation worthwhile.

CHAPTER 20: ACTION, INDICATION AND SANCTIFICATION

120. INTRODUCTION

We come now to approach another facet of the same problem, to study the aims rather than the sources of statistical procedures. Again we shall make a classification, but this time it will be based on the apparent motive which we seem to see as we see statistics used in practice. After discussing this classification, we shall return to its relation to the classification of the previous chapter.

We now divide problems into those of *action*, *indication*, and *sanctification*. Specifically, we use these words as follows:

> A statistical analysis is made for *action*, if its purpose is to affect a single particular given decision, whether to be made now or at some time in the future.
>
> A statistical analysis is made for *indication* (or benchmarking) if its purpose is to make a value (or limits for a value) available for future use in unspecified decisions.
>
> A statistical analysis is made for *sanctification* if its purpose is to throw an aura of statistical reliability about some statement or appearance.

(We have phrased these in terms of decisions in order to avoid argument with strong supporters of the view that "all is decision." We hope by this means to focus attention away from such issues.) These distinctions are often less than perfectly clear. A single analysis can easily fall under more than one heading. Thus a study of the fission cross-section of uranium would be for action if its purpose was to decide between two alternative designs for an atomic bomb, but for indication if its purpose was to provide a reference number to aid the development of theoretical nuclear physics. A census of the United States would be for action if its purpose were the reapportionment of seats in the House of Representatives, and for indication if its purpose were to serve as a basis for many planning decisions by many unspecified planners. And so on.

121. THE ROLE OF *p*%

In the action situation, the relevant assessment of risk is relative. Thus in choosing to accept or reject a lot of raw or manufactured material the proper consideration is the *relative* average economic consequence of each decision. In a program of selecting persons for special training, the proper considerations may be, for example: Will further testing improve the quality of the selected group enough to pay for the added cost? Would more stringent selection according to present tests increase quality enough to make up for reduced quantity and for the loss of potentially excellent persons? In other words, the action situation always resolves itself to — "How should we gamble wisely"? There is no place for 5%, or 1%, or 20%! Even the decision to measure some more before jumping is an attempt to gamble wisely — for the cost of measurement and delay is being balanced against the cost of possible error.

In an indication situation, we have the opposite problem. We cannot try to balance risks — for we know not what decision is to arise nor what the costs will be. We can only say that we have specified confidence that what we are estimating lies in a certain range or area. We may be able to, and should when we can, indicate how this confidence varies with the length of this range or with the size of this area — so that each eventual consumer may be aided in his or her decision problem — but we cannot balance risks for them in their unknown problem. Here is the proper place of 5%, 1% or any *p*%! Here the confidence statement serves as a natural measure of our uncertainty. Here we have the opposite pole from action.

But it is not an unimportant pole — far from it. Most of science is devoted to the accumulation of knowledge for future use — to indication, not action. One of the eventual aims may be action, but there is much indication on the way.

Finally, we have the sanctification situations, where the purpose of the statistical analysis is to throw a cloak of statistical protection around the conclusions already obtained or appearances already noted. Protection against what? Usually only against the charge that "such results could reasonably have arisen by chance." The typical example of a sanctification procedure is the test of significance in its narrowest form. Here, again, 5%, 1% or p% plays a natural and appropriate role — a role somewhat different from that in indication theory. (We shall return to the distinction.) Again balancing of risks is irrelevant.

In summary, then, we may put matters as follows:

(1) in an action situation, risk balancing is the natural and appropriate concept, and p% has no place (except perhaps as a guide to the risks of one sort);

(2) in an indication or sanctification situation, p% is the natural and appropriate concept, and risk balancing has no place.

The nature of the appropriate risk statement is firmly tied to the action-indication-sanctification classification.

122. POWER VS. RISK BALANCING

In the last section we have drawn a clear line between the balancing of risks and the use of p%. At first sight this might seem to be in sharp conflict with the Neyman-Pearson theory of testing hypotheses. The writer feels that this is not the case.

In Neyman's more philosophical accounts of related matters, he lays great stress on "inductive behavior" — words which seem, and which seem intended to seem, to mean action rather than indication, and certainly do not mean sanctification. If it does mean action, then we would judge risk-balancing to be appropriate, and would urge the discussion of both kinds (or types) of error. What we would avoid would be 5%, 1% or p%, feeling that in each individual problem the balancing of risks would be different, and a different type I error would be appropriate. So far as the writer can see, there is nothing in this attitude inconsistent with that of Neyman or Pearson.

The main difference comes in another way. The present writer would assign a large part of statistical analysis to indication rather than to action. By this assignment, if it stands, we have taken much of statistical analysis out of risk balancing and into the domain of error rates.

Such a shift deemphasizes the role of the statistician. He is now not giving the answer — out and out. He is, instead, trying to summarize what the observations say as best he can. (Compare R. A. Fisher, [24, p. 311]). The giving of the answer is being left to someone else, perhaps (or even probably) many answer-givers will use this one summary in answering many questions. Such a shift is more appropriate in many cases than might seem at first glance. The methods of present-day analytical statistics grew around the problems of agriculture and biology. At that time, and in those fields, the average scientist was not quantitatively sophisticated. The growth of statistication among the practitioners in these fields went hand in hand with the growth of quantitative sophistication. Until a large and growing fraction of the scientists in these fields are now both statisticated and quantitatively sophisticated. Today the use of statistics is growing exponentially in chemistry, and in other fields of physical science where quantitative sophistication has been a commonplace for a long time. It is very appropriate here for the statistician to relinquish part of his temporary role as a substitute for the investigator, for here the investigator is far more quantitatively sophisticated, and can better perform his natural function of combining all the evidence, including the present observations, in reaching a decision.

The writer feels that such a position contradicts neither Neyman's emphasis on inductive behavior nor Fisher's emphasis on the summarization of an experiment standing alone. It does contradict the simple-minded way of putting them together, namely "inductive behavior based on this experiment alone." It goes along with a feeling that in very few places have we been able to formulate the risks precisely enough to safely leave the problem of inductive behavior to the statistician (although some aspects of acceptance sampling may have reached this stage). It calls for an effective summary of the individual experiment followed by inductive behavior based on all available knowledge. (All this in action situations.)

Now let us return to 5% (or to p% if you prefer generality) and to indication. Statisticians now well realize that there are many ways of fixing a 95% confidence interval, that some are better than others, and that a natural measure of goodness is given by the analog of the power curve. We are dealing then with something very close to the two types of error. Very close, but not the same! In an indication situation we fix one at 5% (or p%) and then make the other small. Mathematically, we can reach this from risk balancing as a limiting case as the importance of one kind of error becomes arbitrarily large. Conceptually, there is a substantial and important break.

123. SANCTIFICATION VS. INDICATION

It is clearly usually better to express a difference as being so much (within suitable limits) rather than to merely say that it is too large to be reasonably assigned to chance. It is usually better to indicate than to sanctify. There may be exceptions, but what are they? Are they frequent? Important?

There are four sorts of claims for exception that the writer can remember, namely:

(1) We have no scale on which to measure differences, but we can identify direction!

(2) Our accuracy is so poor that "definitely +," "maybe +," "maybe 0," "maybe –," "definitely –" is as many steps as we can use!

(3) This is bad data, and it deserves only the simplest statistical treatment!

(4) There's a test against the null hypothesis, but no statistician has proposed any way of setting confidence limits!

If their assumptions are correct, the first and fourth of these claims are legitimate claims for exception. The first states that the subject-matter field is insufficiently developed to express differences reasonably. The last states that statistical techniques are insufficiently developed. Each claims not only an exception, but also the need for further work, either in the subject-matter field or in the statistical method.

Not every time that they are made are the assumptions of claims (1) and (4) valid. Often a very few minutes will show how to do at least something statistically about rough confidence limits. However, there are really undeveloped areas, and hence real exceptions. A substantial fraction of biology is still quantitatively undeveloped, for example. We must not neglect the need for sanctification, but neither may we overemphasize it.

The second and third claims almost always fail to be grounds for exception. Even when we are comparing only two sets of conditions, it is usually more important to know that A and B agree within 3 units or may disagree by 50 units than to know that they are or are not significantly different. All four combinations are possible, as the following confidence intervals illustrate:

$$(-3,3) \qquad (1,3)$$
$$(-3,50) \qquad (1,50).$$

For many practical purposes, knowledge of the row in this little table is far more important than knowledge of the column. This example bears on most claims (2) and many claims (3).

The rest of the third claims arise in cases where the observations are not just highly variable but are subject to known defects with large and unknown effects. Above all, this is the kind of data we should not *sanctify*.

In conclusion, then, a lack of quantitative development in either subject-matter field or statistics *may* justify stopping with sanctification, but in most cases results should be indicated instead of sanctified.

124. SANCTIFICATION AND ACTION

There is another way in which sanctification can appear. If there is an administrative decision which is going to be taken on the basis of appearances (on the basis of description) it may be very helpful for the administrator to have the action sanctified — to have it shown that the alternative which he was going to adopt anyway is "significantly" better than the others. In our language, not only is his decision presumably wise "for the moment" but it is also reasonable "for the record."

Thus, to some extent, sanctification has contributed to the confusion of scientific and administrative conclusions and statements.

This particular use many be of great administrative convenience. So long as the nature of this use is recognized it should not lead to confusion.

125. EXAMPLES OF COMBINATIONS

It may help at this point to illustrate some of the combinations which arise between the two classifications just introduced.

(1) The procedures of acceptance sampling for sentencing lots of raw or manufactured product form a fine example of decision coupled with action.

(2) The selection of a nominal standard for a physical constant, such as the velocity of light couples indication and description.

(3) The determination of a range of values within which the physical constant is fairly sure to fall couples indication and confidence.

(4) The analysis of a fertilizer trial to show that the yield of crop is rather certainly raised by the use of fertilizer couples sanctification and significance.

These are the four standard combinations. The first is administrative. The second would once have been thought scientific, but is now being recognized as administrative. The last two are scientific. Most nonstandard situations are action situations where description, significance or confidence are being used for decision purposes. Sometimes this is done knowingly and wisely, while at other times it is done unknowingly and unwisely. The prime examples of unwisdom are likely to be interpretations of "nonsignificance" as meaning "no difference."

One other nonstandard situation can be identified, and this arises when a number of determinations are classified into groups. The result is indication, but the methods are those of significance. This is a curious hybrid, one explanation of whose existence is that we were really interested in a much more complete indication of the differences between the determinands, the sort of indication provided by a confidence procedure, specifically by a system of allowances.

CHAPTER 21: SIGNIFICANCE VS. CONFIDENCE

126. THE ENVIRONMENT

When we are concerned with statistical procedures "for the record" we face the distinction between

sanctification-significance

and

indication-confidence.

In many differing situations we must choose one or the other. In this chapter we shall try to set forth some of the pros and cons.

In many cases we set forth a result "for the record" so that it may be used in many and diverse decisions. Thus a statement of the gain in corn yield for a given amount of fertilizer may be used by many farmers in many years to help them decide how much fertilizer to apply. The prices of fertilizer, and the estimated price of corn will vary from case to case. Thus the gain in yield required for no loss or gain — the break-even point for fertilization — which is, after all, the natural null hypothesis — will vary from case to case. A significance procedure may be of no help but a confidence procedure is excellently adapted to the needs of the situation.

As another example, consider a number of high temperature alloys. They are each characterized by values of a number of different characteristics. Different users weight these characteristics differently. To know that alloy 2307 is better than 2313 on characteristics A, C, D, and H but worse on characteristics B, E, F and G is likely to be unhelpful to most users. They need confidence and indication, not significance and sanctification.

Even when the statement "for the record" is intended to be the purest of pure science, in that its aim is only to contribute to the eventual development of new theories and quantitative models, it is still true that the particular measurements being summarized will most frequently be used in quantitative combination with others. Here again indication-confidence is superior to sanctification-significance.

There is another widespread difficulty with significance, one which may be called a fallacy but which recurs too frequently to be neglected. In one type of occurrence, Jones and Smith do the same experiment, measure the same difference, and make tests of significance. One verdict is "significant," the other "not significant." Robinson writes a review paper stating that Smith and Jones found contradictory results. But a confidence analysis may show that *both* experiments are consistent with any value of the difference measured in quite a wide range. Clearly more definiteness is being read into the statements of the significance procedure than belongs there. In another type of occurrence, a single experimenter or observer examines the effects of two variables, finding one "significant" and the other "nonsignificant," and concludes that the first is more important than the second, although there may be far from enough evidence to show this. These are fallacies, and some would not charge them against significance, but they are so frequently and naturally committed that the writer feels that they must be considered.

127. PRESSURE ON PROBABILITY LEVELS

There is another type of consideration in the comparison of confidence and significance procedures. This arises from the simple fact that the significance user usually feels that only "significant" results are meaningful and definite. If this is the case, he may or may not be able to make a definite statement, while his confidence-using colleague is always able to make statements of some definiteness.

As long as "significance" is the hallmark of definiteness, just so long will the users of significance press, consciously or unconsciously, for less extreme probability levels. "You have given a table of 5% levels — we feel that in the XYZ industry 10% or 20% will serve — where could we obtain such tables"? Such questions as this are not only frequent but natural. If only significances count, then let us have more of them!

On the other hand, since some take a "significance" to be conclusive, the careful workers and the skeptical workers tend to push for more extreme probability levels, not 5%, but 1% or even 0.1%. (The writer recently received a letter asking "where can I find values of Student's t for a probability of one in a million." This is clearly going too far.) If we must take a chance of error, let it be a small chance!

These are natural pressures, but neither their mutual tension nor the possibility that one or the other might be overly successful seems a good thing to the writer. They count, in his mind, as a black mark against significance.

There seems to be greater concealed flexibility in confidence procedures. If Jones reports that a certain difference lies between 8 and 12 units (at, say, a 5% error rate) then his readers can easily vary their interpretation. Some will accept 8 to 12. Robinson, that inherent conservative, can say to himself "I know Jones! I had better double the spread and think of 6 to 14," while the dashing young Smith can take the other tack "Jones is overcautious, we can afford to act as if the difference is between 9 and 11." Such mental rescaling of confidence statements is easy — no analogous procedure with significance statements is available.

Thus the pressure on probability levels for confidence statements is far less than for significance. The writer marks this down as a further advantage for confidence procedures.

On the strength of many such examples, it seems to the writer both correct and wise to conclude that sanctification-significance is really just a primitive precursor of indication-confidence and that the discussion of Section 123 can be taken over in very considerable generality.

128. THE OTHER SIDE

In order to be relatively fair to sanctification-significance we must also present the arguments in its favor. There seem to be two such, ease of truth-to-life and availability. We can illustrate them both in the problem of comparing the means of two samples.

If we take an extreme significance point-of-view — in our hypotheses as well as in the forms of our possible conclusions — then our null hypothesis is that the two populations are identical. If they are normal, the two-sample t-test is exact. In a wide variety of cases, as Pitman [64-66] has shown, it is a very close approximation.

If we take a confidence point of view, and admit that the two populations differ in their means, then it is hard to avoid admitting that they may differ in their variances. Even when all populations are normally distributed, this moves us up to the Fisher-Behrens problem which is a number of steps up the difficulty scale from the two sample t-test.

What has been very common in recent years, is to make a compromise, and to proceed as if the variances were the same. Then we can set confidence limits with the two-sample t-procedure. These will be accurate for normal distributions, and good approximations (by Pitman's [64] argument) if the two populations differ only by a translation.

We have three choices in this example, and three similar choices are present in almost all other cases.

We may accept the significance approach completely, and have a greater certainty that our procedures are based on assumptions close to the truth. If we do this, we pay in definiteness of statement. Particularly since we must say "A and B differ in their means," if we wish to nail the probabilities down.

We may accept the confidence approach, but compromise by acting as if all other things remain the same. We now make very definite and useful statements, but our probabilities will sometimes not be what we think they are.

We may accept the confidence approach completely and develop a corresponding procedure which applies when other characteristics (e.g., variances in the example) may differ. We make very definite and useful statements, and our probabilities are much more likely to be what they are supposed to be. We pay in the labor of developing the procedure, and perhaps in the labor of using it.

Except that a scale on which to measure differences is not available, in which case significance is our only recourse, it is almost always possible to find a compromise which provides a (limited) confidence procedure. The argument of availability holds only in these "no scale of difference" cases, where it is final and binding.

When the choice is between significance and limited confidence (i.e., compromise), the writer argues as follows. The purpose of both definiteness of statement and accuracy of probabilities is usefulness. Therefore we are entitled to compare and combine them. Looking back on his experience with practical situations, he feels that the greater definiteness of statement of the limited confidence procedure very noticeably outweighs its smaller accuracy of probability. Therefore he wishes to use (even limited) confidence procedures as a general policy.

The writer thus has a strong general policy in favor of confidence, and feels that the development of, first, limited confidence procedures, and, second, unlimited ones is important.

129. MULTIPLE COMPARISONS

As discussed in more detail later, most of the types of analysis so far suggested for multiple comparison purposes fall into two broad classes:

(1) systems of allowances, and more specifically systems of familywise allowances, which provide individual confidence statements about all differences and other contrasts;

(2) procedures which provide statements of the "significantly different" or "not significantly different" type about groups of determinations — perhaps about pairs, perhaps about larger groups, perhaps only about the whole group.

There are, then, confidence procedures and significance procedures. Both are limited in the sense of tacitly assuming that variances are in known ratios. Based on the arguments and point-of-view set forth in this part, the writer feels strongly that he should use and recommend to his clients confidence procedures rather than significance procedures. Thus he favors some system of allowances.

A further point needs to be made. Many recent multiple-comparison significance procedures (one of them due to the writer) have large error rates familywise — rates which in the writer's present judgment are excessive. As remarked above, these are presumably due to the pressure for "definite" results. No multiple-comparison significance procedure that the writer can imagine will be free from such pressure and, even if the pressure is successfully resisted, there will be a loss in prestige for the procedure since it fails to come up to expectation. This is another general argument against significance procedures in multiple comparisons.

130. STRAGGLER SPOTTING, A SEEMING EXCEPTION

The ingenious reader may have, by this point in the discussion, thought of an example of a multiple comparison situation where significance suffices. This is the "trouble identifying" situation where the data analyst seeks for evidence to say, for example, "number 17 is surely different from "the others" — something is different there." It may properly be argued that only significance is needed here, although it may also be argued that sometimes confidence is a little help.

Such situations are analogous to those in industrial production where the control chart is habitually used. Analogous, but with two differences:

(1) there is no assigned central value, near which all determinands should lie; and

(2) usually there are fewer determinations than points on a control chart.

The typical example is an interlaboratory comparison of measurement techniques, where standard samples are distributed to, and measured by, a number of laboratories, and where the purpose of the analysis is to isolate laboratories that obtain "high" or "low" values, so that the technique of these laboratories can be examined carefully and the causes of discrepancy found. (There are, of course, other sorts of interlaboratory comparisons.)

Such situations only seem to involve multiple comparisons, for the real purpose is *not* comparison *but* repair. Here, significance *is* appropriate, but this is not an exception to the principle that multiple comparisons deserve allowances, since it is not a multiple comparison situation.

131. PERSONALITIES

Innocent readers of the recent periodical literature, might be persuaded, by the way in which the term "interval estimation" is used, that the difference between significance and confidence corresponds in a definite way to the differences in point of view and in mode of development of theory among various well-known statisticians. The writer feels that this is far from being the case.

In Student's days, it was the custom to deal with confidence, somewhat crudely by modern standards, in terms of the "probable error." In addition to developing the t-test of significance, Student spent a substantial part of his published work on "probable errors of correlation coefficients."

While Fisher's name is associated by many with tests of significance, his work on the efficiency of various estimators was fundamental and long continued.

The two terms most closely associated with the name of Neyman are the power of tests and confidence intervals.

The name of Wald is mainly associated with sequential analysis and decision theory, but his work on analysis of variance considered as a significance test and on asymptotically most powerful tests is balanced by his work on tolerance limits and on asymptotical shortest confidence intervals.

The writer sees no evidence that either significance or confidence types of approach have been peculiarly the province of any one well-known statistician or of any school of statistical method. They both have a very long history, just as we might expect if they represent different stages of development in the treatment of almost any problem.

132. SUMMARY

In general, and particularly for the multiple comparison problem, the writer favors confidence procedures to significance procedures. Thus he prefers some kind of familywise allowance.

CHAPTER 22: OPTIMALITY AND QUALITY

133. POWER

While the choices so far discussed are undoubtedly the most basic of all those which will appear in this part, the decision to use some system of familywise allowances or to use some self-consistent significance procedure does not settle our actions. We must still choose among systems of allowances or among significance procedures. We must adopt criteria of relative quality so that we may rank competitive procedures. We must adopt criteria of absolute quality so that we may design experiments with enough replications to give desired quality (i.e., desired accuracy).

In a significance based procedure the criterion furnished by the power function is both natural-seeming and well-accepted. Our only concern with it here is to examine its relation to confidence-based procedures.

The power of a test is often introduced to the reader in a situation where only two hypotheses need be considered, a null hypothesis and a unique alternative. If the test is phrased as a test of hypothesis, rather than as test of significance, the author or lecturer may refer to "accepting one hypothesis or the other" and the situation seems to resemble the confidence approach to the same problem, with its four alternative conclusions:

(1) Both hypotheses are reasonable —

(2) hypothesis A is reasonable —

(3) hypothesis B is reasonable —

(4) neither hypothesis is reasonable —

— in view of the data supplied. But in most realistic situations there are many (and we may as well say infinitely many) possible hypotheses (usually described by the value of one or more parameters. Here the connection of power with confidence is so tenuous as to be negligible.

Yet experience suggests to us that it may not be negligible, for we recall comparing the efficiency of various confidence procedures in terms of the power of the corresponding tests. When we go back and examine these examples, we find the special element. It lies in the fact that these confidence procedures were constructed out of tests of significance by a simple and standard device. This is the device of slippage in some form. The two-sample t-test may be converted into a two-sample difference of means procedure, for example, by testing the difference between the first sample and a dummy sample made up by subtracting the contemplated difference from each observation of the second sample.

A one-parameter confidence procedure which is related to a test of significance by such a device clearly has its characteristics determined by those of the test of significance. But this is not true of confidence procedures in general. Two confidence procedures can be reasonably compared through the powers of the associated test of significance if each is related to the associated test by such a device, and if the slippage devices are the same, and not otherwise. Thus power of the corresponding significance test is not a general criterion for comparing confidence procedures. *A fortiori* it is not a satisfactory absolute criterion for use in designing experiments.

134. TIGHTNESS (parallel of power)

The natural criteria of confidence procedure performance are expressed, for the case of a single parameter, in terms of the length of the confidence interval. Some may like the average length, others the average squared length. Some may prefer the length exceeded at most 50% of the time, others the length exceeded at most 10% of the time. All these seem natural, and we hope that the corresponding notions of optimality are not too different.

In general we have difficulty with anything but the asymptotic case, since there may be no natural scale for the parameter. In many cases, however, there is a scale (and frequently we have used the corresponding slippage device). In such a case we can take any of these criteria as a criterion of absolute quality, as a basis for design of experiment. Such choices as those labelled above by 50% or 10% are perhaps slightly simpler to manage or think about.

The *p% tightness* of a given confidence procedure applied to a given experiment at design we define to be the length of interval exceeded by $(100-p)$% of the confidence intervals arising from this procedure. Thus

90% of the time, the resulting confidence interval is shorter than the 90% tightness.

While other absolute criteria may be equally reasonable, we propose to use the 50% tightness and, more usually, the 90% tightness as absolute criteria of confidence procedure quality. Thus we shall expect to design an experiment to given tightness, not to given power.

135. FURTHER PROBLEMS

Now the selection of tightness as a criterion does not settle our problems. For as is shown in Part E, there are a very large number of systems of allowances which have the following property:

Given any two systems, there are some contrasts for which the first is tighter than the second, and other contrasts for which the second is tighter than the first.

Thus we must decide for which contrasts we want tightness, or else how to weight the relative importance of the tightness attained for various contrasts.

This is another general issue, not as basic as those so far considered, but one that has to be discussed verbally. It is the subject of the next chapter.

CHAPTER 23: IMPORTANCE OF VARIOUS CONTRASTS

136. INTRODUCTION

There are those who wish that statistics were a branch of mathematics rather than a science, for then matters would be neater and tidier, and the diffuse problems of the real world could be shut out and left on the doorstep. For a long time, the question of choice among the procedures of estimation, significance testing and setting confidence intervals was in such a tidy shape. One assumed a criterion of goodness and strove to find a procedure optimum according to the criterion.

The introduction by Wald of the concept of admissibility was a sign of doom for such tidiness. For Wald showed that, in his mathematical models for decision making, there were two kinds of distinctions between procedures.

(1) some procedures were always worse than other definitely corresponding procedures;

(2) some procedures were better or worse than others depending on the choice of a risk or weight function.

In other words, some questions of comparison could be settled once and for all, while others required diffuse considerations from the real world for their solution. Those procedures which could not be ruled out once and for all were ***admissible***, and choice among them was difficult. The mathematically minded blithely (and usefully) started immediately to study classes of admissible procedures. There remained many diffuse and difficult problems for someone, the problem of choosing among the admissible procedures.

It is probable that if we assume normality of fluctuation and homogeneity of variance, all the allowance procedures discussed in Part E of this account are admissible. Thus our problem of choice of allowance procedure is one of these diffuse and difficult ones, a problem of choice among admissible alternatives. It is simple, as such problems go, because the geometric relation of an *A*-allowance to a *B*-allowance is the same in every instance. It is a problem which is now very hard to hide, because *C*-, *F*-, and *W*-allowances:

(1) are all practically available;

(2) have been introduced by different statisticians; and

(3) give different results.

We shall have to tackle this problem, and so, eventually, will many other statisticians.

The problem, however, is difficult for another reason. The twin assumptions of normality of distribution and homogeneity of variance are not ever exactly fulfilled in practice, and often they do not even hold to a good approximation. If we are to grapple with the statistical problems of allowances, we cannot avoid considering the effects of non-normality and inhomogeneity of variance. We shall not be able to settle this problem, but we can advance far enough on it to have some basis for decision. (This is far from surprising, since we are in a somewhat similar situation for the simplest problems — for example when the means of two groups are to be compared.) Almost all the problems discussed in this part are diffuse and correspondingly difficult.

137. SIMPLE COMPARISONS

The writer believes that a vast majority of users of statistical techniques would agree and adhere to the following two propositions:

(A) the usefulness of a result of statistical analysis increases with the "definiteness" of the object or quantity to which it refers;

(B) among the contrasts among a set of determinations, the most "definite" is a simple comparison.

Here "definite" is used in a sense which is clear to the writer, and which he believes is clear to others, but which seems difficult to define. Examples are easy to give. "Bill is taller than Sam" is more definite than "Bill is tall for his age" and this is more definite than "boys of Bill's age are not all of the same height."

If we admit these two propositions, than it is clear that simple comparisons — one determination minus another — have a leading role among contrasts.

Since the writer knows of no one who feels that simple comparisons are unimportant, the argument can be left here.

138. STRAGGLERS

The aim of interest in deviations or one-way discrepancies

$$x_i - x_o \quad \text{or} \quad x_i - x_{(i)}$$

(where $x_{(i)}$ is again the mean of all x_j for $j \neq i$) is clearly a desire to detect stragglers — individuals who are different from the "mass." How much importance shall we give to this desire?

The aim is to isolate individual determinations which are "different." Except for propaganda purposes — which often have a useful role in science, for example in interlaboratory discussions concerning methods of measurement — this bare conclusion is almost useless. We must ask "different from what"? The answer "different from the mean of the others" begs many questions as we shall see.

If it is possible to point to at least certain of the remaining determinations and say "different from this one" or "different from all four of those" then a substantial step forward has been made, both in definiteness and in usefulness. This observation suggests that the real interest in stragglers is an experiment-pinching (allied to penny-pinching) one — an attempt to say something semidefinite when we can't quite say something definite. This is a praiseworthy attempt to stretch experiments as far as possible, and if it cost us nothing elsewhere, it would clearly be a good thing to do. But we know that doing it to excess has a price — and that this price is a weakening of the statements we make about other contrasts, in particular a weakening of the statements we make about comparisons. Is it worth this cost? We must soon consider this question.

Second, we may properly wonder how meaningful is a comparison with a "mass" represented by the remaining determinations. Is this a firm enough reference point, even though somewhat indefinite, to be useful? We shall come to related points in the next section but one and hence leave further discussion aside until we reach that point.

In any event, the most likely areas for the importance of stragglers are those in which *incomplete* finger pointing is both suitable and adequate. The writer would suggest that the most typical such case is that of interlaboratory comparisons.

As the explicit recognition of the inevitability of laboratory-to-laboratory variations in any measurement process increases, there is less odium and more general interest in the fact that laboratory A finds values systematically higher [or lower] than the general run of laboratories considered. Such information is, hopefully, obtained from interlaboratory comparisons of measurements on a set of standards. These comparisons are most often carried on for such purposes as:

(1) investigation of the desirability of a new or modified method of measurement;

(2) improvement of the accuracy of an old method of measurements;

(3) improvement in agreement on one or more measurements among a particular set of laboratories.

In any of these situations, incomplete finger pointing is both suitable and adequate, for, having isolated laboratory A as one of the laboratories which deliver measurements systematically higher (or, alternatively, lower) than the run-of-the-group laboratories involved, the next step is a careful study of differences in actual techniques between laboratory A and other laboratories. It is just incomplete finger pointing which is needed to motivate such further study. But this, as was pointed out in Section 130, calls for significance procedures and not confidence procedures. We need to isolate such cases from the mass of multiple comparison situations, and treat them separately — but not by means of allowances.

It is the writer's opinion that stragglers are of minor importance in comparing allowances, although they warrant the development of special significance procedures for trouble-shooting.

139. INTERACTIONS

Once we come to two or more way tables, we find interactions or, equivalently, h-way discrepancies among the possible contrasts. What can we say of their importance?

Little based on experience, because no effective way of analyzing them has so far been available. We must guess and infer as best we can, and prepare for future alterations. The great importance of having *some* method of going after individual interactions is clear to the writer. But the importance of going after them vigorously, at the expense of other contrasts is not clear.

All the arguments about stragglers seem to apply also to interactions, the main difference being a feeling that special circumstances are probably somewhat less frequent. The writer doubts if interactions rise above intermediate importance, except on rare occasions.

140. SUBGROUP MEANS — THE TRAP

Once introduced to the possibility of setting limits on all contrasts, many statistician's minds jump immediately to the conclusion that the main application of the technique is to differences between subgroup means. We have determined some delicate or obscure characteristic on twenty human subjects — let's compare redheads with blondes, or men with long arms with men with short arms. Perhaps we can show an association between our obscure measurement (perhaps special amino acids in the saliva or double peaked gamma waves in the electroencephalogram) and hair color or arm length — wouldn't that be interesting! There are serious objections to this position which merit our most careful attention.

Let us go back to our model for the multiple determination situation. We took

$$x_i = \mu_i + e_i$$

where the e_i may as well be assumed to be independent with variance σ^2 of which we have one estimate s^2. If subjects 1, 4, 9 and 16 have redhair, while subjects 2, 3, 5, 7, 11, 13, 17 and 19 are blond, then we can set limits on

$$\frac{\mu_1+\mu_4+\mu_9+\mu_{16}}{4}-\frac{\mu_2+\mu_3+\mu_5+\mu_7+\mu_{11}+\mu_{13}+\mu_{17}+\mu_{19}}{8}$$

without difficulty. Suppose, for example, that these limits went from +1 to +14 on the relevant scale. Then we would be entitled to conclude (as always, with an appropriate error rate) that "the means of these four redheads is larger than the mean of these eight blonds." But this is *not* equivalent to, does *not* imply, and is a *far* cry from the statement "The mean of redheads is larger than the mean of blonds." One is a statement about $4+8=12$ individuals which were measured, while the other is a statement about two large populations, largely made up of unmeasured individuals. The first we know how to make, the second is the interesting one.

The nature of the difficulty may not be clear to those unfamiliar with the ideas which lie at the foundation of the analysis of variance. So let us take another example. Suppose that a railroad buys a large quantity of steel rails from each of two suppliers, and suppose that 5 rails from each batch are cut into pieces accurately 12" long and weighed on a high precision scale. Let us suppose that each rail turns out to be of remarkably uniform weight per linear foot, so that the mean weight per foot for any one rail is reliable (with very high probability) to ±0.05 pound. Suppose further that these means run as follows:

Rail No.	*Supplier*	*Weight per foot*
1	A	114.12
2	A	108.76
4	A	110.24
7	A	109.42
10	A	115.21
3	B	111.27
5	B	107.74
6	B	115.32
8	B	112.21
9	B	109.88
mean	A	111.55
mean	B	111.26

Since we can rely on all measurements to ±0.05 pound, the same is surely true of the means, and we are certain that the mean weight per foot of the five rails from supplier A is greater than the mean weight per foot of the five from supplier B. Does this mean that the two lots are different?

Not at all. To see this, let us round our measurement to the nearest pound per foot, when we can neglect errors of measurement altogether. We find, rearranging the values in increasing order:

Supplier A: 109, 109, 110, 114, 115

Supplier B: 108, 110, 111, 112, 115.

No statistician would or should feel that these values indicate a demonstrated difference in the *averages* of the two lots of rails.

In each of these cases, the proper solution is to set up an analysis which recognized the subgroups, as hopefully, *representative* of the populations and asks "what of the populations"? This can always be done, often directly but in every case in terms of an analysis of variance. The relevant lines of the analysis of variance would be

Hair Color
Individuals within hair color
Measurements within individuals

in the first example, and

Suppliers
Rails within suppliers
Measurements within rails

in the second.

If we make such a correct analysis, whether through the formalities of the analysis of variance, or otherwise, the subgroup means will become individual determinations (of a higher order) and the differences in subgroup means will become simple comparisons. Thus in the rail example the 5 measurements on the 5 rails from Supplier A become 5 individual measurements appropriate to the determination of "the average weight of rails received from Supplier A." And similarly for Supplier B. And similarly in every case.

When the subgroups are identified as those parts of two populations which were measured, a proper analysis is required, and it makes the contrast of interest a simple comparison.

141. SUBGROUP MEANS — OTHER CASES

We have dealt with one class of comparisons of subgroup means, there are others yet to be considered. What are they? There seem to be a number of possibilities, and a relatively useful classification seems to be the following:

(1) the subgroups are those *parts* of certain broader classes (populations) which were measured and for which determinations are available;

   (a) these broader classes were identified in advance of the experiment;

   (b) these broader classes were suggested by the experiment;

(2) the subgroups are distinguished by the values of their determinations, and not by any broad classes;

(3) the subgroups each make up *all* of certain broad classes:

   (a) these broad classes were identified in advance of the experiment;

   (b) these broad classes were suggested by the experiment;

(4) the subgroups are meaningless.

We have dealt with (1a) and (1b) in the last section, and have little need to deal with (4).

It would seem to the writer to be a fair summary of scientific policy, as it has evolved to date, to say that:

> "Subgroup means of type (2) should not be neglected, but they should be regarded as of intrinsically lower probable importance than those of type (1)."

Some of the great scientific discoveries may have been found through subgroup means of type (2), but very many more blind alleys have been followed up as the result of apparent differences of this type.

The writer's conclusion, then, is that differences of type (2) subgroup means are of minor importance.

There remain subgroups of type (3). What importance has the difference of their means? The situation here is in part much as it is with stragglers. We have a determination for every relevant individual. We can compare each individual with every other. Why should we give up tightness in these individual comparisons in order to gain tightness in comparing their means? Two reasons seem most likely

(i) the desire to make at least a weak positive statement, when individual differences cannot be pinpointed; or

(ii) a situation where the difference in mean of these subgroups is the key quantity, since we must accept one or the other by our choice or decide to postpone action.

How important are these situations?

All the arguments made against stragglers' importance can be made against (i). They should suffice. But perhaps it is time to point out that this desire to make a positive statement is tightly linked to sanctification rather than to indication (cp. Section 123). If we can say "individual 1 minus individual 3 lies between −2 and +15," we have made at least as definite a statement as "the mean of individuals 1, 4, 9 and 16 is greater than the mean of individuals 2, 3, 5, 7, 11, 13, 17, and 19." By the introduction of quantitative rather than qualitative statements, we have destroyed a major reason for (i).

We come, then, to (ii) and to two reasons against granting it importance. The first is simple. If this is the key quantity, why didn't we set up our analysis so as to spend nearly all of our error rate on this key quantity rather than spreading it over all contrasts among the m determinations? No reason is apparent. The second is a plan of legitimate evasion. If this is a key quantity because of an active decision, then action rather than indication is involved, while allowances are intended for indication. There may be special circumstances where (ii) escapes these reasons, but they must be very special.

In summary, then, differences of subgroup means seem almost certain not to be of the highest importance.

142. REGRESSION COEFFICIENTS

The other class of contrasts which stirs the statistician's interest is that of regression coefficients, whether those corresponding to orthogonal polynomials or to more general functions. We note their existence, and merely say that, while an adequate discussion would take us outside the scope of the present discussion, it seems reasonable to suppose that some segregation (just as in the first case of subgroup means) is going to be part of the solution.

You are welcome to say that cases where regression is important may be regarded, for the present, as exceptions to the discussion we are giving, although we doubt that they will prove to be exceptions.

143. CONCLUSIONS

Our conclusions about the relative importance of various contrasts can now be briefly stated:

(1) simple comparisons are outstandingly important, and their importance is far greater than any other type of contrast;

(2) under special circumstances, stragglers, or even interactions may be equally important;

(3) the importance of regression coefficients is being avoided.

We have concentrated on contrasts, since other linear combinations

(i) can only be discussed if we can compare the importance of relative (about contrasts) and nonrelative (about non-contrasts) statements, which is usually difficult; and

(ii) are a matter of relatively infrequent interest as compared to contrasts. We leave this more difficult and less useful task to other times and places — perhaps to other authors.

PART E
FAMILYWISE ALLOWANCES

CHAPTER 24: GENERAL STRUCTURE

144. INTRODUCTION

Through most of this account we have regarded an allowance as a way of making a large number of individual statements, and have made occasional remarks about treating X-wise allowances as single statements about all determinations at once. We come now to the comparison of systems of allowances, and find it profitable to emphasize the single statement point of view. This change of basis seems natural and significant, for it seems that the user does better with the multiplicity of simple statements, while the theoretical or mathematical statistician finds the single complex statement more useful. We shall concentrate on systems of contrast allowances, mentioning the corresponding systems of universal allowances in passing. (Notice that the PA defines a system of universal allowances.)

Let us revert to the *GSD*, and see how it may be reexpressed as a single complex statement. We are used to its form as the collection of statements:

$$\begin{aligned} | \, x - \mu \, | &= | \, \Sigma \, c_i x_i - \Sigma \, c_i \mu_i \, | \\ &= | \, \Sigma \, c_i e_i \, | \leq (GSD) \, \text{norm}_G \{c_1, c_2, \ldots, c_m\} \end{aligned} \tag{$*$}$$

where

$$\begin{aligned} GSD &= \left\{ \text{upper } p \,\% \text{ point of } \sqrt{2mF} \right\} s \\ &= k_G s, \end{aligned}$$

and

$$\text{norm}_G \{c_1, c_2, \ldots, c_m\} = \left\{ {}^1\!/_2 \, \Sigma \, c_i^2 \right\}^{1/2}.$$

Here

(1) the norm;

(2) The square root of a modified F; and

(3) k_G, the numerical value derived from (2);

stand out in the foreground.

If we shift our emphasis to one complex statement, then we are immediately led to consider those points $(e_1, e_2, \ldots, e_m)$ for which $(*)$ holds for all linear combinations $\Sigma \, c_i x_i$. (By construction, $(100-p)\%$ of all points realized will be of this sort.) If we write out $(*)$ more explicitly, it becomes

$$\begin{aligned} | \, \Sigma \, c_i e_i \, | &\leq (s)(p \,\% \text{ point of } \sqrt{2mF}) \left\{ {}^1\!/_2 \, \Sigma \, c_i^2 \right\}^{1/2} \\ &\leq (s)(p \,\% \text{ point of } \sqrt{mF})(\Sigma \, c_i^2)^{1/2} \\ &\leq A \, (\Sigma \, c_i^2)^{1/2}. \end{aligned}$$

A brief recall of the standard inequalities reminds us that this holds for all c_i if and only if

$$(\Sigma\, e_i^2)^{1/2} \leq A,$$

that is, if

$$\begin{aligned}\Sigma\, e_i^2 \leq A^2 &= (s^2)(p\ \%\ \text{point of } mF)\\ &= (ms^2)(p\ \%\ \text{point of } F)\end{aligned}$$

which we know will occur in $(100-p)\%$ of all experiments.

Now let us regard $(x_1, x_2, \ldots, x_m)$, $(\mu_1, \mu_2, \ldots, \mu_m)$, $(e_i, e_2, \ldots, e_m)$ and so on, as points in the same m-dimensional space. The single complex statement

$$\Sigma\, e_i^2 < A^2$$

can be rewritten

$$\Sigma\,(-e_i)^2 = \Sigma\,(\mu_i - x_i)^2 \leq A^2 \qquad (**)$$

where it becomes a single complex statement about the μ's. We know that $(**)$ is equivalent to the collection of simple statements $(*)$. As a single complex statement it directs our attention to the region of m-dimensional space made up of the points $(\mu_1, \mu_2, \ldots, \mu_m)$ satisfying it.

This region bears the same relation to the point $(x_1, x_2, \ldots, x_m)$ as the region where

$$\Sigma\,(-e_i)^2 = \Sigma\, e_i^2 \leq A^2$$

does to the origin (the point $(0, 0, \ldots, 0)$). Both regions have the same *shape* as the standard region where

$$\Sigma\,(-e_i)^2 = \Sigma\, e_i^2 \leq 1/2$$

and differ from the standard region only by uniform (isotropic) expansion or contraction and (in the case of the μ — region only) translation.

Points $(-e_1, -e_2, \ldots, -e_m)$ which fell in the standard region are, in a well defined sense, more satisfactory and less extended than those which do not. Points $(-e_1, -e_2, \ldots, -e_m)$ for which the points with doubled coordinates $(-2e_1, -2e_2, \ldots, -2e_m)$ lie in the standard region are less extended than those for which this is not true. And so on. It is natural to define the *G*-extent of a point (set of fluctuations) $(e_1, e_2, \ldots, e_m)$ as the least nonnegative number u such that $(-e_1/u, -e_2/u, \ldots, -e_m/u)$ falls in the standard region. Clearly, points of smaller *G*-extent are less extended in the sense just described.

Let us find a direct expression for the *G*-extent. It is the least u such that

$$\Sigma\,(-e_i/u)^2 = \frac{1}{u^2}\,\Sigma\, e_i^2 \leq 1/2$$

but this holds if

$$u^2 > 2\,\Sigma\, e_i^2$$

and hence

$$\text{extent}_G\{e_1, e_2, \ldots, e_m\} = (2\,\Sigma\, e_i^2)^{1/2}$$

a relatively simple expression.

With this definition, the single complex statement becomes

$$\text{extent}_G(\mu_1 - x_1, \mu_2 - x_2, \ldots, \mu_m - x_m) \leq s k_G. \qquad (**)'$$

Here

(1) the extent;

(2) the square root of a modified *F*; and

(3) k_G, the numerical value derived from (2)

stand out in the foreground.

By considering a single complex statement in place of many simple ones, our attention is drawn to an "extent" for fluctuations instead of a "norm" for linear combinations. Each of these determines the other (up to a constant factor). Each determines the same statistic, which we shall call the corresponding *key statistic*, whose upper p % point gives the numerical multiple of s which is to be used. By alternating between these, we can make much progress.

We have earlier considered not only the *GSD* system, but also those corresponding to *WSD, USD, FSD,* and *PA*. Their norms, extents, key statistics, and standard regions are set forth in Table 30.

Table 30

Norms, Extents, Key Statistics, and Standard Regions for Allowances so Far Considered

Allowance	*Norm*	*Extent*	*Key statistic*	*Standard Region* $= R_1$	*Shape*
WSD	$\Sigma^+ c_i = -\Sigma^- c_i$ if $\Sigma c_i = 0$; ∞, otherwise	range $\{e_1, e_2, \ldots, e_m\}$	Studentized range	range $\{e_1, e_2, \ldots, e_m\} \le 1$	W
USD	$\max\{\Sigma^+ c_i, -\Sigma^- c_i\}$	range $(0, e_1, e_2, \ldots, e_m)$	Studentized augmented range	range $(0, e_1, e_2, \ldots, e_m) \le 1$	U
FSD	$(\frac{1}{2} \Sigma c_i^2)^{1/2}$ if $\Sigma c_i = 0$; ∞, otherwise	$(2 \Sigma (e_i - e.)^2)^{1/2}$	$\sqrt{2(m-1)F}$	$\Sigma (e_i - e.)^2 < \frac{1}{2}$	F
GSD	$(\frac{1}{2} \Sigma c_i^2)^{1/2}$	$(2 \Sigma e_i^2)^{1/2}$	$\sqrt{2mF}$	$\Sigma e_i^2 < \frac{1}{2}$	G
PA	$\Sigma \mid c_i \mid$	$2 \max\{\mid e_i \mid\}$	Studentized maximum modulus	$\mid e_i \mid \le \frac{1}{2}$ for all i	P

145. INSTANCES AND SHAPES

We wish now to study the geometry of a single *instance* of a system of allowances, when s is fixed. There are three conditions which we can assume in general, namely:

(A1) The simple statements are of the form

$$\mid \Sigma c_i e_i \mid \le K(c_1, c_2, \ldots, c_m) \tag{+}$$

(A2) whenever not all c_i are zero, then $K(c_1, c_2, \ldots, c_m) > 0$, and

(A3) whenever we wish to make the statement

$$| \Sigma c_i e_i | \leq K(c_1, c_2, \ldots, c_m)$$

we are willing to make the statements

$$| \Sigma c_{\pi(i)} e_i | \leq K(c_1, c_2, \ldots, c_m)$$

where $\pi(1), \pi(2), \ldots, \pi(m)$ is an arbitrary rearrangement (permutation) of $1, 2, \ldots, m$.

These conditions say, respectively, that our individual statements are to be of the general form we have used so far, that none of them will force an exact value on a linear combination of the e_i, and that we are confining our attention to symmetrical situations. These are not novel conditions, but eminently reasonable ones.

Let us suppose that we start with some *subbasic set of statements* of the form (+). According to (A3), we can add all the statements obtained from the subbasic set by permuting the c_i, we shall call the result the *basic set of statements*. There will be points in m-dimensional space $(e_1, e_2, \ldots, e_m)$ which satisfy all these statements, and points which do not. The region, R, of m-dimensional space made up of points which satisfy all these statements will be

(B1) closed;

(B2) convex; and

(B3) symmetrical in the sense that if $(e_1, e_2, \ldots, e_m)$ belongs to the region, so will $(-e_1, -e_2, \ldots, -e_m)$ and $(e_{\pi(1)}, e_{\pi(2)}, \ldots, e_{\pi(m)})$ for any rearrangement (permutation) π; and

(B4) for each point $(e_1, e_2, \ldots, e_m)$ there is a u, $0 < u < \infty$, such that the point $(e_1/u, e_2/u, \ldots, e_m/u)$ lies in R.

Here (B1) and (B2) are consequences of (A1), (B3) is a consequence of our making the collection of statements symmetrical, while (B4) is a consequence of (A2).

If we have a region, R, satisfying (B3) and (B4), we may define a collection of statements satisfying (A1), (A2), and (A3) by choosing

$$K(c_1, c_2, \ldots, c_m) = \max_R \{ | \Sigma c_i e_i | \}$$

where the maximum is taken over all points $(e_1, e_2, \ldots, e_m)$ in R.

As an illustration, consider the case where the subbasic set consists of the single statement

$$|e_1 - e_2| \leq 6$$

which we expand to the symmetrical collection of basic statements

$$|e_i - e_j| \leq 6 \text{ for all } i \neq j$$

and which describes an infinite prismatic region having the lines

$$e_i = t + 3\varepsilon_i$$

where the $\varepsilon_i = \pm 1$ and are fixed for each line, as edges. We have previously described this region, by

$$\text{range}\{e_1, e_2, \ldots, e_m\} \leq 6.$$

Starting from this region, we know that

$$K(c_1, c_2, \ldots, c_m) = \max_R \{ |\Sigma c_i e_i| \} = 6 \max_{R_1} \{ |\Sigma c_i e_i| \}$$

$$= 6 \max \{\Sigma^+ c_i, -\Sigma^- c_i\}$$

where R_1 is the region defined by

$$\text{range}\{e_1, e_2, \ldots, e_m\} \leq 1$$

since we recognize that we are essentially working with the *WSD*.

Going back to the general case, it is easy to prove two results about the result of combining steps from the collection of simple statements to the region and back.

First, if we start from a basic collection of statements, expand by symmetry, find the corresponding region, and then a new collection of statements, we reach exactly all the logical consequences of the basic statements and (A3). As a corollary, if the basic collection was closed under symmetry and logical consequences, as we shall say was a *complete* collection, we come back to exactly the same collection.

Second, if we start from a region satisfying (B1), (B2), (B3), and (B4) and find the corresponding collection of statements, and then treat these as a basic collection and return to a region, we come to exactly the same region with which we started. (If we only assume (B3) and (B4), then the new region is the closed convex cover of the initial region.)

Thus, once we

(a) start with a subbasic collection of statements, expand by symmetry, and pass to a region; or

(b) start with a region satisfying (B1), (B2), (B3) and (B4); or

(c) start with a complete collection of statements;

we can oscillate back and forth between a collection and a region as much as we please, but they will always remain the same. Together they make up an *instance of a system of allowances*, together they are determined by the basic collection of statements.

In our example we already made implicit use of our intuitive feeling that an instance has two aspects, one qualitative and one quantitive, a shape and a scale. Thus, the regions defined by

$$\text{range}\{e_1, e_2, \ldots, e_m\} \leq 0.018$$

$$\text{range}\{e_1, e_2, \ldots, e_m\} \leq 1$$

$$\text{range}\{e_1, e_2, \ldots, e_m\} \leq 6$$

respectively, have the same shape and different scales. We now want to attach a norm for linear combinations and an extent for fluctuations to a shape alone. We can do this by insisting that

$$\text{norm}\{\text{simple comparison}\} = 1$$

and

$$\max\{\, | \, e_i - e_j \, | \text{ for extent } (e_1, e_2, \ldots, e_m) \leq 1\} = 1.$$

We proceed to the construction.

We introduce

$$\text{size}\{R\} = \max_R\{\ |\ e_1 - e_2\ |\ \}$$

as a convenient tool, and place

$$\text{norm}\{c_1, c_2, \ldots, c_m\} \equiv \frac{\max_R\{\ |\ \Sigma c_i e_i\ |\ \}}{\text{size}\{R\}}$$

$$\text{extent}\{e_1, e_2, \ldots, e_m\} \equiv (\text{least } u \text{ for which } \left(\frac{e_1}{u}, \frac{e_2}{u}, \ldots, \frac{e_m}{u}\right) \text{ is in } R)\ (\text{size } R).$$

If we pass from one instance to another by expanding or contracting R by a chosen factor or, as is equivalent, by weakening or strengthening all individual statements by the same factor, it is easily seen that the numerator and denominator of the relation defining the form are altered by the same factor, so that the form is unchanged. A similar compensation is built into the definition of the extent. Form and extent depend only on shape, and not at all on scale.

We shall want to use certain further properties of norm and extent, these we now proceed to derive. Let R_1 be the region made up of points $(e_1, e_2, \ldots, e_m)$ for which extent$\{e_1, e_2, \ldots, e_m\} \leq 1$. Then

$$\text{extent}\{e_1, e_2, \ldots, e_m\} = \text{least } u \text{ for which } \left(\frac{e_1}{u}, \frac{e_2}{u}, \ldots, \frac{e_m}{u}\right) \text{ is in } R_1$$

as is easily seen by studying the points $(e_1/r, e_2/r, \ldots, e_m/r)$ for r near extent$\{e_1, e_2, \ldots, e_m\}$. And, consequently, writing out the definition of "extent" in terms of R_1 and substituting in the last result

$$\text{extent}\{e_1, e_2, \ldots, e_m\} \equiv \text{extent}\{e_1, e_2, \ldots, e_m\}\ \text{size}\{R_1\}$$

so that size$\{R_1\} = 1$, and hence

$$\text{norm}\{c_1, c_2, \ldots, c_m\} = \max_R\{\ |\ \Sigma c_i e_i\ |\ \}.$$

Substituting e_i/r for e_i, where $r = \text{extent}\{e_1, e_2, \ldots, e_m\}$ we find

$$|\Sigma c_i e_i| < \text{norm}\{c_1, c_2, \ldots, c_m\}\ \text{extent}\{e_1, e_2, \ldots, e_m\} \qquad (***)$$

with equality possible for *either* any fixed combination $(c_1, c_2, \ldots, c_m)$ *or* any fixed point $(e_1, e_2, \ldots, e_m)$.

We call shapes satisfying (A1), (A2) and (A3) and consequently (B1), (B2), (B3) and (B4) *regular* shapes and denote them by single capital letters. Thus, the regular shapes corresponding to *WSD, USD, FSD, GSD,* and *PA* are denoted, respectively, *W, U, F, G* and *P*. The corresponding norms and extents have already been given in Table 30. We have just shown that every regular shape is associated with a norm and an extent satisfying (***) such that the extent of any $(e_1, e_2, \ldots, e_m)$ with finite coordinates is finite and the norm of a simple comparison is unity. The norm of any contrast is finite, although the norm of a finite linear combination may be infinite.

146. ALLOWANCES OF GIVEN SHAPE

We come back to statistics, and assume that our m determinations

$$x_1 = \mu_1 + e_1$$

$$x_2 = \mu_2 + e_2$$

$$\ldots \quad \ldots$$

$$x_m = \mu_m + e_m$$

have a symmetrical joint distribution in the sense that the distributions of $(e_1, e_2, \ldots, e_m)$, $(-e_1, -e_2, \ldots, -e_m)$ and $(e_{\pi(1)}, e_{\pi(2)}, \ldots, e_{\pi(m)})$ are the same, where $\pi(1), \pi(2), \ldots, \pi(m)$ is an arbitrary rearrangement (permutation) of $1, 2, \ldots, m$. We further assume that this distribution is known up to a scale factor σ, and that we possess an estimate s^2 with average value σ^2. This means that

$$\left(\frac{e_1}{s}, \frac{e_2}{s}, \ldots, \frac{e_m}{s}\right)$$

has a known distribution.

If we fix a regular shape A, the corresponding extent, $\text{extent}_A\{e_1, e_2, \ldots, e_m\}$ will also be fixed, as will the distribution of the corresponding key statistic

$$\text{key}_A = \text{extent}_A \left\{ \frac{e_1}{s}, \frac{e_2}{s}, \ldots, \frac{e_m}{s} \right\}.$$

Because of (A2) or (B4) there is probability zero of the key statistic being infinite, and there must be a number k_A, the upper p % point of the key statistic, such that

$$\text{Prob}\left\{ \text{extent}_A \left\{ \frac{e_1}{s}, \frac{e_2}{s}, \ldots, \frac{e_m}{s} \right\} \leq k_A \right\}$$

$$= \text{Prob}\{\text{extent}_A\{e_1, e_2, \ldots, e_m\} \leq k_A s\}$$

$$= \text{Prob}\{\text{extent}_A\{-e_1, -e_2, \ldots, -e_m\} \leq k_A s\}$$

$$= \text{Prob}\{\text{extent}_A\{\mu_1 - x_1, \mu_2 - x_2, \ldots, \mu_m - x_m\} \leq k_A s\}$$

$$= (100 - p)\%$$

so that

$$\text{extent}_A\{\mu_1 - x_1, \mu_2 - x_2, \ldots, \mu_m - x_m\} \leq k_A s$$

defines a confidence region for $(\mu_1, \mu_2, \ldots, \mu_m)$ which:

(1) is a function of $s, x_1, x_2, \ldots, x_m$;

(2) is of confidence $(100 - p)\%$;

(3) has shape A (more precisely, is the translation of a region of shape A);

(4) is symmetrical about $(x_1, x_2, \ldots, x_m)$.

It seems probable that Peisakoff's methods [63] would show that among confidence regions satisfying (1), (2), and (3), this is the best, although a few further hypotheses on the distribution of $(e_1, e_2, \ldots, e_m)$ may be needed. At any event it is the only region with properties (1), (2), (3) which we shall consider here. This is the single complex statement of the allowance of shape A.

The collection of simple statements follow from

$$|x - \mu| = |\Sigma c_i x_i - \Sigma c_i \mu_i| = |\Sigma c_i e_i|$$

$$\leq \text{extent}_A \{e_1, e_2, \ldots, e_m\} \text{ norm}_A \{c_1, c_2, \ldots, c_m\}$$

in the form

$$|x - \mu| = |\Sigma c_i x_i - c_i \mu_i| \leq (k_A s) \text{ norm}_A \{c_1, c_2, \ldots, c_m\}$$

where

$$[ASD/s] = k_A = \left| p \text{ \% point of extent}_A \left\{ \frac{e_1}{s}, \frac{e_2}{s}, \ldots, \frac{e_m}{s} \right\} \right| .$$

For $A = W$, we have

$$\text{extent}_W \{e_1, e_2, \ldots, e_m\} = \text{range} \{e_1, e_2, \ldots, e_m\},$$

hence

$$[WSD/s] = \text{upper } p \text{ \% point of range} \left\{ \frac{e_1}{s}, \frac{e_2}{s}, \ldots, \frac{e_m}{s} \right\}$$

$$= \text{upper } p \text{ \% point of studentized range}$$

while for $A = F$, we have

$$\text{extent}_F \{e_1, e_2, \ldots, e_m\} = (2 \Sigma (e_i - e_.)^2)^{1/2}$$

so that

$$[FSD/s] = \text{upper } p\ \% \text{ point of } \left\{ 2\,\Sigma \left(\frac{e_i - e_{.}}{s} \right)^2 \right\}^{1/2}$$

$$= \text{upper } p\ \% \text{ point of } \left\{ 2(m-1) \frac{(\Sigma\,(e_i - e_{.})^2)/m-1}{s^2} \right\}^{1/2}$$

$$= \text{upper } p\ \% \text{ point of } \sqrt{2(m-1)F}\,.$$

In like manner, and in principle, we can obtain $[ASD/s]$ for any regular shape A and thus put a system of allowances of shape A to work if we wish.

CHAPTER 25: THE *C*-SYSTEM AND OTHER EIGENSYSTEMS

147. CORNFIELD, GREENHOUSE, HALPERIN AND ZALOKAR

Cornfield, Greenhouse, Halperin and Zalokar [32A] have independently invented the concept of a confidence procedure with familywise error rates, and have produced an example. Their procedure concentrates on the deviations $x_i - x_{.}$ of each determination from the mean of all, or, as is equivalent, on the discrepancies $x_i - x_{(i)}$, where (i), as before, signifies the formation of the mean over-all values of the subscript except i.

Cornfield, Greenhouse and Halperin discuss both confidence and significance procedures in their paper. In view of the discussion in Sections 130 and 138, we would feel that significance is more likely to replace confidence when *only* deviations or discrepancies are considered, than in any other case. Thus if the Cornfield, Greenhouse and Halperin method were only applicable to deviations and discrepancies, it would not warrant extensive discussion as an allowance.

However, every straggler allowance can be extended to a contrast allowance, just as every comparative allowance can be extended to a contrast allowance. When this is done, the Cornfield, Greenhouse and Halperin procedure generates a *C*-system of allowances.

148. C-ALLOWANCES

The statistic of central importance to Cornfield, Greenhouse and Halperin was

$$\frac{\max \mid x_i - x_. \mid}{s}$$

a bound on which obviously induces a bound, proportional to s, for all contrasts. In terms of the general approach of the last chapter, we reach the corresponding allowances by placing

$$\text{extent}_C \{e_1, e_2, \ldots, e_m\} = 2 \max_i \{e_i - e_.\}$$

where the factor of 2 is required so that the maximum of $\mid e_j - e_k \mid$ shall be 1 over the unit region. The key statistic, in our sense, then becomes

$$2 \frac{\max \mid e_i - e_. \mid}{s} = \frac{2}{s} \max \{\max_i \{e_i\} - e_., e_. - \min_i \{e_i\}\}$$

to whose approximate distribution we shall return in the next section.

The unit region is again a cylinder with generators parallel to the principal diagonal $e_1 = e_2 = \ldots = e_m$, as is the case for all systems of pure contrast allowances, and is, indeed, a prism whose edges are the lines

$$\begin{aligned} e_1 &= t + \delta_1 \\ e_2 &= t + \delta_2 \\ &\qquad\qquad -\infty < t < +\infty \\ &\ldots \ldots \\ e_m &= t + \delta_m \end{aligned}$$

where all $\delta_i = \pm^1/2$ (with one exception with m odd, this exceptional δ_i being $= 0$) exactly as many being + as are −.

In order to determine the norm of a linear combination, we have only to calculate its values on these edges, and take the maximum modulus of the results. We find the values to be

$$\left| \Sigma\, c_i e_i \right| = t \Sigma\, c_i + \Sigma\, c_i \delta_i .$$

If $\Sigma\, c_i$ does not vanish (if we are not concerned with a contrast), then since t takes on all finite values the norm is infinite. If $\Sigma\, c_i = 0$, then we make the modulus of $\Sigma\, c_i \delta_i$ maximum by making the δ_i corresponding to the most positive δ_i positive, and those corresponding to the most negative δ_i negative. This, as is easily seen, yields just one-half the sum of the deviations of the c_i from their median. Hence

$$\text{norm}_C \{c_1, c_2, \ldots, c_m\} = \begin{cases} \infty, & \text{if } \Sigma\, c_i \neq 0 \\ {}^1/2 \text{ (sum of deviations of } c_i \text{ from their} \\ \qquad \text{median)}, & \text{if } \Sigma\, c_i = 0. \end{cases}$$

Once we have some numerical values for the % points of key_C, we are ready to apply the C-allowance.

149. APPROXIMATE % POINTS FOR NORMALITY

While Cornfield, Greenhouse and Halperin [32A] have prepared some tables for the use of C-allowances, those were not available when this account was being drafted, and since comparisons were in order, approximate numerical results were desired. The most useful ones flow from the relations

$$\text{key}_C = 2 \frac{\max \left| e_i - e_. \right|}{s} = 2 \max_i \left\{ \frac{e_i - e_.}{s}, \frac{e_. - e_i}{s} \right\}$$

$$= 2 \sqrt{\frac{m-1}{m}} \max_i \left\{ \sqrt{\frac{m}{m-1}} \frac{e_i - e_.}{s}, \sqrt{\frac{m}{m-1}} \frac{e_. - e_i}{s} \right\}$$

and

$$\text{key}_C = \frac{2\max|e_i - e_.|}{s} = 2\max\left\{\frac{\max\{e_i\} - e_.}{s}, \frac{e_. - \min\{e_i\}}{s}\right\}$$

and the assumption of normality for calibration.

To use the first relation, observe that $e_i - e_.$ is normal with variance $\sigma^2(1-(1/m))$, so that

$$\sqrt{\frac{m}{m-1}}\,\frac{e_i - e_.}{s}$$

follows the t-distribution. Thus we have

$$\text{key}_C = 2\sqrt{\frac{m-1}{m}}\ \max\{2m \text{ correlated values of } t\}.$$

Since some of the values are correlated positively (e.g., $e_1 - e_.$ and $e_2 - e_.$) and others negatively (e.g., $e_1 - e_.$ and $e_. - e_2$) it seems very likely that we can very successfully approximate

$$\max\{2m \text{ correlated } t\text{'s}\} \approx \max\{2m \text{ independent } t\text{'s}\}$$

and derive numerical values as discussed for the studentized maximum modulus in Section 88.

Hence

$$k_C = (\text{upper } 5\% \text{ point of key}_C)$$

$$\approx 2\sqrt{\frac{m-1}{m}}\ (\text{upper } 5/\phi(m)\% \text{ point of } t)$$

$$= 2\sqrt{\frac{m-1}{m}}\ (\text{upper } 5/(^1/_2)\phi(2m)\% \text{ point of } t)$$

where $\phi(2m)$ is roughly $2m$.

To use the second relation, we have only to observe that

$$\frac{\max\{e_i\} - e_.}{s} \quad \text{and} \quad \frac{e_. - \min\{e_i\}}{s}$$

have the same distribution, and to recall that Nair [48] has tabulated some % points of this distribution. We have

$$k_c = (\text{upper } 5\% \text{ of key}_C)$$

$$= 2\left[\text{upper } 5\% \text{ point of max}\left\{\frac{\max\{e_i\} - e_.}{s}, \frac{e_. - \min\{e_i\}}{s}\right\}\right]$$

$$\approx 2\left[\text{upper } 5\% \text{ point of } \frac{\max\{e_i\} - e_.}{s}\right].$$

Applying these two relations, whose results agree quite well in the region of overlap, and choosing constants in the form

$$A + \frac{B}{DDF}$$

to agree with these results for $DDF = 10$ and ∞, we obtain the values in Table 31.

Table 31

Approximate Values Related to C-allowances

m = number of determinations	$[CSD/s]$ ∞	$DDF = 10$	$[CSD/LSD]$	Factor for halving error rate ∞	$DDF = 10$
2	2.77	3.16	$1.00 + 0.0/DDF$		
3	3.90	4.70	$1.41 + 0.8/DDF$		
4	4.32	5.30	$1.56 + 1.2/DDF$		
5	4.60	5.72	$1.66 + 1.4/DDF$	1.091	1.132
6	4.82	6.00	$1.74 + 1.6/DDF$		
7	4.98	6.26	$1.80 + 1.8/DDF$		
8	5.12	6.46	$1.85 + 2.0/DDF$		
9	5.22	6.64	$1.88 + 2.2/DDF$		
10	5.30	6.80	$1.91 + 2.4/DDF$	1.079	1.118
15	5.66	7.44	$2.04 + 3.1/DDF$	1.073	1.109
20	5.90	7.80	$2.13 + 3.5/DDF$	1.066	1.106

$[CSD/s]$ = upper 5% point of k_G

150. EIGENSYSTEMS AND EIGENALLOWANCES

C-, *P*-, and *W*-allowances are all examples of a rather broad class of systems of allowances which we shall call *eigensystems*. These arise when the subbasic collection of statements in the sense of Section 145 consists of one statement — that is to say, when we pick one particular contrast or other linear combination, say $\Sigma\, d_i x_i$, and say that we wish to make as strong a statement about it as possible, subject to the requirement of symmetry. Thus the key statistic becomes

$$(\text{constant}) \cdot \max_{\pi} \left\{ \frac{|\, \Sigma\, d_{\pi(i)} e_i \,|}{s} \right\}$$

where π runs through all rearrangements (permutations) of $1, 2, \ldots, m$.

In the three cases mentioned, we have

$$\Sigma\, d_i x_i = x_1 - x_2$$

and

$$\max_{\pi} \left\{ \frac{|\, \Sigma\, d_{\pi(i)} e_i \,|}{s} \right\} = \frac{\text{range}\{e_i\}}{s}$$

for the *W*-allowance, we have

$$\Sigma\, d_i x_i = x_1$$

and

$$\max_{\pi} \left\{ \frac{|\, \Sigma\, d_{\pi(i)} e_i \,|}{s} \right\} = \frac{\max |\, e_i \,|}{s}$$

for the *P*-allowance, and we have

$$\Sigma\, d_i x_i = x_1 - x_.$$

and

$$\max_{\pi}\left\{\frac{|\Sigma d_{\pi(i)}e_i|}{s}\right\}=\frac{\max|e_i-e_.|}{s}$$

for the C-allowance. We can clearly imitate these examples starting from any contrast or other linear combination for which we can obtain % points for the key statistic. At worst, these can be found by experimental sampling.

Why should there be particular interest in eigensystems? Because the tightest statement about any one chosen contrast or other linear combination is obviously made by the allowance of the corresponding eigensystem! We shall call this the eigenallowance for the contrast. We know that we must balance performance on one kind of contrast or linear combination against performance on another when we come to comparing systems of allowances. It will clearly help to know what is the best possible for a particular contrast or linear combination.

151. APPROXIMATE *k*'s

If we are to use these eigensystems for comparison as just suggested, we need to have some numerical values — and, in particular, numerical values for the upper 5% point of their key statistics. Two ways to obtain approximate values suggest themselves:

(1) to do some experimental sampling;

(2) to make use of the inequality between "X-wise" and "per X."

Both of these have been applied to a single example, namely to $m=10$, $DDF=\infty$, calibration = normality, which was selected as reasonably representative and reasonably accessible. For obvious reasons most of the work was concentrated on more or less symmetrical contrasts.

In order to make use of the inequality between "X-wise" and "per X" we proceed as follows. Consider a particular contrast, say

$$x_1+x_2-2x_3\,.$$

This is one of $m\,(m-1)(m-2)/2$ symmetrical contrasts of the form

$$x_i + x_j - 2x_k$$

where i, j, k are all different. We are thus concerned with $m\,(m-1)(m-2)/2$ expressions of the form

$$\left| \frac{e_i + e_j - 2e_k}{s} \right| .$$

In order to meet an error rate of 5% X-wise, we must find and use the upper 5% point of the distribution of their maximum values. In order to meet an error rate of 5% per X, we must find and use the upper

$$\frac{5}{m\,(m-1)(m-2)/2}\,\%$$

point of the distribution of any one. But each numerator is normal with variance $(1^2 + 1^2 + (-2)^2)\sigma^2 = 6\sigma^2$ and so each is distributed like $\sqrt{6}\;|\,t\,|$. We know that the per X% point will be larger than and near to the X-wise% point, and so we have

$$\text{(upper 5\% point of max} \left| \frac{e_i + e_j - 2e_k}{s} \right| \text{)}$$

$$\le \sqrt{6}\ \text{(upper } \{10/(m)(m-1)(m-2)\}\% \text{ point of } |\,t\,|\text{)}.$$

Thus we can establish an approximate value which is at the same time an upper bound.

We expect the approximation to be relatively good when the symmetrical collection consists of few contrasts, and relatively bad when it consists of many. For when there are many, it is more likely that a point $(e_1, e_2, \ldots, e_m)$ making one contrast unusually large also makes others unusually large.

In general, it is easiest to calculate the ratio

$$\frac{\text{bound for eigenallowance}}{F\text{-allowance}}$$

since the factor $(\Sigma d_i^2)^{1/2}$, which was $\sqrt{6}$ in the example, is just $\sqrt{2}$ norm_F $\{d_1, d_2, \ldots, d_m\}$ so that, using $k_F = [FSD/s] = [FSD/LSD]\, t\sqrt{2}$, this ratio becomes, where r is the number of equivalent contrasts,

$$\frac{\sqrt{2}\ \text{norm}_F\ \{d_1, d_2, \ldots, d_m\}\ (\text{upper } (5/r)\%\ \text{point of } |\, t\,|\,)}{[FSD/LSD]\sqrt{2}\ (\text{upper } 5\%\ \text{point of } |\, t\,|\,)}$$

$$= \frac{(\text{upper } (5/r)\%\ \text{point of } |\, t\,|\,)/(\text{upper } 5\%\ \text{point of } |\, t\,|\,)}{[FSD/LSD]}$$

$$= \frac{\text{error-rate shifting factor for } |\, t\,| \text{ and ratio } r}{[FSD/LSD]}$$

which is a very convenient form. A further upper bound is, of course, given by the fact that this ratio cannot exceed unity.

Values of r, the corresponding factor of shift in error-rate and $[FSD/LSD]$ are given in Table 32. The values of r for all the cases given are special cases of the three general formulas:

all m different, $\qquad r = m!$

k different from each other, others alike, $\qquad r = \dfrac{m!}{(m-k)!}$

k once each +1 and −1, rest zero, $\qquad r = {}^1/_2\ \dfrac{m!}{k!k!(m-2k)!}$.

To make use of experimental sampling, we have only to take down Wold's table of random normal deviates, construct samples of m, calculate the key statistic for each sample, prepare and smooth a frequency distribution and read off an approximate value of the upper 5% point.

The final results of these two processes of approximation are compared with each other, and with calculated values when available in Table 33 for $m = 5$, $DDF = \infty$, and in Table 34 for $m = 10$, $DDF = \infty$.

Because the ratio

$$\frac{\text{eigenallowance for subbasic contrast}}{F\text{-allowance for same contrast}}$$

is 0.77 for a simple contrast and 0.68 for a discrepancy (for $m = 10$, $DDF = \infty$) it is clear that this ratio is not constant. The numerical values indicate clearly that this ratio tends to be larger for contrasts which involve more individual determinations in more distinct ways. (This was mathematically clear for the bound.)

Table 32

Values of r (Factor of Shift in Error-rate) and Bounds for (eigenallowance/F-allowance) for Various Contrasts ($DDF = \infty$)

Coefficients of Subbasic contrast*		Values of r $m = 5$	$m = 10$	Error-rate Shifting Factors $m = 5$	$m = 10$	Bounds** $m = 5$	$m = 10$
$(m-1)^1$	$(-1)^{m-1}$	5	10	1.32	1.43	.84	.68
$(m-2)^2$	$(-2)^{m-2}$	10	45	1.43	1.67	.91	.80
$(1)^1$	$(-1)^1$	10	45	1.43	1.67	.91	.80
$(1)^2$	$(-2)^2$	30	360	1.61	1.94	(1.00)	.92
$(1)^2$	$(-1)^2$	15	630	1.50	2.01	.96	.96
$(1)^3$	$(-1)^3$	—	2100	—	2.15	—	(1.00)
$(1)^4$	$(-1)^4$	—	1575	—	2.13	—	(1.00)
$(1)^5$	$(-1)^5$	—	1260	—	1.81	—	.87
$(3)^1(2)^1(1)^1(-6)^1$		120	5040	1.81	2.25	(1.00)	(1.00)
all different		120	3,628,800	1.81	large	(1.00)	(1.00)
[*FSD*/*LSD*]				(1.57)	(2.10)		

* $(2)^a\ (1)^b\ (-2)^c$ means a 2's, b 1's, c (-2)'s, rest of coefficients zero.

** (1.00) means that bound comes out > 1.00, so that the upper bound of 1.00 has been substituted.

Table 33

Bounds, Experimental Sampling Estimates and Some Calculated Values for the Ratio

$$\frac{\text{eigenallowance}}{F\text{-allowance}}$$

for various subbasic contrasts for $m = 5$, $DDF = \infty$.

Coefficients of contrast	Bound	Exp't'l	Calculated	Reasonable estimate
4, -1, -1, -1, -1	.84		.84(C)	.84
3, 3, -2, -2, -2	.91			
2, 0, 0, -1, -1	(1.00)			
1, 0, 0, 0, -1	.91		.89(W)	.89
1, 1, 0, -1, -1	.96			

Table 34

Bounds, Experimental Sampling Estimates and Some Calculated Values for the Ratio

$$\frac{\text{eigenallowance}}{F\text{-allowance}}$$

for various subbasic contrasts for $m = 10$, $DDF = \infty$.

Coefficients of contrast	Bound	Exp't'l	Calculated	Reasonable estimate
9, -1, -1, -1, -1, -1, -1, -1, -1, -1	0.68	0.66	.68(C)	.68
2, 8, -2, -2, -2, -2, -2, -2, -2, -2,	0.79	0.75	—	.77
2, 0, 0, 0, 0, 0, 0, 0, -1, -1	0.94		—	.88
1, 0, 0, 0, 0, 0, 0, 0, 0, -1	0.79	0.76	0.77(W)	.77
1, 1, 0, 0, 0, 0, 0, 0, -1, -1	0.98	0.84	—	.86
1, 1, 1, 0, 0, 0, 0, -1, -1, -1	(1.00)	0.85	—	.87
1, 1, 1, 1, 0, 0, -1, -1, -1, -1	(1.00)	0.82	—	.84
1, 1, 1, 1, 1, -1, -1, -1, -1, -1	0.68	0.76	—	.74

152. THE MINIMAX REGRET ALLOWANCE

It is natural to assume that the *regret* with which we use any particular allowance on a particular contrast is some increasing function of the ratio

$$\frac{\text{allowance used for contrast}}{\text{eigenallowance for contrast}} .$$

Those statisticians who approve of "minimax regret" will immediately ask if there is a minimax regret allowance. The answer is clearly "yes."

Let us take all possible contrasts, one by one, as subbasic contrasts, and find the corresponding eigenallowances with p% error rate. Let us take all the statements which can be made about any contrasts with any of these allowances, and combine them. The result will be a set of statements where we say of each contrast what its own eigenallowance would say at p%. (Because the eigenallowance makes the sharpest statement of p% of any allowance.) This collection will have an error rate greater than p%, but less than 100%. (For no eigenallowance makes a stronger statement about $(c_1, c_2, \ldots, c_m)$ than $\pm LSD \cdot \text{norm}_F\{c_1, c_2, \ldots, c_m\}$, which has an error rate of p% for that single contrast.) If we weaken every statement in this collection in the ratio $1 + \lambda$ to 1, then the error rate will fall as λ increases, and for a suitable value of λ, say λ_p, the error rate *X*-wise will become p%. The resulting *R*-allowance will have

$$\frac{R\text{-allowance}}{\text{eigenallowance}} = 1 + \lambda_p$$

for all contrasts, and it is easy to see that

$$\begin{matrix}\text{minimum} \\ \text{(allowances)}\end{matrix} \left\{ \begin{matrix}\text{maximum} \\ \text{(contrasts)}\end{matrix} \left\{ \frac{A\text{-allowance}}{\text{eigenallowance}} \right\} \right\} = 1 + \lambda_p$$

so that it is a ***minimax regret*** allowance. We know that it is not the *F*-allowance, since this ratio is not constant for the *F*-allowance. Similar examples show that it is not the *W*-allowance, the *P*-allowance, nor the *C*-allowance. We know that it exists, and is different from the familiar examples, but we do not so far know how to find the numerical values necessary for its use, if it should be advisable to use it.

CHAPTER 26: COMPETITION AND COMPROMISE

153. INTRODUCTORY EXAMPLE

We now have quite a lot of machinery for setting up many systems of allowance, and in the next part we shall come to the comparison of the various systems. We know that one basis for this comparison will be the relative tightness of the various systems for various contrasts (or other linear combinations), and that their relative tightness will depend on the particular contrast. It is time to examine a particular case to see what questions will arise, and, in particular, whether we should consider still further particular systems of allowances because of these questions.

As an example, we take $m = 10$, $DDF = \infty$, calibration = normality, and compare the *C*-, *F*- and *W*-systems as to the tightness of their statements for a discrepancy, a simple comparison, and two differences of means; the mean of three minus the mean of another three, and the mean of five minus the mean of the remaining five. We begin by adopting the *F*-allowance as standard, but we also add the various eigenallowances for comparison, and eventually standardize on them as a revealing alternative. For C- and W-allowances we have the ratios

$$\frac{\text{eigenallowance}}{F\text{-allowance}}$$

for the subbasic contrasts, which we shall term the eigencontrasts of the corresponding systems. To obtain the ratios for another contrast, we have only to multiply by

$$\frac{\text{norm}_A\,\{\text{new contrast}\}}{\text{norm}_F\,\{\text{new contrast}\}} \cdot \frac{\text{norm}_F\,\{\text{eigencontrast}\}}{\text{norm}_A\,\{\text{eigencontrast}\}} .$$

The results are set forth in Table 35.

Clearly the choice between these three systems will depend very greatly on the relative importance of simple comparisons, discrepancies, and differences between subgroup means. (We recall that these relative importances were discussed in Part D.) But the important remark from our present point of view is the size of the differences. It is so large that some sort of compromise seems worth investigation.

Table 35

Relative Widths of Statements Made by Various Allowances ($p\,\% = 5\%$, $m = 10$, $DDF = \infty$, calibration = normality, basic F-allowance = 1.00). (For values with eigenallowance = 1.00, see Table 42.)

Allowance	Discrepancy	Comparison	3 against 3	5 against 5
C	.68	.91	1.58	2.04
F	1.00	1.00	1.00	1.00
W	1.04	.77	1.34	1.72
eigen	.68	.77	.87	.74

154. COMPROMISES

The immediately practical compromise between an A-allowance and a B-allowance, both at a $p\%$ omnibus error rate, is the following:

(1) calculate both A- and B-allowances at $^1/_2\,p\%$ error rate;

(2) apply whichever is narrower; and

(3) observe that the error rate of the result is less than $p\%$.

This is a conservative procedure, and is numerically available whenever we have error rate shifting available for a factor of 2.

The argument showing that the error rate is less than $p\%$ runs as follows: in not more than $^1/_2\,p\%$ of all cases will any $(A, {}^1/_2\,p\%)$ statement be wrong, in not more than $^1/_2\,p\%$ of all cases will any $(B, {}^1/_2\,p\%)$ statement be wrong, in $q\% > 0$ of all cases both an $(A, {}^1/_2\,p\%)$ statement and a $(B, {}^1/_2\,p\%)$ statement will be wrong. But the error rate for the combined statements is now seen to be $^1/_2\,p\,\% + {}^1/_2\,p\,\% - q\,\% < p\,\%$.

155. THE COST OF COMPROMISE

The cost of practical compromise is the weakening of all A statements by the error rate shifting factor for a ratio of 2. We have these factors available for C-, F- and W-allowances. Table 36 compares them, and suggests possible bounds for the corresponding factors for any reasonable kind of allowance. The factors are all small, decreasing as m increases and increasing as DDF decreases.

Table 36

Error Rate Shifting Factors for Various Allowance Procedures and a Ratio of 2
(? = possible bound for all reasonable allowances)

	$DDF = \infty$				$DDF = 10$			
	C	F	W	?	C	F	W	?
m								
2		1.148	1.142	1.16		1.180	1.180	1.19
3		1.108	1.111	1.12		1.152	1.152	1.16
4		1.098	1.098	1.11		1.141	1.140	1.15
5	1.091	1.082	1.089	1.10	1.131	1.134	1.131	1.14
10	1.079	1.063	1.072	1.09	1.118	1.120	1.116	1.13
15	1.073	1.049	1.060	1.08	1.107	1.113	1.109	1.12
20	1.066	1.044	1.058	1.07	1.106	1.102	1.098	1.12

156. COMPROMISE NORMS AND REFERENCE CONTRASTS

If we were to continue with the compromise exactly as indicated, we would have, as the norm for the compromise system

$$\text{minimum}\left\{\text{norm}_A\{\ \},\ \frac{BSD\,(2.5\%,\ DDF)}{ASD\,(2.5\%,\ DDF)}\ \text{norm}_B\{\ \}\right\}$$

and

$$\text{compromise } SD\,(5\%,\ DDF) = ASD\,(2.5\%,\ DDF) \leq BSD\,(2.5\%,\ DDF).$$

In this case, the norm would depend not only on m and 2.5%, but also on DDF, which seems annoying. If we replace

$$\frac{BSD\,(2.5\%,\ DDF)}{ASD\,(2.5\%,\ DDF)} \text{ by } \frac{BSD\,(2.5\%,\ \ )}{ASD\,(2.5\%,\ \ )}$$

and multiply the compromise SD by

$$\text{maximum}\left\{1,\ \frac{BSD\,(2.5\%,\ DDF)}{ASD\,(2.5\%,\ DDF)}\ \frac{ASD\,(2.5\%,\ \infty)}{BSD\,(2.5\%,\ \infty)}\right\}$$

we will have made the situation slightly more conservative. (Very slightly, since this factor does not exceed 1.04 for $DDF \geq 10$, $m \leq 20$ and the allowances considered here.) We denote the resulting modified compromise by $(AB)^*$, the $*$ reminding us of conservatism and numerical applicability.

If we wished to go through the necessary numerical computations, which are feasible in principle and probably feasible in practice, we could find an $(AB)SD$ which, combined with the modified norm would give exactly 5% error rate. The result we call the (AB) system of allowances.

By combining values from Tables 1, 9, 12 and 31 we can calculate the necessary constants, as set forth for $m = 3$, 5, 10 and 20 in Table 37, which gives us the means of making comparisons between the tightness of the various systems. The symbols $\binom{C}{F}$, and so on are defined by

Table 37

Constants for Compromise Allowances for $m = 3, 5, 10$ and 20
(Values in parentheses are rough estimates)

ITEMS	$m = 3$	$m = 5$	$m = 10$	$m = 20$
$(\frac{C}{F})$	1.15	1.07	1.00	1.00
$(\frac{F}{C})$	1.00	1.00	1.09	1.28
$(\frac{C}{W})$	1.19	1.19	1.19	1.19
$(\frac{F}{W})$	1.04	1.11	1.28	1.53
$(CF)^* SD/LSD$	$1.39 + \frac{1.2}{DDF}$	$1.70 + \frac{2.4}{DDF}$	$2.06 + \frac{3.9}{DDF}$	$2.27 + \frac{4.7}{DDF}$
$(CW)^* SD/LSD$	$1.33 + \frac{1.2}{DDF}$	$1.53 + \frac{1.8}{DDF}$	$1.74 + \frac{2.8}{DDF}$	$1.92 + \frac{3.2}{DDF}$
$(FW)^* SD/LSD$	$1.33 + \frac{1.0}{DDF}$	$1.53 + \frac{1.6}{DDF}$	$1.74 + \frac{2.9}{DDF}$	$1.92 + \frac{2.3}{DDF}$
$(CF) SD/LSD$	$(1.35 + \frac{1.0}{DDF})$	$(1.65 + \frac{2.0}{DDF})$	$(2.00 + \frac{3.0}{DDF})$	$(2.20 + \frac{4.5}{DDF})$
$(CW) SD/LSD$	$(1.25 + \frac{0.5}{DDF})$	$(1.45 + \frac{1.0}{DDF})$	$(1.65 + \frac{2.5}{DDF})$	$(1.85 + \frac{3.0}{DDF})$
$(FW) SD/LSD$	$(1.30 + \frac{1.0}{DDF})$	$(1.50 + \frac{1.5}{DDF})$	$(1.70 + \frac{2.5}{DDF})$	$(1.90 + \frac{3.0}{DDF})$

$$\text{norm}_{CF}\{\ \} = \text{minimum}\left[(\tfrac{C}{F})\ \text{norm}_C,\ (\tfrac{F}{C})\ \text{norm}_F\right]$$

$$\text{norm}_{CW}\{\ \} = \text{minimum}\left[(\tfrac{C}{W})\ \text{norm}_C,\ (\tfrac{W}{C})\ \text{norm}_W\right]$$

$$\text{norm}_{FW}\{\ \} = \text{minimum}\left[(\tfrac{F}{W})\ \text{norm}_F,\ (\tfrac{W}{F})\ \text{norm}_W\right].$$

If we examine the contrasts so far considered and to be considered, we find that, with the exception of regression coefficients, they fall into 3 broad categories, as follows:

$$(k \text{ vs. } k)\ \frac{x_1 + x_2 + \ldots + x_k}{k} - \frac{x_{k+1} + x_{k+2} + \ldots + x_{2k}}{k}$$

$$(k \text{ vs. } m-k)\ \frac{x_1 + x_2 + \ldots + x_k}{k} - \frac{x_{k+1} + x_{k+2} + \ldots + x_m}{m-k}$$

$$(1 + (r-1)(c-1)) \text{ vs. } ((r-1) + (c-1))$$

$$x_{ij} - x_{(i)j} - x_{i(j)} + x_{(i)(j)}$$

where in the last case, the 2-way discrepancy, a parenthesis around a subscript indicates a mean for all values of the subscript *except* the one in parenthesis.

When we work out the norms for these cases we find the results in Table 38.

Table 38

Norms of reference contrasts

Norm	k vs. k	k vs. $m-k$	$1+(r-1)(c-1)$ vs. $(r-1)+(c-1)$
C		$(2(1-k/m))^{-1}$	$\frac{3}{2} + \frac{r+c-3}{2(r-1)(c-1)}$
F	$k^{-1/2}$	$(2k\,(1-k/m)\,)^{-1/2}$	$(2(1-\frac{1}{r})(1-\frac{1}{c})\,)^{-1/2}$
W	1	1	2

Combining Tables 19, 12, 31, 38 and the definition of the compromise norms allows us to set forth for comparison the ratios [allowance/LSD] for $m = 3, 5, 10, 20$, various contrasts, and $A = C, F, W, (CF)^*, (CW)^*, (FW)^*$. The results for $DDF = \infty$ are given to one decimal in Table 39.

This table strengthens the general impression that

(1) F-allowances are shorter for most comparisons of means;

(2) W-allowances are shorter for simple comparisons;

(3) C-allowances are shorter for deviations and *one-way* discrepancies.

Clearly the F-allowance is a convenient standard, and Table 41 presents the ratio [A-allowance /F-allowance] for $m = 5$ and 20.

For the case $p\,\% = 5\%$, $m = 10$, $DDF = \infty$, calibration = normality, we can compare various compromises between C-, F-, and W-allowances and these allowances themselves on the basis of the corresponding eigenallowances. This is done in Table 42.

Since the C- and W-allowances have a rather similar pattern to begin with, the compromise $(CW)^*$ shows little improvement over the constituent systems. Both $(CF)^*$ and $(FW)^*$, on the other hand, show substantial

Table 39

Comparisons of allowances for specific contrasts by the values of [allowance/LSD].

Allowance	$m = 3$ 1 vs. 1	$m = 3$ 1 vs. 2	$m = 5$ 1 vs. 1	$m = 5$ 2 vs. 2	$m = 5$ 1 vs. 4	$m = 5$ 2 vs. 3	$m = 10$ 1 vs. 1	$m = 10$ 2 vs. 2	$m = 10$ 3 vs. 2	$m = 10$ 4 vs. 4
C	1.4	1.1	1.7	1.7	1.0	1.4	1.9	1.9	1.9	1.9
F	1.3	1.1	1.6	1.1	1.2	1.0	2.1	1.5	1.2	1.0
W	1.2	1.2	1.4	1.4	1.4	1.4	1.6	1.6	1.6	1.6
$(CF)^*$	1.4	1.2	1.7	1.2	1.1	1.2	2.0	1.6	1.3	1.0
$(CW)^*$	1.3	1.2	1.5	1.5	1.1	1.5	1.7	1.7	1.7	1.7
(FW)	1.3	1.3	1.5	1.2	1.4	1.1	1.7	1.6	1.3	1.0

Allowance	$m = 10$ 5 vs. 5	$m = 10$ 1 vs. 9	$m = 10$ 2 vs. 8	$m = 10$ 3 vs. 7	$m = 10$ 4 vs. 6	$m = 10$ $m = 20$ 1 + 4 vs. 1 + 4	$m = 20$ 1 vs. 1	$m = 20$ 2 vs. 2	$m = 20$ 3 vs. 3
C	1.4	1.0	1.2	1.4	1.6	1.9	2.1	2.1	2.1
F	1.0	1.6	1.2	1.0	1.0	1.2	2.4	2.0	1.6
W	1.4	1.6	1.6	1.6	1.6	1.6	1.8	1.8	1.8
$(CF)^*$	1.1	1.1	1.2	1.1	1.0	1.3	2.3	2.0	1.7
$(CW)^*$	1.5	1.1	1.3	1.5	1.7	1.7	1.9	1.9	1.9
$(FW)^*$	1.1	1.7	1.2	1.1	1.0	1.3	1.9	1.9	1.7

Allowance	$m = 20$ 5 vs. 5	$m = 20$ 10 vs. 10	$m = 20$ 1 vs. 19	$m = 20$ 2 vs. 18	$m = 20$ 3 vs. 17	$m = 20$ 5 vs. 15	$m = 20$ 1 + 9 vs. 1 + 9	$m = 20$ 1 + 12 vs. 3 + 4
C	2.1	2.1	1.1	1.2	1.3	1.4	2.1	1.8
F	1.4	0.9	2.0	1.5	1.2	1.0	1.5	1.4
W	1.8	1.8	1.8	1.8	1.8	1.8	1.8	1.8
$(CF)^*$	1.5	0.9	1.2	1.3	1.3	1.1	1.6	1.5
$(CW)^*$	1.9	1.9	1.2	1.3	1.3	1.5	1.9	1.9
$(FW)^*$	1.5	0.9	1.9	1.6	1.3	1.1	1.6	1.5

The values in this table have been held to one decimal point in order to make rapid examination easy. Some of the results of such an examination (using two-decimal values) are summarized in Table 40.

Table 40

Summary based on Table 39

Allowance	Best for	Within 0.2 of best
C	5 cases (1 vs. $m-1$, 2 vs. 18)	8 cases
F	17 cases	22 cases
W	5 cases (1 vs. 1 for all m, 2 vs. 2 for $m=20$)	8 cases
$(CF)^*$	never	23 cases
$(CW)^*$	never	12 cases
$(FW)^*$	never	22 cases

Table 41

The ratio [A-allowance / F-allowance] for $m = 5, 20$.
($DDF = \infty$. Attached signs show direction of change for finite DDF.)

A	$m=5$ 1 vs. 1	$m=5$ 1 vs. 4	$m=5$ 2 vs. 3	$m=20$ 1 vs. 1	$m=20$ 1 vs. 19	$m=20$ 10 vs. 10
C	1.06^+	0.84^+	1.37^+	0.76	0.55	2.43^-
F	1.00	1.00	1.00	1.00	1.00	1.00
W	0.89^-	1.13^-	1.39^-	0.64^-	0.89^-	2.06^-
$(CF)^*$	1.08^+	0.92^+	1.09^+	0.82^+	0.59^+	1.05
$(CW)^*$	0.98^+	0.92^+	1.50	0.69	0.59^+	2.18
$(FW)^*$	0.98^+	1.08^+	1.09^+	0.69	0.94^+	1.05
$(CFW)^*$	(1.02^+)	(0.97^+)	(1.14^+)	(0.71)	(0.62)	(1.09)

	$m=5$ others	$m=20$ 2 vs. 2	$m=20$ 2 vs. 18	$m=20$ others*
C	1.49^+	1.08	.80	1.01 to 2.28
F	1.00	1.00	1.00	1.00
W	1.26^-	0.92^-	1.22^-	1.12 to 1.90
$(CF)^*$	1.08^+	1.05^+	0.87	1.05^+
$(CW)^*$	1.38^+	0.97^+	0.87	1.08 to 2.00
$(FW)^*$	1.08^+	0.97^+	1.05^+	1.05^+
$(CFW)^*$	(1.13^+)	(1.01^+)	(0.91^+)	(1.09)

(* including cases not in Table 39.)

Table 42

Comparison of Width of Statement for Various Allowances and Compromises for Certain Kinds of Contrasts for $m = 10$.
(Basis = eigenallowance = 1.00. Values in ( ) are rough estimates.)

Allowance	Discrepancy	Simple comparison	Difference in means 3 vs. 3	Difference in means 5 vs. 5
C	1.00	1.18	1.82	2.76
F	1.47	1.30	1.15	1.35
W	1.53	1.00	1.54	2.32
$(CF)^*$	1.08	1.27	1.22	1.44
$(CW)^*$	1.08	1.09	1.65	2.48
$(FW)^*$	1.56	1.07	1.22	1.44
$(CFW)^*$	(1.12)	(1.11)	(1.27)	(1.50)
(CF)	(1.06)	(1.25)	(1.21)	(1.44)
(CW)	(1.04)	(1.05)	(1.59)	(2.40)
(FW)	(1.55)	(1.05)	(1.21)	(1.44)
(CFW)	(1.07)	(1.06)	(1.23)	(1.47)

improvements for some contrasts and slight losses for others. Of the two, $(CF)^*$ appears to come closest to being minimax regret, however $(FW)^*$ does noticeably better for the important case of simple comparisons.

It is not our immediate purpose to argue the merits of these various allowances, but only to point out that compromise is possible and substantially rewarding, as the following extract from Table 42 shows.

Allowance	Comparison	5 vs. 5
F	1.30	1.35
W	1.00	2.32
$(FW)^*$	1.07	1.44

Thus compromise systems must be considered.

157. CONSEQUENCES

As a consequence of our discussion in Part D (Chapter 23) we have a preference order for the various systems of allowances just considered. This order is found by looking first and foremost at simple comparisons, next at deviations (stragglers), next at interactions, and very much last at all others. We shall now apply this procedure, naming only x'd compromises, with the convention that each is to be replaced by the corresponding exact compromise when numerically available.

If we examine the behavior for simple comparisons (in Section 128, for example,) we find:

(1) the W-allowance far ahead;

(2) $(CW)^*$, $(FW)^*$, and $(CFW)^*$, within a reasonable distance;

(3) C- and (CF)-allowances next;

(4) the F-allowance last for $m = 10$ or 20 but ahead of those in (3) for $m = 5$.

If we accept the overwhelming importance of simple comparisons, the W-allowance is a clear first choice, and we continue only to pick the second and third choices.

If we examine the behavior for deviations, we find

(1) the C-allowance first;

(2) $(CF)^*$, $(CW)^*$, $(CFW)^*$ within a reasonable distance;

(3) F, W, and $(FW)^*$ last.

The leading candidates for second choice are $(CW)^*$ then (C), and then $(CF)^*$.

If we examine interactions, we find:

(1) F first;

(2) $(CF)^*$, $(FW)^*$, and $(CFW)^*$, within a reasonable distance;

(3) W, and $(CW)^*$ next;

(4) C last.

This suggests that $(CFW)^*$ is now slightly ahead of $(CW)^*$ for second choice with C further to the rear, just ahead of $(CF)^*$.

When we consider the difference of two means, we find:

(1) F first;

(2) $(CF)^*$, $(FW)^*$, $(CFW)^*$, within a reasonable distance;

(3) W, and $(CW)^*$ next;

(4) C last.

This seems to make it quite clear that $(CFW)^*$ is the second choice.

We conclude that as far as giving narrow limits and making tight statements are concerned, we have the following order (where all $*$'s are to be deleted as soon as the corresponding exact compromise becomes available):

first choice — W-allowance

second choice — $(CFW)^*$-allowance

next choice — $(CW)^*$-allowance

next choice — $(FW)^*$-allowance

next choice — C- and $(CF)^*$-allowances

last choice — F-allowances.

These are the conclusions based on tightness.

CHAPTER 27: INTRODUCTION TO NON-NORMALITY OF FLUCTUATION

158. INTRODUCTION

Most of the detailed discussion of allowances in this account has been restricted, explicitly or implicitly, by necessary or apparent hypotheses of normality of distribution and homogeneity of variance. It is important for us to know to what extent these hypotheses are necessary and to what extent they are apparent. For both of these hypotheses will fail to a greater or lesser degree in practice. This is not a simple problem, and we shall not reach final answers. Yet we can go far enough to provide a further basis for comparing different sorts of allowances.

We shall begin by discussing the effects of non-normality on the % points of $|t|$, usual and extreme. This is relevant to all "per X" allowances, and to a considerable degree to parallel allowances. Second, we shall consider the case of a pure comparative allowance which has distinctive properties. Finally, we shall discuss contrast allowances, beginning with a description of the Pitman-Welch results about the stability of F, and continuing to find a crude method of indicating sensitivity to non-normality which interpolates known results.

The problems of inhomogeneity of variance are quite similar to those of non-normality of distribution, and are treated in Chapter 29.

159. THE % POINTS OF $|t|$

The best available ***numerical*** results on the percentage points of $|t|$ when the underlying distribution is not normal are due to A. K. Gayen [31] who assumed that the population followed an Edgeworth series, retaining terms in the 3rd, 4th, and 5th relative cumulants. If we observe that we deal only with $|t|$, the terms in λ_3 and λ_5 drop out, yielding

$$\text{Prob}\{\,|t| > t_0\} = P_0(t_0) + \lambda_4 P_4(t_0) + \lambda_3^2 P_{33}(t_0)$$

where we have used subscripts "4" and "33" in place of Gayen's "λ_4" and "λ_5." If we examine Gayen's tables, we see that $P_4(t_0)$ and $P_{33}(t_0)$ have zeros for $p\,(t_0)$ in or near the usual significance range. In particular, we find the following:

Degrees of freedom:	4	6	12	24
$P\,(t_0)$ when $P_4(t_0) = 0$:	11%	9%	6%	5%
$P\,(t_0)$ when $P_{33}(t_0) = 0$:	18%	17%	17%	17%
$P_4(t_0)/P\,(t_0)$ for $P\,(t_0) = 1\%$:	.1	.15	.1	.05
$P_{33}(t)/P\,(t_0)$ for $P\,(t_0) = 1\%$:	.8	.7	.6	.4
$P_4(t_0)/P\,(t_0)$ for $P\,(t_0) = 0.1\%$:	?	?	.4	.2
$P_{33}(t_0)/P\,(t_0)$ for $P\,(t_0) = 0.1\%$:	?	?	2.0	1.5

Notice that the correction for λ_4 vanishes close to the 5% point, while that for λ_3^2 vanishes almost at 17% point, while for more extreme % points the ratio of correction to basic probability increases steadily.

These results are for a one-sample $|t|$, while our "per X" or parallel applications are most frequently to a pooled denominator. Thus, an unknown amount of adjustment has to be made for the effect of pooling, which almost inevitably will increase the effects of non-normality, since this will reduce the correlation between numerator and denominator due to extreme values and thus reduce the desirable compensation for the effect of such values. (Our t's with pooled denominators deviate from one-sample t's in the opposite sense from two-sample t's.)

It seems reasonable, however, that we can carry over the qualitative results to the pooled case, and say:

(1) the 5% point of $|t|$ is only moderately sensitive to non-normality; but

(2) this sensitivity rises rapidly as we go to more and more extreme % points.

Thus, "per *X*" and parallel allowances are expected to become increasingly sensitive to non-normality as the number of determinations increases.

160. COMPARATIVE ALLOWANCES

Let us next consider the case where we are only interested in simple comparisons, each the difference of two determinations. Only one allowance has been discussed in this whole account which can reasonably be regarded purely as a comparative allowance — an allowance spending an error rate of 5% familywise on simple comparison ***alone***. This is the *W*-allowance.

For, if we consider any of the other contrast allowances, say the *F*-allowances, we will find that it is spending part of its error rate on other contrasts, and so has an error rate of less than 5% for comparisons. We can approximate the extent of this effect, using Tables 1 and 12. We find for $DDF = \infty$ ($DDF = 10$) the following values:

$m =$	5	10	20
[*WSD*/*LSD*]	1.40 (1.48)	1.62 (1.78)	1.81 (2.06)
[*FSD*/*LSD*]	1.57 (1.67)	2.10 (2.34)	2.80 (3.26)
equivalent %*	1.82% (2.6%)	0.17% (0.8%)	0.0001% (0.18%)

(* % level, where $WSD_{\infty}\% = FSD_{?}\%$)

None of the equivalent %'s resemble 5%, and some are strikingly different.

If *only* a comparative allowance is desired, then changing to an *F*-allowance gives up the 5% error rate completely. Thus, even if the error rate of the *F*-allowance as a contrast allowance is stable near 5%, it is not thereby a satisfactory ***comparative*** allowance.

The only way in which a comparative allowance, which spends a 5% error rate on comparisons in the presence of normality, can have its susceptibility to non-normality varied is by changing the estimate of σ^2 which is used, by changing its "denominator." Before we discuss the possibilities here, we need to discuss the Pitman-Welch results.

161. THE WORK OF PITMAN AND WELCH

We are preparing to consider the relative stability against non-normality of the key statistics corresponding to various allowances. We shall confine our detailed consideration to the case of a balanced single classification, where the appropriate model is: $y_{ij} = \mu + v_i + w_{ij}$.

y_{ij} is the jth observation under ith set of conditions, where i runs from 1 to m and j runs from 1 to r and the w_{ij} are the individual fluctuations and are a sample from some distribution. The ith determination is

$$x_i = y_{i.} = \mu + v_i + w_{i.}$$

(where a "." represents the *mean* over all values of the subscript replaced) and its fluctuation is $w_{i.}$ the mean of r individual fluctuations. The underlying σ^2 is to be estimated from the scatter of the r observations under each of the m sets of conditions.

The Pitman-Welch results are confined to the key statistic for the F-allowance, the classical F-statistic, which in the single classification reduces to

$$F = \frac{r\,\Sigma\,(w_{i.} - w_{..})^2}{\Sigma\,\Sigma\,(w_{ij} - w_{i.})^2}\,.$$

If the distribution of individual fluctuations is exactly normal, then

(1) the numerator follows a multiple of a chi-square distribution;

(2) the denominator follows a multiple of a chi-square distribution;

(3) the numerator and denominator are independently distributed.

When the distribution of the individual fluctuations is not normal, it is easy to show that all three of these fail. For many years the general opinion was that since these three points fail, and fail rather badly, there was no hope for any stability of the F-distribution toward non-normality.

In 1939, Pitman [64,65,66] and Welch [81] discovered that this was not the case, that there was very considerable stability. The explanation lay, as it must, in the fact that (1), (2) and (3) fail in compensating ways. The basic observation is that, since

$$\Sigma\,\Sigma\,(w_{ij} - w_{..})^2 = \Sigma\,\Sigma\,(w_{ij} - w_{i.})^2 + r\Sigma\,(w_{i.} - w_{..})^2$$

we have

$$\frac{F}{F+r-1} = \frac{r\Sigma(w_{i.}-w_{..})^2}{\Sigma\Sigma(w_{ij}-w_{..})^2}$$

where the denominator is unaffected by permutations of the *mr* *w*'s. By studying the effects of permutations on a convenient multiple of this new ratio, Pitman was able to show that its low moments were nearly the same for any reasonable set of *mr* *w*'s and hence that its distribution was very nearly the same for samples from any reasonable population.

This device is clearly peculiar to this one ratio. Indeed it seems unlikely that any similar device will serve to teach us about any other ratio of interest. We have learned that *F*-allowances are highly immune to non-normality, and that is all.

162. THE SUGGESTED INDICATOR

There is, however, some help to be gained from the Pitman-Welch results for other key statistics. The basic idea which led Pitman to the first of these results was simple. He sought a key statistic in terms of what happened as the w_{ij} were permuted but the set were kept fixed, hoping to have the characteristics of the set of *mr* *w*'s disappear from the answer. We are not going this far, having particular numerators in mind and being prepared to tailor denominators to fit them. We want an indicator of mismatch between numerator and denominator, where this indicator refers to a single set of *mr* fluctuations, or to a simple family of such sets. If we can work on such a basis, we can avoid considering the detailed nature of the non-normality of the distribution from which the w_{ij} are, indeed, a sample.

We are reasonably convinced that the type of distribution which will bother us most will be one with long tails (e.g., with a slowly decreasing density function). Such distributions will affect us through samples $\{w_{ij}\}$ where a very few of the values are unusually deviant. We will probably get a good indication of the situation if we limit ourselves to one or two deviant fluctuations. The suggestion then is to study the behavior of numerator and denominator of the various key statistics when one or two fluctuations are deviant.

Let us start with the *F*-statistic, in order to build a basis for comparisons, and with one deviant fluctuation (which by symmetry may as well be in the 11 position), for simplicity. Thus, we replace w_{11} by $w_{11}+u$ and consider the effects on numerator and denominator. We find that

$$r\Sigma(w_{i.}-w_{..})^2 \to r\Sigma(w_{i.}-w_{..})^2 + 2(w_{i.}-w_{..})u + \frac{1}{r}\left(\frac{m-1}{m}\right)u^2$$

$$\Sigma\,\Sigma(w_{ij}-w_{i.})^2 \to \Sigma\,\Sigma(w_{ij}-w_{i.})^2 + \frac{r-1}{r}u^2$$

and on the average over all permutations of the w_{ij} the right-hand sides become

$$(m-1)\sigma^2 + \frac{1}{r}\left(\frac{m-1}{m}\right)u^2 = (m-1)\sigma^2\left(1+\frac{1}{mr}\frac{u^2}{\sigma^2}\right)$$

$$m\,(r-1)\sigma^2 + \frac{r-1}{r}u^2 = m\,(r-1)\sigma^2\left(1+\frac{1}{mr}\frac{u^2}{\sigma^2}\right)$$

$$(\text{where } (mr-1)\sigma^2 = \Sigma\,\Sigma(w_{ij}-w_{..})^2)$$

and we note that the *average* of *both* denominator and numerator, which are increased by factors which are quadratic polynomials in u/σ, are increased by the *same* factor.

Since the resistance of F to non-normality is very high, there is some reason to think that failure of such factors to be equal is an indication of sensitivity to non-normality. We shall try to apply this indicator to other numerators and denominators.

163. INDICATORS AND RANGES

We introduce

$$W_+ = \text{range}\{w_{i+}\}$$

and

$$W_i = \underset{j}{\text{range}}\{w_{ij}\}$$

and proceed to consider the effects of $w_{11} \to w_{11} + u$. Clearly W_i for $i \neq 1$ is unaffected, but W_1 and W_+ change. By how much?

To save subscripts, consider $W = \text{range}\{w_i\}$ while $w_1 \to w_1 + u$. We have

$$W \to \begin{cases} W + |u| - (w_d - w_1), & \text{for } u \le w_d - w_1 \\ W, & \text{for } w_d - w_1 \le u \le w_u - w_1 \\ W + u - (w_u - w_1), & \text{for } w_u - w_1 \le u \end{cases}$$

where w_d is the algebraically lowest and w_u is the algebraically highest w (d for "down" and u for "up"). If we average over all permutations of the mr w's (including w_1) we find that

$$\text{ave}\{W\} \to \text{ave}\{W\} + uf(u) \approx \text{ave}\{w\} + |u|$$

where $f(0) = 0$, and $f(u) \to 1$ as $u \to \pm\infty$.

On the other hand, for $u \to \pm\infty$

$$W \to W + u - (w_u - w_1) = (w_1 - w_d) + u$$

$$W^2 \to (w_1 - w_d)^2 + 2u(w_1 - w_d) + u^2$$

and since $W = w_u - w_d$

$$\text{ave}\{W - (w_d - w_1)\} = \text{ave}\{w_1 - w_d\} = 1/2 \text{ ave}\{w\}$$

and

$$1/4 \text{ ave}\{W^2\} \le \text{ave}\{(w_1 - w_d)^2\} \le 1/3 \{W^2\}$$

(the last inequality holding for unimodal distributions) this leads to

$$\text{ave}\{W\} \to 1/2 \text{ ave}\{W\} + u$$

$$\text{ave}\{W^2\} \to 0.3 \text{ ave}\{W^2\} + u \text{ ave}\{W\} + u^2$$

where 0.3 is approximately correct.

Applying these results to W_i and W_+, we have, for $|u|$ large

$$\text{ave}\{W_i\} \to 0.5 \text{ ave}\{W_i\} + |u|$$

$$\text{ave}\{W_i^2\} \to 0.3 \text{ ave}\{W_i^2\} + |u| \text{ ave}\{W_i\} + u^2$$

$$\text{ave}\{W_+\} \to 0.5 \text{ ave}\{W_+\} + |u|$$

$$\text{ave}\{W_+^2\} = 0.3 \text{ ave}\{W_+^2\} + |u| \text{ ave}\{W_+\} + u^2$$

$$\text{ave}\{\Sigma W_i^2\} = \frac{m - 0.7}{m} \text{ ave}\{W_i^2\} + |u| \text{ ave}\{W_i\} + u^2.$$

We also wish to deal with $\text{ave}\{(\Sigma W_i)^2\}$ which is seemingly a little harder because of the correlation between the W_i. But

$$(\Sigma W_i)^2 \to (\Sigma W_i)^2 + 2u \Sigma' W_i + u^2$$

where Σ' indicate a sum over $m - 1$ values of $i \neq 1$, and

$$W_1 \to W_1 + u$$

and hence

$$\text{ave}\{(\Sigma W_i)^2\} \to \text{ave}\{(\Sigma W_i)^2\} + 2\left(\frac{1}{r} |u| - 0.5 \text{ ave}\{W_i\}\right)$$

$$(m-1) \text{ ave}\{W_i\} + u^2 + |u| \text{ ave}\{W_i\} - 0.7 \text{ ave}\{W_i^2\}$$

$$= \text{ave}\{(\Sigma W_i)^2\} - (m-1)(\text{ave}\{W_i\})^2 - 0.7 \text{ ave}\{W_i^2\}$$

$$+ \left(1 + \frac{2(m-1)}{r}\right) |u| \text{ ave}\{W_i\} + u^2.$$

164. AVERAGE VALUES

We introduce the abbreviations A_k, B_k, C_k and D_k by

$$\text{ave}\{W_i^2\} = rC_r\,\sigma^2$$

$$\text{ave}\{W_+^2\} = mC_m\,r\sigma^2$$

$$(\text{ave}\{W_i\})^2 = rD_r\,\sigma^2$$

$$\text{ave}\{W_i W_j\} = rA_r\,\sigma^2$$

$$\text{ave}\{(\max\{\mid w_i - w_{..}\mid\})^2\} = mB_m\frac{\sigma^2}{r}\left(1-\frac{1}{m}\right).$$

We can get an idea of the size of these quantities, which depend on the particular set of mr w's involved, since the averaging is over rearrangements of a fixed set, by examining a table of their values for samples from a normal distribution (after all, $w_{11} \to w_{11} + u$ represents the main part of our non-normality). These run as follows:

$r =$	2	3	4	5	6	8	10	15	20
$C_r =$	1.04	1.22	1.25	1.23	1.19	1.10	1.01	.84	.72
$A_r = D_r$	.63	.96	1.06	1.08	1.07	1.01	0.96	.81	.70
$B_r =$	(1.6)						(1.1)		(.75)

Thus, it is reasonable to treat these quantities as being near unity, and to treat $C_r - A_r$ as rather small.

The changes in the denominators and numerators so far considered are (when we take $\sigma^2 = 1$ as we may do):

(ΣW_i^2) $$mrC_r \to \left(1-\frac{0.7}{m}\right) mrC_r + \mid u \mid \sqrt{rD_r} + u^2$$

$((\Sigma W_i)^2)$ $$mrC_r + m(m-1)\,rA_r \to mrC_r + m(m-1)\,rA_r - (m-1)rC_r$$
$$-0.7\,rD_r + (1+2(m-1)/r)\sqrt{rD_r}\mid u \mid + u^2$$

$(\Sigma\Sigma(w_{ij} - w_{i.})^2)$ $$m(r-1) \to m(r-1) + \frac{r-1}{r}u^2$$

(W_+^2) $$rmC_m \to 0.3\,rmC_m + \mid u \mid \sqrt{rD_r} + u^2$$

$$((\max\{\,|\,w_{i.}-w_{..}\,|\,\})^2) \qquad mB_m\frac{1}{r} \to \frac{1}{r}\left(1-\frac{1}{m}\right) + {}^{1}/_{2}\left(1-\frac{1}{m}\right)^2 u^2$$

$$(\Sigma\,(w_{i.}-w_{..})^2) \qquad (m-1) \to m-1+\frac{1}{r}\left(\frac{m-1}{m}\right)u^2$$

and the factors of increase are most simply expressed in the form

$$\alpha + \beta\,\frac{u}{\sigma}\sqrt{\frac{rD_r}{mr}} + \gamma\,\frac{1}{mr}\,\frac{u^2}{\sigma^2}$$

where α, β and γ take the following values:

For	α	β	γ
$\Sigma\, W_i^2$	$\left(1-\frac{0.7}{m}\right)$	C_r^{-1}	C_r^{-1}
$(\Sigma\, W_i)^2$	$1-\frac{\left(1-\frac{C_r}{m}+0.7\frac{D_r}{m}\right)}{C_r+(m-1)A_r}$	$\frac{1+2(m-1)/r}{C_r+(m-1)A_r}$	$\frac{1}{C_r+(m-1)A_r}$
$\Sigma\,\Sigma\,(w_{ij}-w_{i.})^2$	1	0	1
W_+^2	0.3	$C_m{-}1$	C_m^{-1}
$(\max\{\,|\,w_{i.}-w_{..}\,|\,\})^2$	$\frac{1}{mB_m}$	0	$(1-\frac{1}{m})B_m^{-1}$
$\Sigma\,(w_{i.}-w_{..})^2$	1	0	1

165. TWO EXTREME FLUCTUATIONS

If we suppose two different fluctuations are shifted, one by u and the other by $\pm v$, where $u > v > 0$ (or $-u > -v > 0$), then we can follow through a similar calculation. The quadratic terms are relatively easily found, and are shown in Table 43. Again the numerator and denominator of F show perfect balance. If we look for a denominator to match W_+^2, we fail to find one either for like signs or for unlike signs. We are driven toward $(\Sigma\, W_i)^2$ until we realize that while this denominator treats u^2+v^2 and $2uv$ similarly, the coefficient γ (see previous table) is ~$1/mA_r$ for this denominator which is far from both the 1 and the C_m^{-1} for the numerators.

Table 43

Factors Corresponding to u^2 when Two Fluctuations are Shifted

	for $(u, +v)$	for $(u, -v)$
ΣW_i^2	$u^2 + \left(1 - \frac{r-1}{r-1}\right) v^2$	$u^2 + v^2 + 2\,uv \left(\frac{r-1}{r-1}\right)$
$(\Sigma W_i)^2$	$u^2 + v^2 + 2uv$	$u^2 + v^2 + 2uv \left(1 - \frac{2\,(r-1)}{r-1}\right)$
$\Sigma\,\Sigma\,(w_{ij} - w_{i.})^2$	$u^2 + v^2 - \frac{2uv}{r-1}$	$u^2 + v^2 + \frac{2uv}{r-1}$
W_+^2	$u^2 + \frac{r-1}{r-1}\,(2uv + v^2)$	$u^2 + v^2 + 2uv \left(1 - \frac{2(r-1)}{2}\right)$
$(\max\{\mid w_{i.} - w_{..} \mid\})^2$	$u^2 + \frac{1}{(m-1)}\,2\,v^2 - \frac{2}{m-1}\,uv$	$u^2 + \frac{1}{m-1}\,2\,v^2 + \frac{r-1}{m-1}\,uv$
$\Sigma\,(w_{i.} - w_{..})^2$	$u^2 + v^2 - \frac{2uv}{r-1}$	$u^2 + v^2 + \frac{2uv}{r-1}$

The conclusion is that, except for F-allowances, all the other systems so far considered go to pot when *two* or more really large fluctuations are likely to occur among the set of mr fluctuations. If we must have stability against this much non-normality, we must go to an F-allowance with a conventional denominator.

CHAPTER 28: FURTHER NON-NORMALITY

166. INTRODUCTION

We have so far carried out our computations as if what we wished was to have the effect on the averages of numerator and denominator the same, where the averaging is over all permutations of a particular set of mr w's and all permutations of a particular additional shift u, or two additional shifts u and $\pm v$. Such very highgrade balancing turned out to be possible for the F-statistic, but not for any others. Perhaps we may still receive useful guidance if we slacken our requirements a little.

One natural approach would be as follows:

(1) agree to normalize both numerator and denominator to average unity for sample of mr from normality;

(2) seek to have the effect of a large shift on the normalized numerator and denominator such as will bring the ratio back to unity.

In this case, the A_k, B_k, C_k, and D_k to be inserted in the formulas are the values appropriate to normality, rather than to some particular set of mr fluctuations. We can use the known numerical values, and see a little further.

167. BACK TO COMPARISONS

Let us go back to the comparative allowances, where we started. Here we wish to combine the W-numerator, W_+^2, with some suitable denominator, of which three are being considered so far:

(short cut)	$(\Sigma\, W_i)^2$
(half cut)	$\Sigma\, W_i^2$
(conventional)	$\Sigma\,\Sigma\,(w_{ij} - w_{i.})^2$

Let us begin by considering the quadratic terms for specific values of m and r. The case $m = 2$ is uninteresting. So we take the following cases:

$$(m = 3,\ r = 10)$$
$$(m = 10,\ r = 5)$$
$$(m = 20,\ r = 3)$$

which seem to reflect likely practice fairly well. Substituting in previous results, we find the coefficients of $1/mr\sigma^2$ set forth in Table 44.

The bottom panel is appropriate for symmetrical but non-normal fluctuations. It indicates the following approximate percentages of compensation of denominator for numerator.

short cut	5% to 60%
half cut	45% to 120%
conventional	55% to 125%

Clearly the short cut denominator is nearly useless in balancing out the effects of symmetrical non-normality. The difference between the half cut

Table 44

Coefficients of $1/mr\sigma^2$ for two random shifts and a comparative allowance

	$m=3, r=10$	$m=10, r=5$	$m=20, r=3$
(u and $+v$, with $u > v > o$, or their negatives)			
numerator	$.82u^2+.50uv+.25v^2$	$.99u^2+.16uv+.08v^2$	$1.39u^2+.10uv+.05v^2$
short cut	$.34u^2+.68uv+.34v^2$	$.09u^2+.18uv+.09v^2$	$.05u^2+.10uv+.05v^2$
half cut	$.99u^2+.00uv+.68v^2$	$.81u^2+.00uv+.07v^2$	$.82u^2+.00uv+.02v^2$
conventional	$u^2+.07uv+v^2$	$u^2-.04uv+v^2$	$u^2-.03uv+v^2$
(u and $-v$, with $u > v > 0$, or their negatives)			
numerator	$.82u^2+.62uv+.82v^2$	$.99u^2+1.62uv+.99v^2$	$1.39u^2+2.58uv+1.39v^2$
short cut	$.34u^2+.26uv+.34v^2$	$.09u^2+.15uv+.09v^2$	$.05u^2+.09uv+.05v^2$
half cut	$.99u^2+.61uv+.99v^2$	$.81u^2+.13uv+.81v^2$	$.82u^2+.06uv+.82v^2$
conventional	$u^2+.07uv+v^2$	$u^2+.04uv+v^2$	$u^2+.03uv+v^2$
(average of two panels above)			
numerator	$.82u^2+.56uv+.54v^2$	$.99u^2+.09uv+.54v^2$	$1.39u^2+1.34uv+.72v^2$
short cut	$.34u^2+.47uv+.34v^2$	$.09u^2+.16uv+.09v^2$	$.05u^2+.10uv+.05v^2$
half cut	$.99u^2+.31uv+.84v^2$	$.81u^2+.06uv+.81v^2$	$.82u^2+.03uv+.82v^2$
conventional	$u^2+.00uv+v^2$	$u^2+.00uv+v^2$	$u^2+.00uv+v^2$

and the conventional is smaller, actually varying from 8% to 20%. The half cut is not as good as the conventional by a considerable margin, but it provides substantial protection.

The top panel is appropriate for very unsymmetrical cases of non-normality, and here the following approximate percentages of compensation are indicated.

short cut	5% to 90%
half cut	45% to 120%
conventional	70% to 15%

Again the sort cut fails us, but the other two are considerably improved, the advantage of the conventional denominator being somewhat more marked.

The compensation for the latter two denominators is good enough to make it worthwhile at least to look at the coefficients of $|u|\sqrt{rD_r}/mr$ — to look at the linear terms. These coefficients are given in Table 45.

Table 45

Coefficients of $|u|\sqrt{rD_r}/mr\sigma$ for a Single Random Shift and a Comparative Allowance

-ator	$m=3, r=10$	$m=10, r=5$	$m=20, r=3$
numerator	0.82	0.99	1.39
half cut	0.99	0.81	0.82
conventional	0	0	0

Thus a certain amount of compensation for moderate non-normality is given quite well by the half cut and not at all by the conventional denominator. Against moderate non-normality then, the half cut may perform on the average as well as the conventional denominator.

Our conclusions for comparative allowances then, run as follows:

(1) there is only one numerator to consider;

(2) for violent non-normality the conventional denominator is superior, and the half cut may be usable;

(3) for moderate non-normality, these two denominators are about on a par;

(4) the short cut should only be used when somewhere near normality;

(5) in general, these allowances will be more sensitive to non-normality than those based on the F-statistic.

168. A WORD ABOUT TAILORING

There remains a real possibility that other denominators can be found which do a better job of balancing the comparative allowance numerator. These new denominators would be selected to reflect the effects of extreme fluctuations into the denominator with greater amplification where needed.

Such a denominator as, for a single classification

"the sum of the squares of the r largest W_i (or of all if $r > m$)"

would clearly tend to magnify the effects of extreme fluctuations more than does the half cut. Perhaps such denominators will be developed in time. Their use must meet two major difficulties, however. First, the distributional problems will not be easy. Second, and more important, the investigation of sensitivity to non-normality, the selection of a denominator, and the development of quantitative distributional information will have to be repeated for each of many situations. We have in this account considered only two situations explicitly — balanced single classifications and balanced double classifications. But there are many more situations in which we find estimates of σ^2. The proper comparison-compensating denominators may be very different. Thus, developing good comparison compensating denominators would be a major problem.

In this chapter we have only considered single classifications, and only by an indicator of uncertain merit, and on this small foundation is based our advice to the user. The small size of this foundation is realized, but so is the fact that it is the only foundation so far available. When more evidence comes in, the recommendations will be gladly changed. But for the moment, this is the best that we can do.

169. CONTRAST ALLOWANCES

It is now time to consider the various possible contrast allowances from the point of view of non-normality. Both numerators and denominators are at our disposition. It may be well to begin by disposing of comparison numerators at once.

The nature of a compromise allowance is, of course, this: we calculate several allowances (at a suitable fraction of 5%) and take the narrowest. The compromise key statistic is then the maximum of a number of key statistics. Its sensitivity to normality will be controlled by the sensitivities of the more sensitive key statistics which enter into its formation. Roughly speaking, the compromise allowance will be nearly as sensitive to non-normality as the most sensitive of the allowances involved in the compromise. Having made this argument, we can be content to confine our considerations to allowances which are not explicit compromises.

There are, then, three numerators C, F, and W, and three denominators, short cut, half cut and conventional. We have already tabulated the coefficients in the indicator and the modified indicator, and have compared the denominators in some detail. The corresponding comparison for numerators is made in Table 46 and can be summarized as follows:

$m = 3, r = 10$: C much less affected than others which are nearly the same,

$m = 10, r = 5$: C somewhat less than F, which is a little less than W,

$m = 20, r = 3$: F somewhat less than C, which is a little less than W.

Since the conventional denominator balances F exactly, it will overcompensate C for the first two cases and undercompensate for the last. W is compensated in the first case and undercompensated in the last two. Thus, C is somewhat better behaved than W in so far as these results indicate stability against non-normality.

The comments about the short and half cut denominators, made in connection with comparative allowances carry over without essential change to C or W contrast allowances (the case where the half cut fails to compensate the most is the case $m = 20$, $r = 3$ where the C-numerator is most like the W-numerator.)

Table 46

Coefficients of $1/mr\sigma^2$ in the modified indicator for two random shifts and contrast allowances.

numerator	$m = 3, r = 10$	$m = 10, r = 5$	$m = 20, r = 3$
	(u and $+v$, with $u > v > 0$ or $u < v < 0$)		
C	$.4u^2 - .4uv + .1v^2$	$.8u^2 - .2uv + .01v^2$	$1.3u^2 - .13uv + .00v^2$
F	$u^2 - .07uv + v^2$	$u^2 - .04uv + v^2$	$u^2 - .03uv + v^2$
W	$.8u^2 + .50uv + .25v^2$	$.99u^2 + .16uv + .08v^2$	$1.39u^2 + .10uv + .05v^2$
	(u and $-v$, and $u > v > 0$ or $u < v < 0$)		
C	$.4u^2 + .4uv + .1v^2$	$.8u^2 + .2uv + .01v^2$	$1.3u^2 + .13uv + .00v^2$
F	$u^2 + .07uv + v^2$	$u^2 + .04uv + v^2$	$u^2 + .03uv + v^2$
W	$.82u^2 + .62uv + .82v^2$	$.99u^2 + 1.62uv + .99v^2$	$1.39u^2 + 2.58uv + v^2$
	(average of two panels above)		
C	$.4u^2 + .00uv + .1v^2$	$.8u^2 + .00uv + .01v^2$	$1.3u^2 + .00uv + .00v^2$
F	$u^2 + .00uv + v^2$	$u^2 + .00uv + v^2$	$u^2 + .00uv + v^2$
W	$.82u^2 + .56uv + .82v^2$	$.99u^2 + .89uv + .99v^2$	$1.39u^2 + 1.34uv + .72v^2$

170. CONCLUSIONS

In summary, we have come to the following main conclusions:

(1) avoid the short cut denominator in the presence of substantial non-normality;

(2) avoid the half cut in the presence of extreme non-normality;

(3) the *F*-numerator — conventional denominator combination — is least sensitive to non-normality;

(4) the *C*-numerator is a little less sensitive than the *W*-numerator;

(5) compromises are nearly as sensitive as the most sensitive of their components.

CHAPTER 29: CONCLUSION OF COMPARISON

171. INTRODUCTION

We have dealt with the two most important considerations in comparing allowances — tightness of statement and sensitivity to non-normality — and we have found them diametrically opposite. For desirable tightness of statement we want a system of allowances that grips comparisons and deviations tightly. To avoid sensitivity to normality we want a system of allowances that grips comparisons and deviations loosely. The eventual choice of allowances for use will depend mainly on a judgment of the balance between these two forces, but there are other considerations which need to be brought in. These we consider next.

172. INHOMOGENEITY

The most important of these is the other standard bugaboo of all analytical procedures, inhomogeneity of variance. We shall continue to confine our discussion to the balanced single classification. As before, let y_{ij} represent the jth observation under the ith set of conditions.

We have $y_{ij} = \mu + v_i + \sigma_i w_{ij}$ where σ_i^2 is the variance appropriate to the ith set of conditions, and the w_{ij} are now a sample from a distribution with unit variance. Since we do not expect any great interaction between non-normality and inhomogeneity, and for reasons of simplicity, we shall usually assume that the w_{ij} are a sample from the unit normal distribution.

It is natural to inquire concerning the average values of numerator: and denominators. We find:

$$\text{ave}\{\Sigma\, w_{i^+}^2\} = r\Sigma\sigma_i^2 = rm\sigma^2$$

$$\text{ave}\{\Sigma\,(w_{i^+} - \frac{1}{m}w_{++})^2\} = r\,(m-1)\sigma_{\cdot}^2$$

$$\text{ave}\{W_+^2\} = ?$$

$$\text{ave}\{(\max\{\mid w_{i.} - w_{..}\mid\})^2\} = ?$$

$$\text{ave}\{\Sigma\,(w_{ij} - \frac{1}{r}w_{i+})^2\} = (r-1)\,\Sigma\,\sigma_i^2 = (r-1)m\sigma_{\cdot}^2$$

$$\text{ave}\{\Sigma\, W_i^2\} = rC_r \Sigma\, \sigma_i^2 = rmC_r \sigma_{\cdot}^2$$

$$\text{ave}\{(\Sigma\, W_i)^2\} = rC_r \sum_i \sigma_i^2 + rD_r \sum_{i \neq j}\sum \sigma_i \sigma_j$$

$$= rD_r\, (\Sigma\, \sigma_i)^2 + mr\, (C_m - D_m)\sigma_{\cdot}^2$$

where $m\sigma^2 = \Sigma\, \sigma_i^2$, so that $\sigma_{\cdot}^2$ is the average variance.

Note that the average values of (i) the G-numerator, (ii) the F-numerator, (iii) the conventional numerator, and (iv) the half cut denominator remain proportional to one another. So far as averages go, there is no effect due to inhomogeneity. The average value of the short cut denominator involves $\bar{\sigma}^2$ rather than $\overline{\sigma^2}$, and what the average value of the W-numerator may be is very hard to say. We can go little further forward in this type of examination.

We are, of course, really interested in upper percentage points and not in moments. There is one case — a rather extreme case — where we can examine upper percentage points. This arises when the first k values of σ_i^2 are equal, say to unity, while the other $m - k$ values vanish. This means that the F-numerator has fewer degrees of freedom, that the W-numerator is the augmented range of k rather than the range of m, and that the denominators need involve only summation over k columns instead of m. The F-numerator changes from $m - 1$ to an equivalent number of degrees of freedom of

$$k - \frac{(k-1)^2 + (k-1)}{(m-1)^2 + (k-1)}$$

which is slightly greater than and nearly equal to $k - 1$. We have thus to deal with the dependence on k of

(for C- and C_h-allowances)	???
(for F- and F_h-allowances)	$\sqrt{F}$ on $(k-1)$ and kn DF
(for W- and W_h-allowances)	studentized range of k for kn DF

where n is a factor expressing the number of equivalent denominator degrees of freedom. For the conventional denominator $n = r - 1$. For the half cut denominator, n is somewhat less than this.

Numerical investigation shows that the dependence on k is almost, if not quite, that of a fractional power, and the exponent fitting exactly for

$k=4$ and $k=20$ seems an adequate indicator of the overall behavior of the correct width of allowance.

When the short cut is used, another factor of (k/c) enters the denominator, and the dependence of allowance width on k is increased by an additional factor $k^{-0.5}$.

Further inquiry shows that the fractional power dependence on k can be put in the form

$$k^{-A-.20/n}$$

to adequate accuracy for $n \geq 1$, where the value of k depends on the allowance in question. Since this means that when the extreme inhomogeneity represented by $k < m$ is present, the variance allowances are too narrow in the ratio

$$\left(\frac{k}{m}\right)^{-A-.20/n}$$

it is clear that small values of A are desirable. The resulting values of A are as follows:

Numerator	Denominator		
	conventional	half cut	short cut
C	?	?	?
F	0.15	0.15	0.65
W	0.29	0.29	0.79

Compromises suitably intermediate

The general direction of choice suggested by the desire for insensitivity to inhomogeneity of variance is thus the same as that suggested by the desire for insensitivity to non-normality.

173. SIMPLICITY OF USE

It is easy to rank the various allowances on simplicity of use. This can be done separately in the numerator and denominator.

In the numerator, from simplest to least simple we run as follows:

$$W > C > F >> \text{compromises} >>> R$$

the last being unavailable. The differences between W and C and between C and F are small, being essentially the difference between finding $\text{norm}_W\{c_i\}$, $\text{norm}_C\{c_i\}$, and $\text{norm}_F\{c_i\}$ for the contrasts of interest. This avoidance of squaring and square-rooting is a small matter, but sometimes worthy of consideration. Between F and the compromises there is a greater difference, since the use of a compromise requires the calculation of at least two norms and the selection of one for further use.

In the denominator, the order is obviously

$$\text{short cut} > \text{half cut} >> \text{conventional}$$

for those designs where convenient and adequate short cuts have been prepared. The differences between short cut and half cut seems to depend on circumstances. When observations are being analyzed in a hurry, far from computing equipment, the short cut is notably more convenient that the half cut. When analysis is routine, a few squares and one square root do not amount to a great difference. In any case, either is far ahead of the conventional analysis in simplicity for those designs like balanced single and double classifications where it can be easily applied.

174. EASE OF INTERPRETATION

Here we can repeat the remarks made about numerators in the last section, making necessary changes, and strengthening them somewhat. It is very easy to tell anyone how the *WSD* is used, and what it means, less easy for the *CSD*, still less easy for the *FSD* and far less easy for compromise *SD*'s.

175. FLEXIBILITY OF APPLICATION

All systems of allowances can be easily extended to the following cases:

(1) the m determinations have a common variance and a common covariance, and only contrasts are to be considered;

(2) the m determinations have variances in known ratio and their covariances vanish.

These two cases have reasonably wide scope.

In case (1) we have only to introduce constant factors into the definition of the determinations chosen so as to equalize the variance. In case (1), if σ^2 and $\rho\sigma^2$ are the common variance and common covariance, then

$$\operatorname{var}\{\Sigma c_i x_i\} = (\Sigma c_i^2)\sigma^2 + ((\Sigma c_i)^2 - \Sigma c_i^2)\rho\sigma^2 = \Sigma c_i^2(\sigma^2 - \rho\sigma^2)$$

where we have used $\Sigma c_i = 0$. If we arrange for the average value of s^2 to be $(1-\rho)\sigma^2$ in place of σ^2, as is easily and automatically done, we are again reduced to the standard situation.

In case (2), we have two choices, an unsatisfactory exact solution and a rather satisfactory approximate one. We can reduce case (2) to the standard case by introducing constant factors into the definition of the determinations, so that the new determinations have constant variance. The reduction is exact, but has serious disadvantages. A simple comparison of two old determinations is not necessarily even a contrast among the new ones.

The approximate solution, which we adopted in Chapter 4, is to calculate allowances as if the determinations were of equal variance and then adjust them at the last moment by the ratios indicated by the known ratio of variances. Here simple comparisons remain simple comparisons, and the approximation consists of neglecting the fact that the corresponding covariances are not what they would be for the case of equal variances over all determinations. The effect of this approximation on error rates will depend on the system of allowances in an unknown manner. An important virtue of this approximate solution is its applicability when the common covariance is not zero. All that is really required is:

(a) a common covariance; and

(b) knowledge of the ratios of (variance minus common covariance) for the various determinations.

This last fact makes the approximate solution available for use in complex analyses of variance.

In principle, all allowances can be extended exactly to the general case where

$$\operatorname{cov}\{x_i, x_j\} = g_{ij}\sigma^2$$

and the g_{ij} are known. If we define a specific transformation $\{f_{ij}\}$ for each $\{g_{ij}\}$ so that

$$x_i = \Sigma\, f_{ij} y_j$$

implies that

$$\operatorname{cov}\{y_i, y_j\} = \begin{cases} \sigma^2, & i = j \\ 0, & i \neq j \end{cases}$$

then

$$\Sigma\, c_i x_i = \sum_j (\Sigma\, c_i f_{ij})\, y_j$$

and if we define a generalized norm, norm_g, by

$$\text{norm}_g \{c_i\} = \text{norm}\{\sum_k c_k f_{ki}\}$$

we have, in principle, extended any allowance to the general case.

In practice, the definition of the $\{f_{ij}\}$ is not easy, and the properties of norm_g are not likely to be what we want. In certain special cases we can obtain a usable limited extension in this way. For example, if there is a common variance and a common covariance, so that

$$\operatorname{cov}\{x_i, x_j\} = g_{ij}\sigma^2 = \begin{cases} \sigma^2, & i = j \\ \rho\sigma^2, & i \neq j \end{cases}$$

then we may choose

$$f_{ij} = \begin{cases} 1 + \lambda, & i = j \\ \lambda, & i \neq j \end{cases}$$

where

$$(1 - \rho)\,(1 + 2\lambda + m\lambda^2) = 1$$

and obtain such an extension. Since

$$\Sigma\, c_k f_{ki} = c_i + \lambda(\Sigma\, c_i)$$

the new norm coincides with the old norm for contrasts. However, this is but a special case.

As a reasonable approximation, we can copy the approximate solution for case (2) in the general case. The properties of the result here are even less clear.

Finally, for the F- and G-allowances alone, because of their symmetry, we may define an extension by

$$\text{norm}_h \{c_i\} = \left[{}^1\!/_2\, \Sigma\, \Sigma\, c_i c_j g_{ij} \right]^{1/2}.$$

This seems to be the only practical extension to the general case and is, of course, the generalized F-allowance introduced by Scheffé [68-69].

We have then, two alternatives:

(1) to use the F-allowance for the standard case and the generalized F-allowance for the general case; or

(2) to use some other allowance for the standard case, and change to the generalized F-allowance for the general case.

Such a change could only be pedagogically important, since the "equal accuracy" and "substantially unequal accuracy" cases are well distinguished from one another in practice. This pedagogical importance would lie only in the apparent continuity in method between one case and another. This apparent continuity is, the writer feels, less important in practice than might seem to the teacher of mathematical statistics. There is a considerable and necessary increase in complexity when we go to the generalized case, and this introduces enough of a break so that the additional change, for example, from a W-type norm to an F-type norm does not, in the writer's opinion, increase the break enough to matter.

176. AN UNIMPORTANT PROPERTY

We should also mention a property of the F-allowance which seems entirely unimportant from an overall point of view. This is the property that the error rate of every individual statement is the same. Since the unimportance of this may not be clear, we shall discuss the point a little.

If we were *solely* interested in comparisons, we would, I think, feel that the error rates for individual statements about other contrasts were irrelevant, and we would want to spend our error rate on comparisons alone. If we were *equally* interested in all contrasts, we would, I think, be willing and eager to give each individual statement the same error rate. If our state of mind *varied* slowly and continuously from one of these situations to the other, I am sure our desirable allocation of error rates would vary slowly and continuously also. We are, in fact, near the "solely in comparisons" end of this scale. Our optimum allocation is far from equality. Hence, equality is irrelevant.

177. SUMMARY OF OTHER CONSIDERATIONS

The result of trying to summarize the minor considerations is set forth in Table 47. These are one man's attempt to put down in mournful numbers what he has earlier said in words. These can be summarized as

(1) a strong vote against compromises and minimax regret; and

(2) a weaker vote against both forms of short cutting.

So much for the minor considerations.

Table 47

Subjective Scores (9 = best) for Minor Considerations

	Numerators							
	C	*F*	*W*	compromises	*R*	conv.	half	short
Ease of interpretation and perspicuity	8	6	9	4	1	9	7	7
Flexibility of application	5	9	5	5	5	9	5	5
Mathematical elegance	7	7	7	3	9	9	8	7
Conformance to tradition	6	9	7	4	1	9	7	5
Total	26	31	28	16	16	36	27	24

When we bring in "simplicity of use" as a matter of great importance, we strengthen the vote against compromises and minimax regret, and reverse the vote on short cutting, so that we now have

(1) a very strong vote against compromises and minimax regret; and

(2) a moderate vote in favor of short cutting (when available).

Considerations of availability are of equal importance, but are adequately taken care of by the parenthetical statement in (2) above.

178. BASIC CONSIDERATIONS

We are now ready to examine the relation between the basic characteristics of tightness (of statement) and insensitivity (to non-normality and inhomogeneity). Figure 9 represents an attempt to represent graphically the conclusions that we have reached verbally earlier. The "admissible"' allowances from these two points of view will be those on the upper right boundary of the diagram, namely F, $(CF)^*$, $(CFW)^*$ and W, with the others close behind. In view of the strong vote against compromises from other considerations, it seems appropriate to replace $(CF)^*$ by (C), at a slight cost in sensitivity, and to consider $(CFW)^*$ as out of the picture. Thus, we are finally back to C, F and W again.

In view of this return to simplicity, the role of the compromises warrants some discussion. If the writer's point of view about the relative importance of various contrasts is accepted, then the compromises fall out of the picture. But if a less extreme view is taken on this point, and insensitivity is not regarded as highly important, then such compromises as $(FW)^*$ become highly attractive. Under some conditions, we dare avoid them, but under others not.

179. THE ULTIMATE DECISION

We must now choose between C, F and W on the basis of all considerations. If tightness and insensitivity were always equally important, then we might consider choosing C as a compromise. But, in actuality, tightness is always important, but insensitivity is of a very varying importance. Therefore, it is reasonable to choose at least two allowances as standard and to vary their use in accordance with the judged amount of danger from non-normality or inhomogeneity. In view of the extreme insensitivity of F, it must

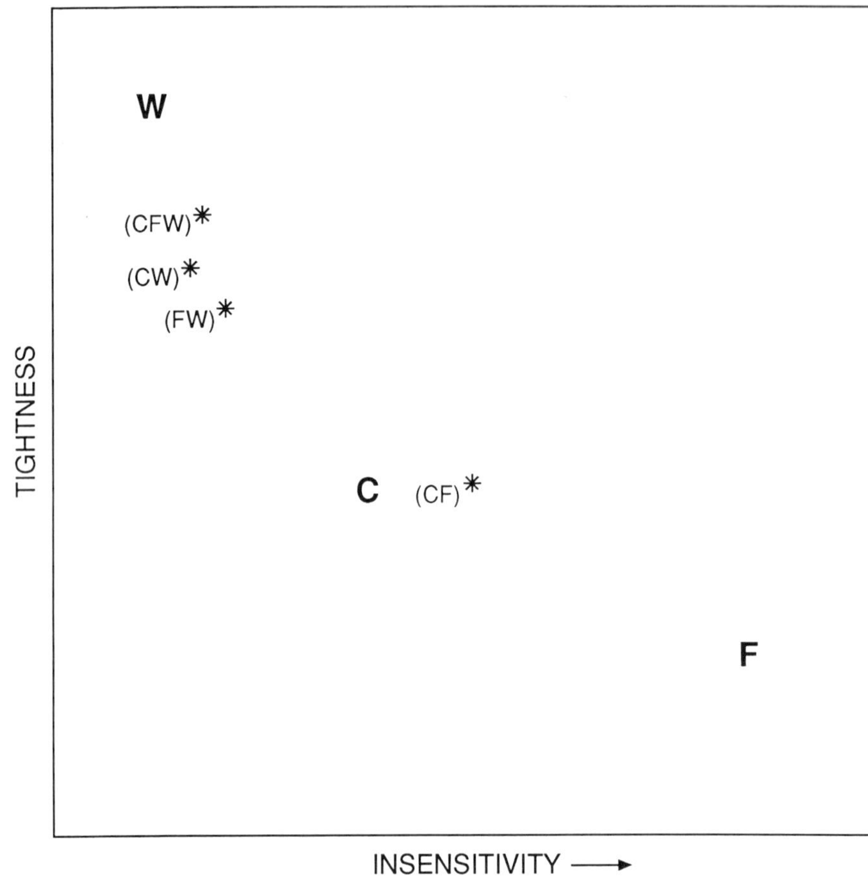

Figure 9. Tightness vs. insensitivity.

clearly be included. In the writer's opinion the tightness of W exceeds the tightness of *C* — where it matters most — by enough to make its inclusion essential.

Since *C* seems to fail to reach the (subjectively defined) halfway point between *F* and *W* in either tightness or insensitivity, and since two standards are to be preferred to three, the writer would eliminate *C*.

Since *F* is selected because of insensitivity, which is only present when the conventional denominator is used, only the conventional denominator is appropriate with the *F*-numerator.

On the contrary, it seems appropriate to use any denominator, chosen according to the needs of the situation, with the *W*-numerator.

These are our ultimate decisions.

PART F
SELF CONSISTENT SIGNIFICANCE PROCEDURES

CHAPTER 30: THE GENERAL FRAMEWORK

180. INTRODUCTION

Classically, there was only one type of significance procedure for a family of determinations. There the spread-outness of the family was compared with s, the square-root of the estimate of variance, and then the whole family was declared either "significant" or "nonsignificant." No attempt was made to focus significance more sharply on parts of the family. Such procedures, from the point of view about to be developed, are naturally described as *rudimentary* significance procedures.

The significance procedures with which we shall be mainly concerned go far beyond the rudimentary level and declare "significant" or "nonsignificant" *each* group of 2 or more determinations in the family. Thus much more detailed conclusions are put to explicit test. We shall find it possible to set up many such procedures.

The "natural" way to begin would seem at first glance to be clear: just test each group of k as if it were a whole family of determinations, and state the results. A simple example will show the unsatisfactory nature of such a procedure. Let us suppose that DDF is very large, that $s = 1$, that $m = 50$, and that $x_1 = 98$ and $x_{50} = 103$. We wish to show that the values of the other 48 determinations affects the common sense significance of the group of 2 determinations made up of x_1 and x_{50}.

Let us first suppose that the other 48 determinations all lie somewhere between 200 and 250. Then they are all very different from x_1 and x_{50}, and the same is almost certainly true of the corresponding determinands. The difference $x_{50} - x_1$ is $5/\sqrt{2} = 3.5^+$ times its standard error, so that we have no ground for doubting the significance of this group of two determinations.

On the other hand, if we suppose that the other 48 determinations all lie between 98 and 102, our attitude and conclusions are quite different. The values of all 50 determinations, taken together, are quite likely to be consistent with the hypothesis that all 50 determinands have the same value, and that the observed variations in the determinations, which range from 98 to 103, are solely due to fluctuations. It is common sense to assert that x_1 and x_{50} are not significantly different.

Comparing then two situations, we see that common sense significance depends on the values of the determinations outside the group considered, and we expect that such dependence will also appear in systematic significance procedures. Having come to this conclusion, we are led to formulate the three natural criteria of monotony, scaling, and extreme recognition as follows:

(M) If the estimated standard deviation, s of the determinations is decreased, without any change in the determinations x_i, then any group previously "significant" remains "significant."

(S) The significance of a group depends only on the ratio of the differences in the whole family to s.

(X) If the range of the determinations in any group becomes arbitrarily large in comparison with s, that group is declared significant.

Since we choose to be concerned only with symmetrical situations, we shall also require that our procedures treat all the determinations symmetrically.

181. SELF-CONSISTENCY

Besides the three natural criteria just mentioned, which express how we feel that procedure should behave for any one particular group, there is a further criterion involving two or more groups. One that does not spring to mind as rapidly and directly as the other three criteria, but one which becomes important when we consider how we might use our declarations of significance. What could we do with the ***combination*** of the two statements

(1) x_1 and x_2 are "significant" (= significantly different);

(2) x_1, x_2 and x_3 are "not significant" (= not significantly different)?

Can we draw any useful conclusions or guidance? The writer believes not, and so adds a fourth criterion

(C) If a given group of determinations is called significant, so is every group that contains the given group.

The writer feels that monotony, scaling, extreme recognition, symmetry and self-consistency are requirements which do not exclude any interesting or useful significance procedures, and which thus do not really restrict our freedom of choice. We shall call procedures which satisfy all these conditions merely *self-consistent* procedures for brevity.

Our next step will be to focus our attention on a special class of self-consistent procedures which are obtainable in a particular manner. This class includes almost all the multiple comparison significance procedures so far proposed, and in particular contains all rudimentary procedures. It does not include *all* self-consistent significance procedures, and at some later time some of us might be interested in significance procedures outside this special class.

182. APPARENT SIGNIFICANCE

In order to define the class of multilayer significance procedures, we shall need a few more ideas and techniques. These center around notions of "apparent significance," which is to be thought of as a tool on the way to a notion of "significance" and never as an end in itself, not as in any sense equivalent or closely related to "significance."

One simple way in which symmetrical self-consistent procedures can be set up is the following:

(1) For each size of group we have a notion of apparent significance.

(2) A group of determinations is called significant if it is apparently significant, and so is every group which contains it as a part.

If B is a group which contains a significant group A, then every C containing B also contains A and is therefore apparently significant. Hence B is significant. Thus self-consistency is guaranteed.

If we require monotony, scaling, and extreme recognition of "apparent significance," we retrieve them for the corresponding notion of "significance." Symmetry is already included, since the notion of apparent significance was required to depend only on the size of the group involved. Thus a notion of apparent significance, suitably restricted, will generate a self-consistent significance procedure. As we shall see briefly later, any self-consistent significance procedure can be generated by a notion of apparent significance, but not by one that is *local* in the sense of the rather natural condition:

(L) The apparent significance of any group of determinations does not depend on the values of the determinations of the family which are outside the group.

In conjunction with the other conditions, this implies that the notion of apparent significance may only depend on:

(1) the ratio $(x_i - x_j)/s$ of differences of determinations in the *group* to the square-root of the estimate of their common variance;

(2) the number of degrees of freedom, *DDF*, associated with s; and

(3) the total number of determinations, m, in the *family*.

It is only through m that the determinations outside the group can make themselves felt at all in determining local ***apparent*** significance.

If a self-consistent significance procedure can be derived from a local notion of apparent significance, it is a ***multilayer*** significance procedure.

183. DISCUSSION

We might, for example, take as the notion of apparent significance for a group merely a significant F-test at the 5% level. If we revert to our 50 determinations example, then x_1 and x_{50} would always be an apparently significant pair.

In the first form, where the other determinations all exceeded 200, every group including x_1 and x_{50} would also be apparently significant, and so the group consisting of x_1 and x_{50} would be significant.

In the second form, where the other determinations were concentrated between 98 and 102, they might well be so concentrated that the F-test for all 50 determinations would not reach the 5% level. Thus the whole family would not constitute an apparently significant group. Since the group of 2 is part of this not apparently significant group of 50, the group of 2 would *not* be declared significant. Thus such a procedure can agree with common sense in this example.

While we may not wish to use just this procedure in practice, for reasons which will appear, it does illustrate the basic workings of a local notion of apparent significance — of the workings of the sort of apparent significance in which we are interested.

If we give up the local requirement for a moment, we can exhibit any self-consistent significance procedure in terms of a notion of apparent significance. For if we define "apparent significance" to be the same as the original notion of "significance," and consider the new notion of "significance," based on this "apparent significance," then we shall find all three notions the same. This remark has interest but no particular application or usefulness.

184. APPARENT SIGNIFICANCE LEVELS

Having limited our selection of significance procedures on very general grounds, we come next to the definition of their significance levels. Levels in the plural, because regretfully there must be several, for the situation is unfortunately and irremediably complicated. In the simple case of two determinations there is no difficulty, as we know, since the significance level is just the probability of false significance. In the multiple determination case there are many different kinds of false significance, and the probabilities of their occurrence depend on many things.

Thus in an example with the 50 determinations, the probability of declaring the group consisting of x_1 and x_2 significant when $\mu_1 = \mu_2$ will depend on the values of $x_3, x_4, \ldots, x_{50}$ and hence on the values of μ_3, $\mu_4, \ldots, \mu_{50}$. There is no unique probability of false significance for this pair of determinations. (Attempts to ensure a unique probability produce obviously bad procedures.) The only reasonable solution is to take as the significance level the greatest probability of false significance, whether attained for some specific situation or as a limit.

Thus we define the *apparent significance level* of a procedure at a particular group of determinations as the greatest probability of false significance for that group. For a multilayer procedure, this is the same as the probability of false apparent significance — of apparent significance when all the determinands corresponding to the group are equal. To show that this is so, we may argue as follows: Take all the determinands corresponding to the given group equal to one value, and all the other determinands equal to a second value. If these two common values differ by very many multiples of s, then any group containing the given group and any other determination is almost certainly apparently significant. This follows from the property of extreme recognition for apparent significance. Hence, almost certainly, the significance of the given group of determinations will be the same as its apparent significance. Since the given group is never significant without being apparently significant, the greatest probability of false significance is the same as the probability of apparent significance, and this is what we wished to prove.

By symmetry, the apparent significance level depends only on the size of the group. We denote it by $\alpha(k)$.

185. NULL HYPOTHESES

The hypothesis that a given group of determinands are equal, for example that

$$\mu_1 = \mu_2 = \mu_4$$

will be called an *apparent null hypotheses*. The term "apparent" is used for two reasons first, because an apparent significance level corresponds to an apparent null hypothesis. Second, and more importantly, because several apparent null hypotheses may hold together. Thus if $m = 7$, and

$$\mu_1 = \mu_2 = \mu_4 << \mu_3 = \mu_6 << \mu_5 = \mu_7.$$

We have an *overall null hypothesis* made up of three apparent null hypotheses, namely $\mu_1 = \mu_2 = \mu_4$ and $\mu_3 = \mu_6$ and $\mu_5 = \mu_7$. We refer to the three corresponding groups of determinands or determinations as *coherent bunches* and refer to the common value of the determinands of a coherent bunch as its common value.

This particular overall null hypothesis is subject to false significance in three ways, since any or all of the coherent bunches may be falsely assigned significance. The corresponding apparent significance levels are

$\alpha(3)$ for the group (1, 2, 4)

$\alpha(2)$ for the group (3, 6)

$\alpha(2)$ for the group (5, 7).

The actual probabilities of false significance increase and approach $\alpha(3)$, $\alpha(2)$ and $\alpha(2)$ as closely as we please, when the common values of the three bunches are all separated by larger and larger multiples of s. Even in the limit, however, the occurrences of false significance in the three bunches do not become independent unless $DDF = \infty$, since the same s is used in assessing of apparent significance. This association is positive, and the probability of one or more false apparent significances, and hence of one or more false significances is less than for independence. If $\alpha(k)$ does not depend on DDF, then, we maximize the probability of one or more false significances by choosing $DDF = \infty$ and separating the common values as widely as possible.

186. MORE SIGNIFICANCE LEVELS

The *overall* significance level, for an overall null hypothesis, is defined to be the chance of one or more false significances when this null hypothesis holds. Its actual value will depend on the separations between the groups of equal determinands and, for $\alpha(2)$, $\alpha(3)$, ... fixed, on the number of degrees of freedom, *DDF*, associated with s^2. It will increase as the separations and the number of degrees of freedom increase, and its limiting value corresponds to independent occurrence of false significance in each coherent bunch.

Thus if we have a null hypothesis with coherent bunches of 3, 2 and 2, we shall have (overall significance level) less than and tending to $\alpha(3, 2, 2)$ where $\alpha(3, 2, 2)$ is the probability of occurrence of one or more of the independent events with respective probabilities $\alpha(3)$, $\alpha(2)$, $\alpha(2)$, which is given by

$$(1-\alpha(3, 2, 2)) = (1-\alpha(3))(1-\alpha(2))(1-\alpha(2)).$$

We shall adopt a similar notation for any system of coherent bunches.

We note that since $\alpha(1) = 0$, and $\alpha(3) \geq \alpha(2)$, we must have

$$\alpha(3) \geq \alpha(2, 1)$$

and, in general, that

$$\alpha(k+1) \geq \alpha(k, 1).$$

Thus we may reasonably define as the *standard* significance levels, the $\alpha(k_1, k_2, \ldots, k_p)$ for which

(1) each $k_i \geq 2$; and

(2) $k_1 + k_2 + \ldots + k_p = m$;

that is for which each coherent bunch contains at least two determinands. Every overall significance level is less than some standard level. Every standard level is the limit of overall levels, and hence of the probability of finding one or more false significances under the corresponding null hypothesis.

The α's for which

$$k_1 + k_2 + \ldots + k_p = m - 1,$$

that is, those with a *single* exceptional coherent bunch made up of only *one* determination, are *near standard* significance levels.

If all coherent bunches but one consist of one determination, and if all their common values are widely separated, then the only false significances are those for this exceptional coherent bunch containing k determinations, and the probability of this occurring approaches $\alpha(k)$. Thus under exceptional circumstances, any apparent significance level is an overall significance level. But conversely, every standard significance level is also, under suitable circumstances, an overall significance level.

Thus if we wish to summarize the significance level behavior of a multilayer significance procedure it is natural to give three significance levels, namely:

(1) the minimum apparent significance level;

(2) the maximum apparent significance level;

(3) the maximum standard significance level;

since every overall significance level lies between the first and the third. These may be widely different, and stating only one may be misleading.

187. DIFFERENCES ONLY

If we are only concerned with the "significance" or "nonsignificance" of the simple comparisons, $x_i - x_j$, then we only need consider groups of size 2. The extreme relevant significance levels for m even will always be the apparent significance level $\alpha(2)$ and the standard significance level.

$$\alpha(2, 2, \ldots, 2) = 1 - [1 - \alpha(2)]^{m/2}.$$

Notice that the relation between these levels does not depend on anything except m. Thus the difficulties caused by their inequality are irremediable among multilayer significance procedures.

The fact that a procedure with $\alpha(2) = 5\%$ has a much higher overall significance level when the determinands are equal in pairs, and the common values are widely separated was first pointed out to the writer by John Mandel of the National Bureau of Standards in connection with a procedure suggested by the writer in BIOMETRICS [73] and discussed in Chapter 34. As noted above, this is an essential, and unfortunate, feature of all multilayer significance procedures.

188. PLOTTING FOR $m = 3$

If we take the simplest meaningful case, that of three determinations, we can set forth the situation graphically. It is easy to plot the differences of three values in a plane. The procedure can be described in at least three alternative ways.

We may say that we are taking $(x_2 - x_3)/\sqrt{2}$ and $(2x_1 - x_2 - x_3)/\sqrt{6}$ as rectangular coordinates, and that, since any contrast can be expressed in terms of these two, each point in the plane determines x_1, x_2 and x_3 up to an additive constant.

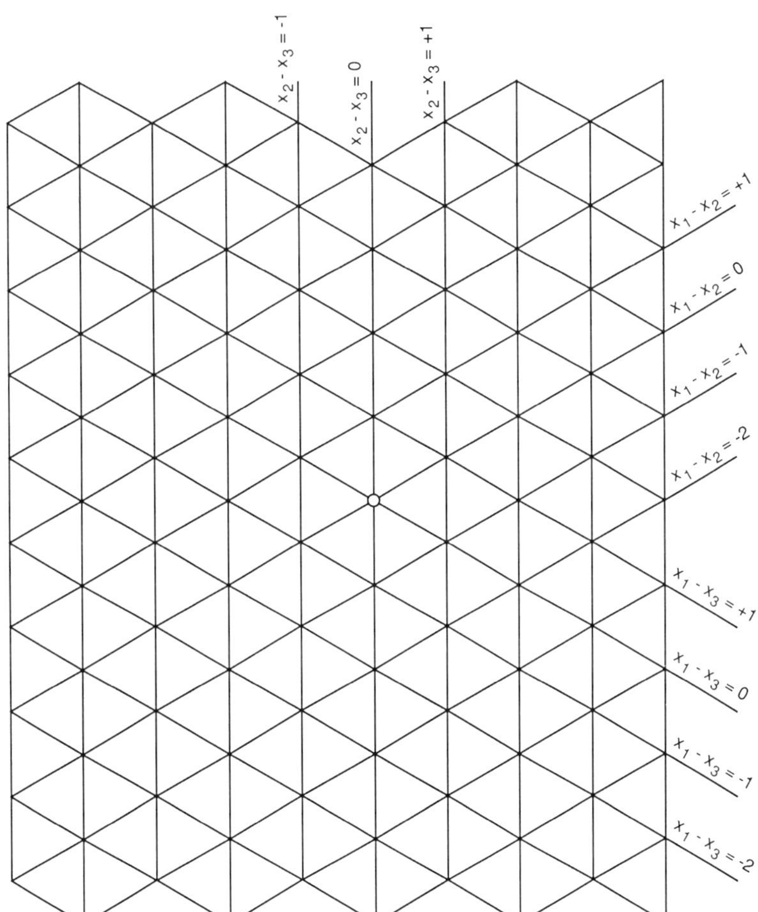

Figure 10. Coordinate system for differences of three determinations.

We may say that we are taking (x_1, x_2, x_3) as rectangular coordinates in space, and projecting the corresponding point parallel to the line $x_1 = x_2 = x_3$ onto a plane perpendicular to that line.

We may say that we are taking $x_1 - x_2$, $x_2 - x_3$ and $x_3 - x_1$ as trilinear coordinates in the plane.

x_2, x_3 *or* of μ_1, μ_2, μ_3 in the plane, and the coordinate grid for the differences appears as in Figure 10. The regions in which

(a) $| x_1 - x_2 | \leq A$;

(b) $\text{range}\{x_1, x_2, x_3\} \leq B$;

(c) $\Sigma (x_i - x_.)^2 \leq C$;

(d) there is a gap between x_1 and x_2 of length greater than D;

are shown in Figure 11.

If we take the unit of our scale as a fixed multiple of s, and fix our attention on a particular group (of which there are four (1,2), (1,3), (2,3) and (1,2,3), then the plane will be divided into two regions. If the point representing x_1, x_2, x_3 falls into one region, the given group will be declared significant, if falls into the other, the given group will not be declared significant. Thus a diagram showing all four such regions makes it possible to set forth the details of any particular multilayer significance procedure for $m = 3$.

In the examples with which we shall be concerned, there will be no loss of generality in assuming σ known and equal to our unit, so we shall do this.

189. THE FIRST EXAMPLE

We choose to begin with apparent significance of a group of two defined by

$$| x_i - x_j | \geq 2.77$$

(that is, ≥ 2.77) and the apparent significance of a group of three defined by

$$\text{range}\{x_1, x_2, x_3\} \geq 3.27.$$

The resulting regions are shown in Figure 12. The shaded strips, one vertical and two slanting, each consists of the points where a particular

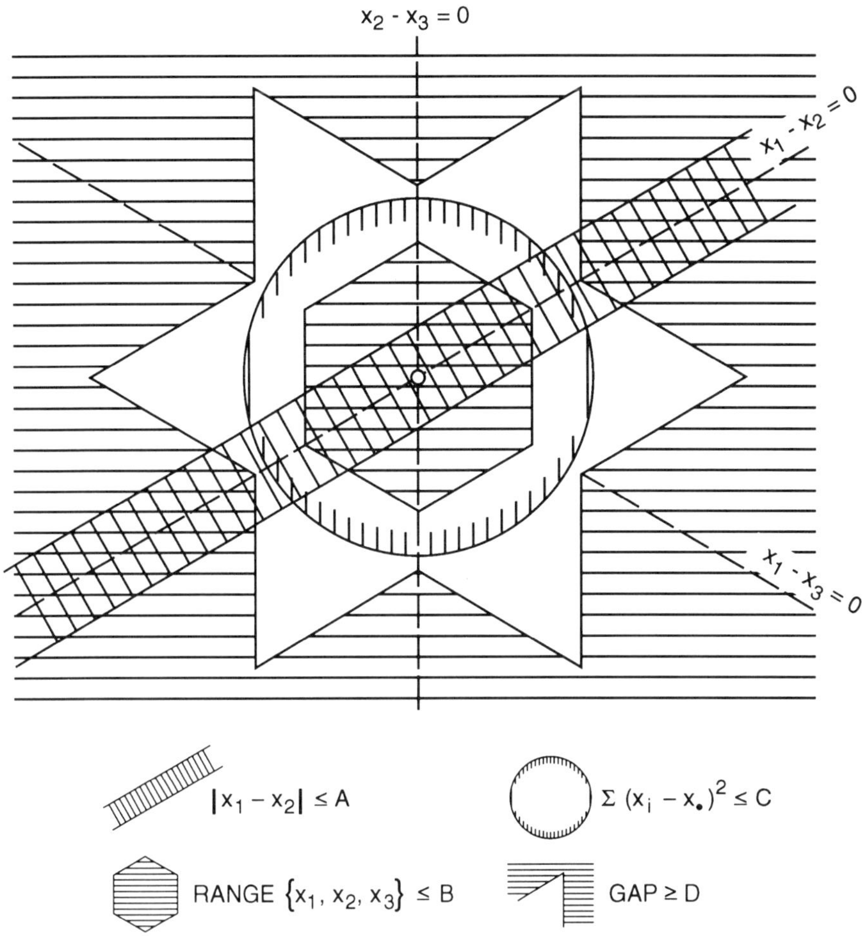

Figure 11. Shapes of four types of region.

group of two is apparently not significant. Their common part, a hexagon, is contained in a larger hexagon, which consists of the points where the group of 3 is not apparently significant. The various detailed possibilities are as follows:

In the corners A_1 to A_6, all four groups are apparently significant. Hence all four groups are declared significant.

In the strip-ends B_1 to B_6, two of the groups of 2 and the group of three are apparently significant. Hence these three groups are declared significant.

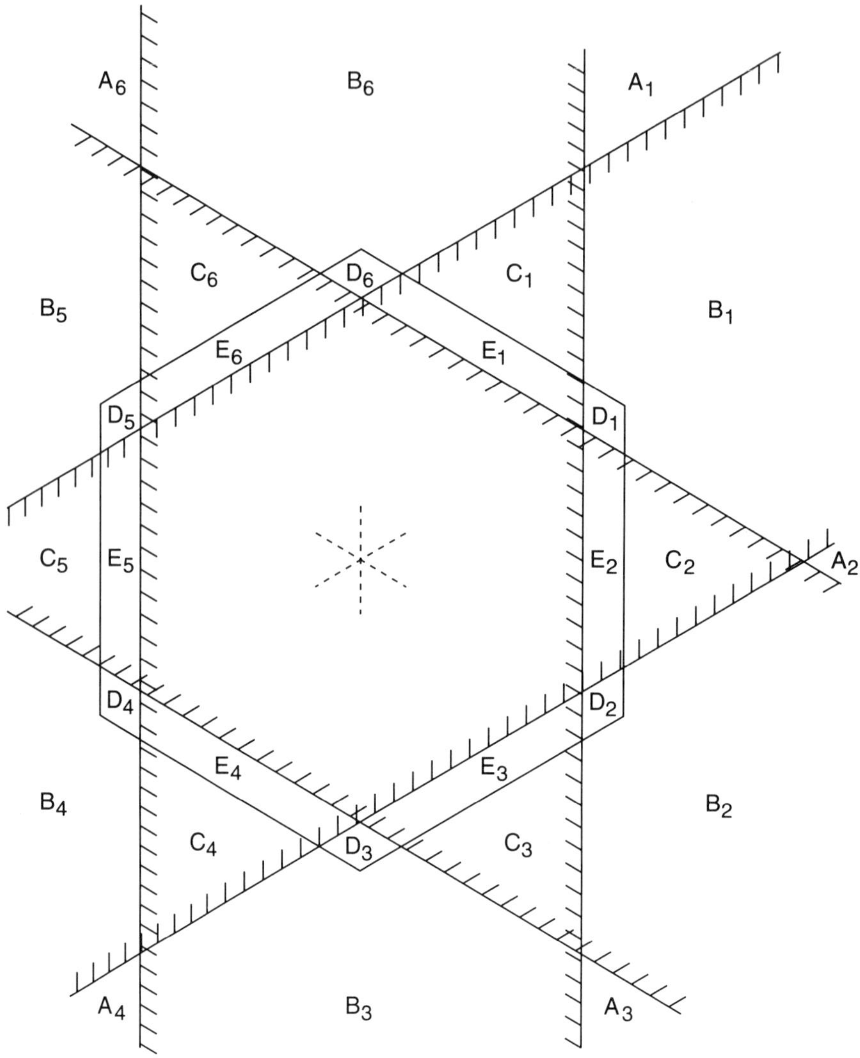

Figure 12

In the triangles C_1 to C_6, one of the groups of 2 and the group of 3 are apparently significant. Hence these two groups are declared significant.

In the small parallograms D_1 to D_6 two of the groups of two are apparently significant, but not the group of 3. Hence no group is declared significant.

In the narrow trapezoids E_1 to E_6, one group of 2 is all that is apparently significant. Hence no group is declared significant.

In the central hexagon F, no group is apparently significant, hence no group is declared significant.

The apparent significance levels for this example are easily found from a glance at the percentage points of the range to be $\alpha(2) = 5\%$, $\alpha(3) = 5\%$.

190. OTHER EXAMPLES

The second example differs from the first in only one respect, the criterion for apparent significance for a group of three is taken as

$$\text{range}\{x_1, x_2, x_3\} \geq 2.40.$$

The resulting diagram is Figure 13. Because of the shrinking of the hexagon in which the group of 3 is not significant, regions D_1 to D_6 and E_1 to E_6 have disappeared and have been replaced by a hexagonal ring labelled G.

In the hexagonal ring G, no group of 2 is apparently significant, but the group of 3 is apparently significant. Hence the group of 3 is declared significant.

In all other regions the situation is the same as in Figure 12. Reference to the % points of the range shows that $\alpha(2) = 5\%$, $\alpha(3) = 20\%$ for this example.

If instead of shrinking the hexagon, we were to expand it by using

$$\text{range}\{x_1, x_2, x_3\} \geq 5.55$$

as the criterion of apparent significance for the group of 3, regions D_1 to D_6 and E_1 to E_6 would expand, while regions C_1 to C_6 would disappear. The resulting procedure would have $\alpha(2) = 5\%$, $\alpha(3) \leq 0.05\%$.

As a different alternative, we might replace the hexagon by a circle by taking

$$\Sigma (x_i - x_.)^2 \geq C$$

as the criterion of apparent significance for the group of 3. If C were small, the regions would be much as in the second example. If C is made larger and larger, the circle will eventually touch the inner side of regions C_1 to C_6 and then bite into them, cutting out narrow regions E_1 to E_6 with one

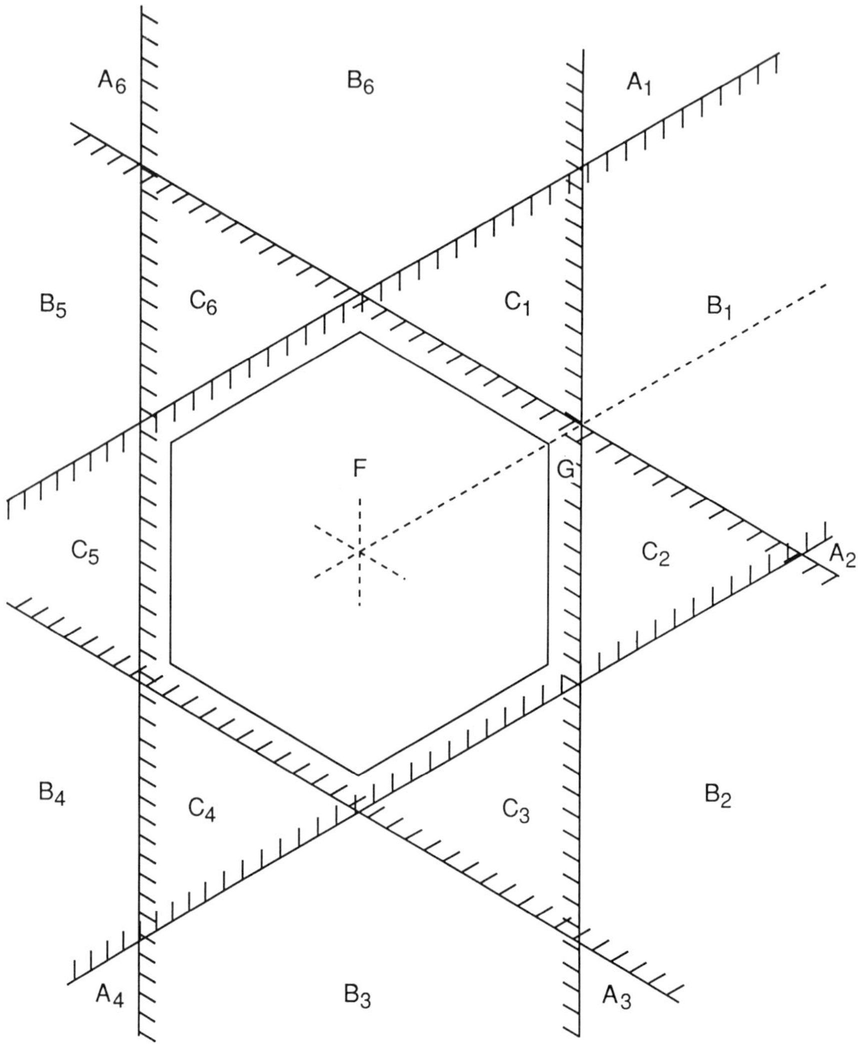

Figure 13

circular side and one straight side, and breaking up region G into six parts G_1 to G_6. As the circle expands further, G_1 to G_6 will shrink, to disappear when the circle first touches B_1 to B_6.

No new types of region will appear with a circular central region than appear with a hexagon. The same is true for the regions resulting from any other definitions of apparent significance.

191. ACKNOWLEDGEMENT

The writer wishes to express his indebtedness to David B. Duncan, and in particular to the paper [22] presented by him before the American Statistical Association at the 1952 annual meeting. While the point of view expressed in these pages is quite different from that of Duncan, it owes many of the essential insights to his work. He seems to have been the first to have a clear idea of what he was doing in setting up a multilayer significance procedure.

CHAPTER 31: SYSTEMS OF SIGNIFICANCE LEVELS

192. THE γ's

We are going to deal with standard significance levels, and since

$$1-\alpha(k_1, k_2) = [1-\alpha(k_1)]\,[1-\alpha(k_2)]$$

it is convenient to introduce

$$\gamma = -\ln(1-\alpha)$$

so that

$$\gamma(k_1, k_2) = \gamma(k_1) + \gamma(k_2),$$

and in general

$$\gamma(k_1, k_2, \ldots, k_p) = \sum_i \gamma(k_i).$$

We shall work to a considerable extent in terms of the γ's and therefore give a short table of corresponding values:

α	γ	:	α	γ
1%	.0101	·	39.4%	.5
5%	.0514	·	63.5%	1.0
10%	.105	·	86.5%	2.0
20%	.223	·	99.3%	5.0

193. CONSTANT APPARENT LEVELS

One natural proposal is to have constant apparent significance levels, so that

$$\gamma(k) = -\log_e(1-\alpha)(k)) = -\log_e(1-\alpha) = \gamma .$$

In this case

$$\alpha(k_1, k_2, \ldots, k_p) = 1 - e^{-p\gamma}$$

which attains its maximum when p is as large as possible. This occurs when all k_i are equal to w, and $m = 2p_{\max}$ or $m = 2p_{\max} + 1$. For a number of reasonable cases the summary levels are given in Table 48. The maximum standard significance levels are moderately to very substantially larger than the nominal levels.

194. CONSTANT STANDARD LEVELS

Exactly constant standard levels would be a natural and conservative choice. For $m = 5$ this would require

$$\alpha(5) = \alpha(2, 3)$$

and we might consider letting $\alpha(4)$ and $\alpha(2, 2k)$, which cannot be greater than this common value, also equal it. Thus we investigate

$$\alpha(5) = \alpha(4) = \alpha(2, 3) = \alpha(2, 2)$$

which corresponds to

$$\gamma(5) = \gamma(4) = \gamma(2, 3) = \gamma(2, 2)$$

and since

$$\gamma(2, 3) = \gamma(2) + \gamma(3),$$

$$\gamma(2, 2) = 2\gamma(2),$$

we must have

$$\gamma(5) = \gamma(4) = 2\gamma(3) = 2\gamma(2).$$

If we take the nominal α as 5%, we find

$$\alpha(2) = \alpha(3) = 2.53\%,$$

$$\alpha(4) = \alpha(5) = 5\%.$$

Table 48

Summary significance levels for nominal levels of 5% and 1%
(a) = constant standard (c) = independent comparisons
(b) = constant apparent (d) = extreme case

	$\alpha = 5\%$		$m = 5$		$\alpha = 1\%$		$m = 5$	
	(a)	(b)	(c)	(d)	(a)	(b)	(c)	(d)
minimum apparent	2.53%	5%	5%	5%	0.5%	1%	1%	1%
maximum apparent	5%	5%	18.6%	29%	1%	1%	3.9%	7.3%
maximum standard	5%	9.8%	18.6%	20%	1%	2%	3.9%	7.3%

	$\alpha = 5\%$		$m = 10$		$\alpha = 1\%$		$m = 10$	
	(a)	(b)	(c)	(d)	(a)	(b)	(c)	(d)
minimum apparent	1.0%	5%	5%	5%	0.2%	1%	1%	1%
maximum apparent	5%	5%	37%	63%	1%	1%	8.6%	23%
maximum standard	5%	23%	37%	63%	1%	4.9%	8.6%	23%

	$\alpha = 5\%$		$m = 20$		$\alpha = 1\%$		$m = 20$	
	(a)	(b)	(c)	(d)	(a)	(b)	(c)	(d)
minimum apparent	0.5%	5%	5%	5%	0.1%	1%	1%	1%
maximum apparent	5%	5%	62%	92%	1%	1%	15%	53%
maximum standard	5%	40%	62%	92%	1%	10%	15%	53%

Thus there is no difficulty in obtaining constant standard and near standard levels for $m = 5$. (And for $m < 5$.)

If we try to carry out the same process for $m = 6$, however, we find

$$2\gamma(3) = \gamma(3) + \gamma(2) = 3\gamma(2)\ ,$$

whose only solution is $\gamma(2) = \gamma(3) = 0$. Thus exactly constant standard *and* near standard levels are not available for $m = 6$.

If we relax our condition a little and require that all standard levels are equal, while all near standard levels are as large as possible, then we can find a solution for $m = 6$, namely

$$\gamma(5) = \gamma(6) = \gamma\,,$$

$$\gamma(4) = {}^2/_3\gamma\,,$$

$$\gamma(3) = {}^1/_2\gamma\,,$$

$$\gamma(2) = {}^1/_3\gamma\,,$$

which differs from

$$\gamma(k) = \frac{k}{m}\gamma$$

for $k = 5$ but not otherwise.

It is easy to show that the corresponding general solution for $m > 5$ is

$$\gamma(k) = \begin{cases} \gamma, & k = m - 1 \\ \dfrac{k}{m}\gamma, & \text{otherwise.} \end{cases}$$

For convenience we shall adopt $\gamma(k) = k\gamma/m$ as our standard. The resulting summary significance levels are determined by

$$\gamma(\text{appar. min.}) = \frac{2}{m}\gamma\,,$$

$$\gamma(\text{appar. max.}) = \gamma,$$

$$\gamma(\text{std. max.}) = \gamma,$$

For a few interesting cases these are given in Table 48. Here all but the minimum apparent significance levels cluster around the nominal level.

195. INDEPENDENT COMPARISONS BASIS

Duncan [17], in the procedure discussed in Chapter 33, makes use of a sequence of individual significance values obtainable from the following argument.

All differences among $x_1, x_2, \ldots, x_k$ can be computed from $k-1$ independent comparisons. We could measure these independent comparisons in $k-1$ separate experiments. If we were entitled to an error rate of $\alpha = 1 - e^{-\gamma}$ in each experiment, and set confidence limits accordingly, then all $k-1$ confidence statements would be correct in all but $1 - e^{-(k-1)\gamma}$ of all situations. If we picked these $k-1$ independent comparisons in a particular way, one or more of our resulting statements about the differences would be wrong if any of the independent comparisons were wrong — $1 - e^{-(k-1)\gamma}$ of all situations. We didn't do the separate experiments, but perhaps $\gamma_k = (k-1)\gamma$ indicates a reasonable choice of significance levels.

In this case, the summary significance levels are determined by

$$\gamma(\text{appar.min.}) = \gamma,$$

$$\gamma(\text{appar. max.}) = (m-1)\gamma,$$

$$\gamma(\text{std. max.}) = (m-1)\gamma,$$

For a few interesting cases these are given in Table 48. Here the maximum levels run very substantially above the nominal levels.

196. THE EXTREME CASE

For completeness we consider the extreme case where a group of size k is called apparently significant if one or more differences among its determinations is apparently significant. That is, the criterion of apparent significance takes the form

$$\text{range} \geq As$$

for a suitable constant A, which does not depend on the size of the group. We can adjust any one apparent significance level by the choice of A.

For convenience and simplicity, we shall confine our numerical example to the case where σ is known. For three choices of A we find the following results by interpolation in a table of % points of the range:

	$A = 2.77$		$A = 3.64$		$A = 4.47$	
k	$\alpha(k)$	$\gamma(k)$	$\alpha(k)$	$\gamma(k)$	$\alpha(k)$	$\gamma(k)$
2	5%	.051	.1%	.010	0.16%	.0016
3	13%	.14	2.8%	.028	0.44%	.0044
4	20%	.22	4.9%	.050	0.9%	.009
5	29%	.34	7.3%	.076	1.4%	.014
7	44%	.82	13%	.144	2.6%	.026
10	63%	.90	23%	.26	5%	.026
15	83%	1.77	39%	.49	10%	.051
20	92%	2.5	53%	.64	16%	.17

197. COMPARISON

In order to make comparison easy, Table 48 (p. 267) shows the four summary significance levels for nominal significance levels of 5% and 1% and 5, 10, or 20 determinations.

We may also relate these various systems to various types of error rates for confidence procedures. The natural correspondences are easily seen to be:

constant standard levels	error-rate familywise
constant apparent levels	error-rate per pair of determinations
independent comparisons basis	error-rate degree-of-freedomwise
extreme case	error-rate per comparison

Two of the error rate bases are not familiar in connection with specific confidence procedures. Of these, "per pair of determinations" can always be expressed in terms of "per determination." At a time when the distinction between "per X" and "X wise" was not clear to him, the writer suggested to Duncan [19], that "error rate per degree of freedom" was appropriate for the independent comparisons basis. Today he prefers to say "degree-of-freedomwise."

The graduation of the four systems of significance levels is clear, with the relation of actual significance levels to nominal levels becoming more and more dangerous as we move from near constant standard levels toward the extreme case.

No serious writer on multiple comparison has favored the extreme case, and so we owe it little discussion. Its high maximum significance levels have looked very dangerous to all.

The independent comparisons basis has so far appealed only to Duncan, and we shall discuss its possibilities and dangers in connection with Duncan's procedure.

The "constant apparent" system has been proposed by authors like Newman and Keuls, usually with a word of hedging or warning.

The "constant standard" system seems not to have been explicitly proposed before, and appeals most of all to the present writer, who is apparently conservative about error rates.

We shall meet the last three systems in connection with various specific procedures. This emphasizes the fact that the considerations of this chapter are independent of the particular test statistic used to determine apparent significance.

CHAPTER 32: PROCEDURES INVOLVING RANGES

198. INTRODUCTION

The procedures involving ranges are among the simplest of the multilayer significance procedures. This is especially true since we need to make tests of apparent significance in any multilayer significance procedure. Usually nowhere near all the $2^m - m - 1$ possible tests need to be made, but even after ingenious short cuts have been exploited to the full, there will remain a goodly number of tests. Thus the ease of computation of the range is particularly helpful here.

199. NEWMAN'S PROCEDURE

The first more or less detailed multiple comparisons procedure was proposed by Newman in 1939 [51] in connection with the first tabulation of % points of the studentized range. This involved constant apparent significance levels implicitly.

If, as in earlier parts, the *WSD* for j determinations is defined by

$$WSD_j = (s) \text{ (upper } p \text{ \% point of studentized range of } j)$$

so that

$$\text{Prob}\{\text{range}\{x_1 - \mu_1, x_2 - \mu_2, \ldots, x_j - \mu_j\} > WSD_j\} = p\ \%,$$

Newman's procedure could be defined as follows:

(1) calculate the *WSD* for m determinations;

(2) look to see if it is exceeded by the range of the observed determinations;

(3) if not, stop;

(4) if exceeded, drop k determinations from one end, recalculate the *WSD* for $m - k$ determinations and recompare;

(5) repeat step (4) for various choices of k from one or both ends; and

(6) make a rather general statement about the conclusions.

The degree of judgment required was indicated by Newman as follows (p. 27) "the method followed . . . must be employed with discretion, as is always the case when observations are rejected and a hypothesis tested using the selected data that remain." It seems likely that this diffuseness of proposed procedure hindered the spread of the use of the studentized range in such problems.

200. KEULS' FORMULATION

Very recently, a similar method has been expounded by Keuls' [39] in a paper in *Euphytica*. The steps here are the following:

(1) calculate the *WSD*'s for $m, m-1, m-2, \ldots$ determinations;

(2) compare the range with the *WSD* for m;

(3) if significant drop the highest observed determination;

(4) apply the *WSD* for the number of determinations remaining;

(5) repeat steps (3) and (4) until significance ceases;

(6) repeat steps (3), (4) and (5) with "highest" replaced by "lowest";

(7) assign to each determination the ranks not contradicted by steps (2) to (6).

The author comments (p. 121) "Mathematical study about a good test remains to be done."

This procedure may have been what Newman had in mind, and is explicitly a constant apparent significance level procedure, based on the studentized range as a test statistic.

201. THE *X*-PROCEDURE

The constant standard level procedure based on the studentized range attracts the writer more than any of the other procedures discussed in this chapter and the next. On the one hand, it guarantees the total absence of false significances in all buy 5% of the situations to which it is applied at a nominal 5% level. On the other hand, it is as simple to apply as any multilayer significance procedure. No other proposal combines these merits.

The critical range, which we denote by *XSD*, depends both on the total number of determinations, m, and on the number of determinations in the group considered, k. It will be shown in the next section that, to a working approximation, we may take

$$XSD_{m,k} = {}^1/_2(WSD_m + WSD_k),$$

so that no further tabulation is required before starting to use the procedure.

This alternative is somewhat more conservative than the Newman-Keuls procedure, as it must be if it is to avoid standard significance levels greater than the nominal level. But the degree of conservatism is smaller than the writer would have thought possible, being 50% in the sense that is suggested by the following comparison of critical ranges.

W-allowance	WSD_m
X-procedure	$1/2(WSD_m + WSD_k)$
NK-procedure	WSD_k

This consideration has forced the writer to change ground rapidly. For he has often argued that the desires, frequently expressed by friends and colleagues, to use a critical range which decreased with k was an attempt to get something for nothing, and was thus doomed to failure. These words have now to be taken back. Those who wish a conservative and effective significance procedure now may use the *X*-procedure.

202. VALIDATION OF APPROXIMATION

The constant standard level procedure is defined (for $k \leq m - 2$) by

$$\gamma(k, m) = \frac{k}{m}$$

where γ corresponds to the nominal level. For a nominal level of 5%, $\gamma = .051$ and we find that u (see Table 9) is given by

$$u = -\log t - .025(1-t) + .0004(1-t^2)t \ldots$$

where $t = k/m$. Thus we have

t =	0.9	0.8	0.7	0.6	0.5	0.4	0.3	0.2	0.1
u =	.045	.093	.048	.212	.290	.383	.506	.697	.978

and we may use Tables 1 and 9 to find, approximately, $\alpha(k, m)$ points of the studentized range which we denote by $X_{k,m}$. These are compared in Table 49 with an empirical approximation consisting of the average of the 5% points of the studentized range for groups of k and m respectively.

The agreement for $DDF = \infty$ (σ known) and for $DDF = 10$ and $m = 5$ is only to be described as surprisingly good. The differences are without

Table 49

Comparison of Computed and Empirical Approximate Values of $X_{k,m}$.

	$m = 5$						$m = 20$				
	$DDF = \infty$			$DDF = 10$			$DDF = \infty$			$DDF = 10$	
k	comp.	emp.	k	comp.	emp.	k	comp.	emp.	k	comp.	emp.
4	3.73	3.74	4	4.51	4.50	18	4.97	4.97	18	6.44	6.40
3	3.54	3.59	3	4.26	4.27	16	4.93	4.92	16	6.41	6.34
2	3.26	3.31	2	3.89	3.90	14	4.88	4.88	14	6.36	6.25
						12	4.83	4.82	12	6.30	6.15
	$m = 10$										
9	4.44	4.44	9	5.56	5.53	10	4.76	4.76	10	6.22	6.04
8	4.39	4.38	8	5.50	5.45	8	4.69	4.65	8	6.16	5.88
7	4.32	4.32	7	5.43	5.36	6	4.56	4.52	6	6.00	5.79
6	4.26	4.25	6	5.36	5.26	4	4.37	4.34	4	5.74	5.40
						2	3.95	3.89	2	5.05	4.81
5	4.17	4.16	5	5.26	5.13						
4	4.06	4.05	4	5.11	4.96						
3	3.91	3.90	3	4.90	4.74						
2	3.61	3.62	2	4.47	4.38						

doubt no larger than the uncertainties in the table of % points of the studentized range and its extrapolation. For $DDF = 10$ and $m = 10$ or 20, there are differences of as much as 2 to 4% between the computed and approximate values. For a moment, this might seem to be a large enough difference to make the empirical values unusable. However, there is a compensating consideration. For $DDF = 10$ the standard significance level overestimates the actual rate of false significances somewhat (because of the positive association arising from the fluctuations of s, which occurs in all tests). If we were to correct the computed values to make this nearly constant, and this is presumably what we would like to do, we would reduce the discrepancy. On the whole, then the empirical approximation should give working accuracy. Whenever the situation has been given a complete numerical going-over, it may be desirable to re-examine the quality of this approximation.

203. ANOTHER POSSIBILITY

We point out the possibility of a procedure using significance levels on the independent comparisons basis, as is done in Duncan's procedure, but using the studentized range as the test statistic. We shall not expand on this possibility here, since the independent comparisons basis does not attract the writer.

204. INCOMPLETE PROCEDURES

The other multiple-comparison significance procedures based on the ranges of groups of determinations are appropriately mentioned at this point. They fall into two broad classes; the work of Paulsen on comparison with the top determination or a control, and the work initiated at the University of London on short-cut versions of conventional rudimentary procedures.

In 1949, Paulsen [57] recognized that the studentized range could be used to compare the apparently highest determination with all others. Unfortunately, his published paper gives far more attention to the difficulty of obtaining numerical values for the power function of the test than to the real advantages of his technique. If he had considered it as an estimation procedure, it seems likely that he would have discovered the *W* allowance, since he was the first to realize the possibility of simultaneous comparisons.

More recently [58], Paulsen has discussed the problem of comparison with a control from a point of new "roughly analogous to the type I and type II errors in the Neyman-Pearson theory of testing a hypothesis." The connection with the present approach is unclear, and may be tenuous.

The work on short-cut versions of classical rudimentary significance tests began with Lord's work [42,43] on a range-in-the-denominator version of Student's *t*. The initial multiple comparison versions were by Patnaik [56], who treated the double classification. The computational procedure was simplified by the writer [74-78] who observed that it could be extended to a system of allowances and extensive tables, based on a slightly modified approximation were prepared by Kurtz, Link, Tukey and Wallace [40].

These methods were extended to more complex designs by H. A. David [15,1951] who covered a substantial number of the standard designs.

CHAPTER 33: PROCEDURES INVOLVING *F*

205. DUNCAN'S PROCEDURE

In 1947, D. B. Duncan wrote a Ph.D. thesis at Iowa State College on the multiple comparison problem [17]. This is the first work known to the writer which really attacked the simultaneous comparison problem, and any deficiencies which it shows from the standpoint of 1953 are, the writer feels, more than compensated for by its path-breaking character. The thesis set forth the theoretical approach and did not provide the numerical values required to make the method immediately usable. The necessary tables were computed, and expositions of practice and theory appeared in 1951 [19] and 1952 [20]. This procedure was the first to explicitly use a multilayer significance procedure.

In principle, each group of k is to be F tested for apparent significance at the level suggested by the independent comparisons basis. Since this would involve 1013 F-tests for $m = 10$, just as the Neuman-Keuls or X-procedures would in principle involve | 0 | 3 range tests, some short cuts are necessary. Duncan has worked out a scheme of preliminary range tests and a well-chosen order of F-tests which reduce the computation to a minimum. It has to be admitted, however, that the computation is still substantial in amount.

The "restricted" form of the procedure has the objective [Duncan 1951, p. 184] "to test only the differences between single means taken in all possible pairs," and is the result of paying attention only to groups of two determinations. The "complete" form of the test also deals with groups of size greater than 2, and does this is a specific, and very individual manner.

206. GROUPS OF SIZE GREATER THAN 2

Given a group of determinations which is "significant" but of which no subgroup is "significant" the complete procedure searches among pairs of subgroups of the group satisfying the following conditions:

(1) all determinations in the first subgroup are below the mean of all determinations in the group;

(2) all determinations in the second subgroup are above this mean;

(3) the two subgroups, taken together as a group are "significant";

for pairs of subgroups which cannot be made smaller without losing the last property. Duncan then asserts that "the difference between the subgroups is significant." What interpretation the user of the test is to put on this statement is not made clear.

If this statement means that at least one determinand in one subgroup is different from determinand in the other group, then the statement is of some mild use. But the meaning of the statement should be more clearly signposted, for otherwise far too many users will feel that "the groups B and CH are significantly different" [1951, p. 186] means that B is different from *both* C and H, rather than from *at least one* of C and H.

The author's intent is, however, still unclear to the writer, particularly since has has indicated on the previous page [1951, p. 185] a particular contrast which can surely be considered "significant" when a group of k is found to be "significant."

All these questions arise in connection with regions analogous to the hexagonal ring G in Figure 13, where a group of k is "significant" although none of its subgroups of 2 are significant. We note in passing that if Duncan had chosen to use the studentized range instead of F as his measure of apparent significance, such questions would never have arisen, even if the apparent significance levels were still expanded at the radical rate corresponding to the independent comparisons basis.

207. GUIDING CONSIDERATIONS

It is taken as a working axiom that [1951, p.177] "For a given test T to be satisfactory, (a) it must not reject any p-treatment null hypothesis whenever the latter would not be rejected by an α_p-level likelihood-ratio test, and (b) it must reject each p-treatment null hypothesis as often as it is possibly consistent with (a)." To the writer this axiom seems to carry in itself the seeds of conceptual inconsistency, as we shall see.

The likelihood-ratio test (in this case an F-test) is given a central position, and the likelihood itself is emphasized as a controlling factor. Thus [1951, p. 177] ". . . for the range test. For that test, the null hypothesis is sometimes rejected in cases when it has a larger likelihood than in other cases when it is accepted. This is a decided intuitive weakness of any test of a null hypothesis that does not conform to the likelihood-ratio criterion." This gives the likelihood a central place. But when the underlying fluctuations are normal, the likelihood is invariant under orthogonal transformations of the determinations. Thus

$$^1/_2\,(\mu_1+\mu_2+\mu_3+\mu_4),$$

$$^1/_2\,(\mu_1-\mu_2),$$

$$^1/_6\,(\mu_1+\mu_2-2\mu_3),$$

$$^1/_{12}\,(\mu_1+\mu_2+\mu_3-3\mu_4),$$

stand before the likelihood and the likelihood-ratio criterion with the same right and prestige as $\mu_1, \mu_2, \mu_3, \mu_4$. Thus (a) is symmetrical, and treats all contrasts alike.

On the other hand (b) refers specifically and essentially to comparisons, and treats them differently from other comparisons. If it were to read "(b′) it must reject each $(m-p)$ dimensional linear null hypothesis as often as is possible consistent with (a)" it would be giving symmetrical treatment to all contrasts. This it does not do.

Thus we have a peculiar combination of orthogonal symmetry for (a) and the much weaker symmetry of the permutation group for (b). Without a clear conceptual reason for this change in symmetry, the combination of (a) and (b) seems a strange requirement. Anyone is permitted to use such a combination, but others are permitted to doubt its usefulness.

208. MINIMUM WEIGHTED RISK

The latest paper [1952] presents a plausibility argument to show that the procedure resulting from the criterion just discussed approximately minimizes a certain class of weighted risks. In so doing, Duncan operates in the frame work of that one of Wald's theories of statistical inference [80] which retains the type I error of the Neyman-Pearson theory but replaces their type II error by a weighted risk, weighted by an *a priori* distribution in the parameter space.

A particular class of risk functions is defined, which is far from implausible. On the observation that its optimum procedures will not satisfy (a) above, it is suitably modified, so that a plausibility argument can be given for the approximate optimality of the Duncan procedure. No discussion is given of the rationale behind the modified risk function.

Under these circumstances, the writer feels that the basic criterion came first, and the weighted risks second. Until a discussion of the rationale behind the modified risk function is available, he will direct his own critical appraisal to the basic criterion (discussed in the last section) and let his opinions of Duncan's procedure rise and fall with his opinions of this basic criterion.

209. SUMMARY FOR DUNCAN'S PROCEDURE

Duncan's procedure fits into the framework of multilayer significance procedures. The tests of apparent significance are *F*- (or *t*-) tests. The apparent significance levels are on the independent comparisons basis, whose use Duncan admits [1951, p. 188] is a radical step.

As is to be expected with path-breaking work, the criteria of selection for a particular procedure seem to have been incompletely formulated and analyzed. The initial considerations based on likelihood seem more fundamental than an attempted justification *via* minimum weighted risk.

A minor problem arises in dealing with significant groups which contain no significant difference.

210. THE PROTECTED *LSD*

One of the few multiple comparison procedures of respectable antiquity, though probably not of respectable error rate, consists of the following steps:

(1) a preliminary application of an F-test to all m determinations; and

(2) if the F-test proves significant, application of t-tests to each difference.

This fits into the pattern of multilayer significance procedures when we agree that:

(1) a group of size $< m$ is apparently significant if it contains an apparently significant difference;

(2) the whole group of m is apparently significant if it passes an F-test.

It is reasonably described as a two-layer procedure. Examples of its use may be found, for example, in the book of Davies [15A, Section 5.28, p. 80].

The properties of this procedure have been criticized by Nandi [49] and by Duncan [19 p. 176]. The basic difficulty lies in the large apparent significance levels for $k = m - 1, m - 2, \ldots$.

211. OTHER POSSIBILITIES

Analogs of Newman-Keuls and X-procedures based on F rather than on the studentized range are possible, but seem to offer no advantages.

212. THE CLASSICAL *F*-TEST

No discussion of multiple comparisons would be possible without mention of the earliest and most used of all multiple comparison procedures, the simple *F*-test. Both procedure and distribution theory are of course due to R. A. Fisher [25].

Its main merits seem to be these:

(1) If we have a components-of-variance situation rather than a multiple comparison situation, and if all is normally distributed, then the *F*-test is:

   (a) the test based on sufficient statistics;

   (b) the uniformly most powerful test, and so on.

(2) If we believe in likelihood-ratio tests, the *F*-test is one.

(3) Among all tests whose power is a function of the mean square deviation of the determinands, and of this alone, the *F*-test is the most powerful.

These are sometimes convincing reasons for its use, and sometimes not.

CHAPTER 34: PROCEDURES INVOLVING GAPS

213. THE BIOMETRICS PROCEDURE

The writer introduced [73] in 1949 a composite procedure for comparing individual determinations. This procedure operated as follows:

(1) calculate a critical gap length (or as we shall say a gap-guide);

(2) arrange the observed determinations in order from lowest to highest;

(3) seek for places where the difference between adjacent observed determinations exceeded the gap guide; and

(4) declare such differences gaps, and thus divide the observations into groups;

(5) apply straggler tests within groups;

(6) *F*-test whatever was not segregated by straggler testing.

The results of this procedure were, then

(1) a subdivision into groups;

(2) a further pointing out of stragglers;

(3) a possibility of pointing to groups where some further differences appeared to exist.

Considered as a test procedure for identifying differences, this procedure, with its first emphasis on division into groups (surely the most definite sort of finger pointing), its lesser emphasis on stragglers (where one end of the difference was identified) and its minor emphasis on unidentified differences, was very attractive. It endeavored to put first things first.

It had a number of major defects, some acknowledged at the time and others discovered later. Some of these were the facts that:

(1) the apparent significance levels were not known, because of the use of multiple tests, whose properties were only known for the separate tests;

(2) the overall and standard significance levels were not the same as the apparent levels;

(3) the straggler procedure was set up for a 10% significance level instead of a 5% level as stated in the paper (by neglecting the distinction between expressions (**) and (*) of Section 220 of the next chapter).

Of these major defects, the first was discussed carefully in the paper which presented the procedure. As noted earlier, defect (2), which is shared by all multilayer significance procedure, was pointed out to the author, in connection with the BIOMETRICS procedure, by John Mandel. Defect (3), which was due to a careless oversight by the writer, was pointed out to the author by Jerome C. R. Li.

214. DISCUSSION

Even with these defects known, the BIOMETRICS procedure was, and is, far better than no procedure at all or the protected *LSD*. (Since even these defects, all leaning away from conservatism, do not keep the procedure from remaining more conservative than Duncan's procedure, it is safer for the user than Duncan's.) But today, with a variety of systems of allowances available (and with the *X*-procedure for the confirmed and unredeemable significance tester) there seems no doubt that the BIOMETRICS procedure is *entirely obsolete*, together with various gap procedures proposed in the interim by the writer [74,75,76] and accepted in good faith by other writers [7,16].

We need go no further with the BIOMETRICS procedure than to point out that it *almost* falls into the pattern of multilayer significance procedures, once we define apparent significance as follows:

(1) a set of k determinations is apparently significant if it contains a gap greater than the gap guide;

(2) a set of k determinations is also apparently significant if indicated by straggler or F-tests;

(3) a set of k determinations is not apparently significant except by the action of (1) or (2).

The multilayer significance procedure defined by (1), (2) and (3) differs from the BIOMETRICS procedure in just one way. If driven to an F-test in the BIOMETRICS procedure, the analysis stopped there, even if the result was "significant." The procedure just specified would allow and require further tests on subgroups of a group found significant by F, just as on subgroups found significant by a straggler test.

215. SIMPLE GAP PROCEDURES

In addition to providing a definite, usable procedure, the BIOMETRICS procedure introduced the systematic use of gaps. Although the BIOMETRICS procedure is obsolete, the use of gaps may not be, and further study of gap procedures is indicated.

We begin by remarking that numerical computation has shown [73] that the apparent significance level corresponding to a gap guide of constant length becomes more extreme as k increases. This is contrary to the behavior of all the systems of apparent significance levels discussed in Chapter 31. Thus it is natural to try to reduce the gap-guide as k increases. There are, as we shall immediately see, difficulties in such reduction.

Let y_k denote the length of the gap-guide for a group of k and suppose that $m = 3$ and

$$y_3 < x_2 - x_1,\ x_3 - x_2 < y_2.$$

Then we must accept $x_2 - x_1$ as large enough to mean a gap. Having admitted $\mu_1 \leq \mu_2, \mu_3$, we have only a group of $k = 2$ left and $x_3 - x_2$ is not large enough to mean a gap. Conclusion: μ_1 less than μ_2 and μ_3, which may or may not be equal.

But x_1, x_2, x_3 are essentially symmetric, so that if we look at $x_3 - x_2$ *first*, we come to the conclusion: μ_1 and μ_2, which may or may not be equal, less than μ_3.

The irrelevant choice of picking up the gaps first from one end or first from the other has led us to different conclusions. This is clearly unbearable. And so we cannot follow a rule of following form (for 3 determinations):

> Use $y_3 < y_2$ to localize the first gap, and if a first gap is found,
> use y_2 to localize any further gap.

Other devices for avoiding conservatism run into the same general sort of difficulty.

216. POSSIBILITIES FOR THE FUTURE

If we are to use a gap guide of constant length, and apparent significance levels which increase with k, we must declare groups apparently significant for other reasons than for gaps. There are of course many possible test statistics, but we naturally turn to the studentized range.

The general sort of procedure which we might well consider is the following:

(1) a gap is significant if it is longer than $Y_{m,2}$;

(2) a gap of k is apparently significant if its range exceeds $Y_{m,k}$.

In setting this notation, we have made use of the fact that gap and range criteria for groups of 2 may as well coincide.

There will be a different version of such a procedure for each choice of a system of apparent significance levels. The most attractive of these to the writer will be those with constant standard levels. By analogy with the X-procedure, it is natural to call this the Y-procedure. It would have the good properties of the X-procedure, except for a slight increase in computational work, and since it makes use of the extremely definite concept of a gap in its conclusions it would give very useful statements for a significance procedure. Unfortunately, before we can reduce this procedure to numerical practice, we need to know more about the joint distribution of range and largest gap in random samples of k.

For the present, the writer will probably succumb to temptation and use the approximation

$$Y_{n,k} \approx X_{m,k} ,$$

with the realization that the maximum standard significance level will be somewhat larger than $p\%$, in those few cases where a significance procedure directed toward a separation of determinands seems appropriate.

CHAPTER 35: STRAGGLER SPOTTING

217. INTRODUCTION

As noted in Section 147, Cornfield, Greenhouse and Halperin's procedures were introduced with an emphasis on stragglers, in particular by concentration of attention on deviations and discrepancies. The argument of Sections 130 and 138 made it clear that significance procedures are particularly fitting in such cases. In this chapter we shall discuss various significance procedures focused on stragglers.

218. THE ORIGINAL *CGH*-PROCEDURE

As originally proposed by Cornfield, Greenhouse and Halperin, the simple (not localizable, self-consistent) significance testing procedure based on key_C would operate about as follows:

(1) Calculate $| x_i - x_. | /s$ for all i;

(2) compare each with the appropriate % point of (∗); and

(3) declare all x_i "significant" for which $| x_i - x_. | /s$ is extreme by comparison with this critical value.

As a straggler spotting procedure, this is simple and apparently straightforward.

As usual, however, in a multiple comparison situation, there is more than meets the eye. Let us consider a simple example, where

$$\mu_1 = \mu_2 = \ldots = \mu_{m-1} = \mu,$$

$$\mu_m = \mu + m\phi,$$

where ϕ is relatively large. We have

$$x_{.} = \mu_{.} + e_{.} = \mu + \phi + e_{.}$$

thus

$$\frac{x_i - x_.}{s} = \frac{\mu_i - \mu_. = \phi}{s} + \frac{e_i - e_.}{s}$$

and for $i = 1, 2, \ldots, m - 1$ this becomes

$$\frac{x_i - x_.}{s} = -\frac{\phi}{s} + \frac{e_i - e_.}{s_.}.$$

If ϕ is very large, then not only will x_m be declared "significant" in almost every case, but so will $x_1, x_2, \ldots, x_{m-1}$. But for almost all purposes that the writer can imagine, we wish to declare only x_m "significant." Thus there are serious complications.

If we were sure that there would never be more than one real straggler per family, we could avoid these complications very easily. We should only have to modify the procedure to run as follows:

(1) Calculate $| x_i - x_. | /s$ for all i;

(2) compare the largest of these with the appropriate % points of (*); and

(3) declare all x_i for which $| x_i - x_. | /s$ attains its largest value to be "significant."

With one straggler such a modified procedure operates nicely, but in the presence of 2 or more stragglers it fails to do what we wish. Thus, for example, if $s = 1$ with *DDF* very large, and

$$x_1 = 3.2,\ x_2 = 1.9,\ x_3 = 2.4,\ x_4 = 27.3,\ x_5 = 128.2,$$

then this procedure would declare x_5 "significant," as it should, but would fail to announce the obvious significance of x_4. We must be more ingenious.

219. MULTILAYER PROCEDURES

If we take (*) as the key statistic for determining the apparent significance of any group of determinations, we can construct a multilayer significance procedure for stragglers corresponding to any desired choice of apparent significance levels. The reader will not be surprised if we investigate only the constant standard rate version, which we shall call the *D*-procedure.

We do not have either detailed or reliable tables to facilitate or permit the calculation of *CSD*'s. The approximate values available suggest, however, the approximation

$$DSD_{k,m} = 0.7\ CSD_m + 0.3\ CSD_k$$

an approximation analogous in form to, but differing in weights from, the approximation for the *XSD*. The practical use of such a procedure must wait upon better and more extensive tables.

Actually, however, we may modify the form of standard multilayer type of procedure, and obtain equivalent results as far as straggler spotting goes by using the following *iterative* procedure, starting with $k = m$:

(1) Calculate the largest of the k values of $|\ x_i - x_.\ |$ and compare with $^1/_2 DSD_{k,m}$. If not significant, stop here. If significant, determine the signs of $x_i - x_.$ for these extreme determinations.

(2) If a majority of the extreme determinations give the same sign to $x_i - x_.$, set this majority aside and continue.

(3) If an equal number of extreme determinations give each sign to $x_i - x_.$, calculate $|\ \underline{x_j - x_0}\ |$ where

(a) x_j is from one extreme group;

(b) x_0 is the mean of all x_i except those in the other extreme group;

and compare with $\frac{1}{2}DSD_{K,m}$, where $K = k -$ (number per extreme group). If significant, set aside both extreme groups and continue. If not significant

(4) Repeat (1) to (3) with the remaining determinations and the new *reduced* value of k.

(5) Declare all set-aside x_i to be significant.

Ordinarily, only one x_i will be set aside at a time, and only a few steps will be required.

This seems to be the best straggler spotting procedure proposed to date. Its only lack is an adequate set of tables. Presumably tables for short cut or half out DSD's would also be useful.

220. RELATED PROCEDURES

The key statistic of the C-allowance is, of course,

$$\frac{\max\{\,|\,e_i - e_.\,|\,\}}{s}. \qquad (*)$$

A close relative and predecessor,

$$\frac{\max\{e_i - e_.\}}{s} \qquad (**)$$

was introduced, and its distribution tabled by K. R. Nair [48] who indicated its use as a straggler test, without considering possible extensions.

The $p\%$ point of $(**)$ corresponds to *about* the $2p\%$ point of $(*)$. Reference to [35] indicates that $\max\{e_i - e_.\}$ and $\max\{e_. - e_i\}$ are positively correlated when the e's are normally distributed. Thus the error rate familywise of $(*)$, where used at the $p\%$ point of $(**)$, is slightly *less* than $2p\%$.

Nair's numerical results were taken up by the writer [73] in connection with the BIOMETRICS procedure (described in the last chapter).

The numerical results were also taken up by Moshman [46]. Since it is based on (**) rather than (*), its error rate familywise is slightly less than $2p\%$, where $p\%$ is the nominal value.

CHAPTER 36: DISCUSSION

221. RANGE VS. *F*

In this chapter we are concerned with the various choices explicitly or implicitly indicated by the earlier chapters of this part. We shall try to set out at least some of the pros and cons for a number of these choices.

First, let us discuss the choice of test statistic, which is mainly a choice between F and the studentized range. We have considered a similar problem, for allowances rather than multilayer significance procedures in Chapter 29, but there are some novel features.

A point of view which seems to underlie Duncan's reliance on the likelihood-ratio may be phrased as follows: the rules of significance testing should conform to the principles of abstract justice — to the principles of abstract justice assuming the corresponding null hypothesis. We should not declare any particular sample "significant" unless we are willing to do the same for any other sample of equal likelihood on the null hypothesis. This is an internally consistent principle, but neither an obvious principle nor one the writer wishes to accept. (It bears a strange resemblance to the simple property of Section 176.)

The main issue, in the writer's mind, is still that definiteness of statement implies usefulness of statement. With a range criterion, if anything is declared "significant," at least one difference of two determinations — at least one simple comparison — is so declared. Thus there is always a considerable amount of definiteness in a positive result with a range criterion. This is not necessarily the case when an F criterion is used, as the complete form of Duncan's procedure shows. The range method also provides more power for simple comparisons at the expense of power for more complex situations. This too seems good. Thus even without consideration of relative case of computation, the writer feels that range based multilayer significance procedures are preferable to F based ones.

The matter of relative ease of computation is not a small one, and it too favors range based methods.

222. SYSTEMS OF SIGNIFICANCE LEVELS

In Chapter 31 we discussed four systems of significance levels, one of which was very extreme, and was merely for contrast. The remaining three can be characterized with reasonable fidelity as follows:

constant standard levels — "conservative"

constant apparent levels — "hidden radical"

independent comparisons basis — "radical."

We do not hesitate to call the last "radical" in view of Duncan's remark [1951, p. 188]. We hasten to point out that "radical" and "conservative" have no political meaning here, referring only to the relation of the maximum probability of false significance to the ***nominal*** significance level.

When the conservative system is used, the maximum probability that any false significance shall be found in a family of determinations is equal to the nominal significance level. This is just the standard we have been accustomed to in the routine use of the F-test in the analysis of variance.

When the hidden radical system is used, the probability of any false significance in a coherent bunch of determinations is equal to the nominal value. Here a coherent bunch of determinations is a group all of whose determinands are equal. Since a family of determinations may break up into many coherent bunches, the possible rate of false significance among families will rise far above the nominal significance level. This is the effect noted by Mandel. It seems very appropriate to call it "hidden" radicalism.

When the "radical" system is used, the only probabilities of false significance which are equal to the nominal level are those for an individual difference. To the writer, a probability of 62% of false significance in some difference in a family of 20 determinations seems a long, long way from the nominal level of 5%. This system seems well termed "radical."

Now that the pressures for definite statements, which the writer feels are responsible for the suggestion of the radical system (cp. Chapter 21), can be met through a confidence procedure (i.e., the W system of allowances) there seems little excuse for the radical system of significance levels.

223. CONCOMITANT USE OF GAPS

Not now, but assuredly in the relatively near future, we could use procedures combining gaps and ranges. Are we to consider these procedures better than those which use range alone? The writer holds to the affirmative.

Given a family of determinations, almost the most definite qualitative statement that can be made is that they fall into two groups, the determinands in the first being individually lower than the determinands in the second. (The only *more* definite qualitative statements the writer can think of are those which specify two or more gaps.) Since we require definiteness above almost everything else, the use of gaps seems definitely very desirable.

Just how desirable will depend on how much we have to open out the critical ranges to permit the use of the critical gap while maintaining apparent significance levels. We do not know how much, but it seems probable that the answer is "very little." The writer would bet on the combined procedure including gaps.

224. IF A SIGNIFICANCE PROCEDURE IS NECESSARY

In Part D we admitted two valid sorts of reasons for using a significance procedure rather than a confidence procedure. These were

(1) the absence of a scale on which to measure differences; or

(2) the absence of an appropriate statistical technique.

Neither of these can arise where *any* of the multiple comparisons procedures discussed in Parts E or F can be used. How then could it be necessary to use a significance procedure? Only for nonstatistical reasons.

Sometimes such reasons are binding. Thus if

(1) the director of the project "understands" significance procedure and not confidence procedures; or if

(2) the results of a test need to be expressed, for administrative reasons, in a form stressing that "we looked and found nothing";

it may be necessary to use a significance procedure. In each such case there are hidden dangers of misinterpretation, but it *may* seem wise to risk these dangers.

If we must test significance, which multilayer significance procedure should we use? Based on the arguments already set forth in this chapter, the writer concludes that he would

(1) use the Y-procedure if it were available; and

(2) in its absence, use the X-procedure if sticking to $p\%$ were essential;

(3) otherwise use the approximate Y-procedure based on $X_{m,k}$.

These procedures are simple, powerful and conservative.

225. STRAGGLER SPOTTING PROCEDURES

Here significance seems justified (cp. Sections 130 and 138). When tables are available, the modified D-procedure of Section 220 would seem to be most appropriate.

PART G
REFERENCE MATERIAL

CHAPTER 37: BIBLIOGRAPHY

226. GENERAL REFERENCES

The following list of references includes

(1) all items referred to in the text, except those concerning numerical examples (which follow in Section 227); and

(2) items on the multiple comparison problem not necessarily referred to in the text.

The items involved include books, papers, abstracts, and addresses before societies. The last two classifications are identified by one of the letters "p," "q," "r," ... attached to the year. Particularly since handouts have been distributed by various speakers, so that much information on multiple comparisons have circulated in this manner, it has seemed advisable to include items consisting of talks or addresses, whether abstracted or not.

[1] (1952). Various papers including: *American Potato Journal* **29**(3), 55-61.

[2] Bartlett, M. S. (1939). Complete simultaneous fiducial distributions, *Ann. Math. Stat.* **10**, 129-137.

[3] Bechhofer, Robert E. (1951p). The probability of a correct ranking (Preliminary Report), abstracted in (1952). *Ann. Math. Stat.* **23**, 139-140.

[4] Bechhofer, Robert E. (1952p). On a multiple decision procedure associated with certain ranking problems. Paper before the Institute of Mathematical Statistics and the Biometric Society, Blacksburg, 19 March 1952.

[5] Bechhofer, Robert E., Dunnett, Charles W. and Sobel, Milton, (1952p). A two-sample multiple decision procedure for ranking means of normal populations with unknown variances, abstracted in (1953). *Ann. Math. Stat.* **24**, 136.

[6] Bechhofer, Robert E. and Sobel, Milton, (1952p). A sequential multiple decision procedure for ranking means of normal populations with known variances (Preliminary Report), abstracted in (1952). *Ann. Math. Stat.* **24**, 136-7.

[7] Bliss, C. I. and Calhoun, W. W. (1951p). An outline of biometry, — privately circulated, especially 94ff and Tables XI and XII.

[8a] Bose, R. C. and Roy, S. N. (1952p). Simultaneous confidence interval estimation, abstracted in (1953). *Ann. Math. Stat.* **24**, 144.

[8b] Bose, R. C. and Roy, S. N. (1952q). On a set of simultaneous confidence interval statements in multivariate analysis of variance, abstracted in (1953). *Ann. Math. Stat.* **24**, 144-145.

[9] Bose, R. C. (1958p). The General Theory and Application to Analysis of Variance and Covariance, Address before the *Inst. Math. Stat.* and *Biom. Soc.*, Washington, April 1953.

[10] Brookner, Ralph J. (1945). Choice of one among several statistical hypotheses, *Ann. Math. Stat.* **16**, 221-242.

[11] Clopper, C. J., and Pearson, E. S. (1934). The use of confidence or fiducial limits illustrated in the case of the binomial, *Biometrika* **26**, 404-413.

[12] Cochran, W. G. (1941-42). The distribution of the largest of a set of estimated variances as a fraction of their total, *Ann. Eugenics (London)* **11**, 47-52.

[13] Cochran, W. G. (1951). Improvement by means of selection. *Proc. 2nd Berkeley Symp Math Statist* + Prob 449-470.

[14] (blank)

[15] David, H. A. (1951). Further applications of range to the analysis of variance, *Biometrika* **38**, 393-409. For [15A] see end of section.

[16] Duncan, Acheson J. (1952). Quality control and industrial statistics, Richard D. Irwin, Chicago, especially 440ff, 449ff, tables L1, L2a, L2b.

[17] Duncan, D. B. (1947). Significance tests for differences between ranked variates drawn from normal populations. Iowa State College Thesis, iii and 117pp.

[18] Duncan, D. B. (1950p). A significance test for differences among ranked treatments in an analysis of variance, abstracted in (1951). *Ann. Math. Stat.* **22**, 142.

[19] Duncan, D. B. (1951). A significance test for differences between ranked treatments in an analysis of variance, *Virginia J. of Sci.* **2**, 172-189.

[20] Duncan, D. B. (1952a). On the properties of the multiple comparisons test, *Virginia J. of Sci.* **3**, 49-67.

[21] Duncan, D. B. (1952p). The multiple comparisons test for separating ranked treatments in an analysis of variance. Paper before the Institute of Mathematical Statistics and the Biometric Society, 19 March 1952.

[22] Duncan, D. B. (1952q). Testing the homogeneity of treatment means in an analysis of variance of engineering data. Paper presented at Annual Statistical Meetings, Chicago, Dec. 1952 with 7 page handout.

[23] Feller, William (1950). *An introduction to probability theory and its applications*. xii and 419 pp. Wiley, New York.

[24] Fisher, R. A. (1922). On the mathematical foundations of theoretical statistics *Phil. Trans. (Roy. Soc. London)* **A222**, 309-368. (Reprinted in his *Contrib. to Math. Statist*. as paper 10).

[25] Fisher, R. A. (1925). *Statistical Methods for Research Workers*. Oliver and Boyd, Edinburgh, 1st edition 1925, 11th edition 1950.

[26] Fisher, R. A. (1929). Tests of significance in harmonic analysis, *Proc Roy Soc (London)* **A125**, 54-59. (Reprinted in his *Contrib. to Math. Statist*. as paper 16).

[27] Fisher, R. A. (1929-30). Inverse probability, *Proc. Comb. Phil. Soc. 326*, 528-535. (Reprinted in his *Contrib. to Math. Statist.* as paper 22).

[28] Fisher, R. A. (1933). The concepts of inverse probability and fiducial probability referring to unknown parameters, *Proc Roy Soc (London)* **A139**, 343-348.

[29] Fisher, R. A. (1935-1960). *The Design of Experiments.* Oliver and Boyd, Edinburgh. 1st edition 1935, 7th edition 1960.

[30] Fraser, D. A. S. (1952p). Confidence bounds for a set of means, abstract in *Ann. Math. Stat.* **23**, 303.

[31] Gayen, A. K. (1950). The distribution of 'student's *t* in random samples of any size drawn from non-normal universes, *Biometrika* **36**, 353-369.

[32] Hald, A. (1952). *Statistical tables and formulas.* 97pp. John Wiley, New York; Chapman and Hall, London.

[32A] Halperin, M., Greenhouse, S. W., Cornfield, J., and Zalokar, J. (1955). Tables of percentage points for the studentized maximum absolute deviate in normal samples, *J. Am. Stat. Assoc.* **50**, 185-195.

[33] Harris, Marylin, Horvitz, D. G., and Mood, A. M. (1948). On the determination of sample sizes in designing experiments, *J. Am. Stat. Assoc.* **43**, 391-402.

[34] Hartley, H. O. (1950). The use of range in analysis of variance, *Biometrika* **37**, 271-280.

[35] Hastings, C. Jr., Mosteller, F., Tukey, J. W., and Winsor, C. P. (1947). Low moments for small samples: A comparative study of order statistics, *Ann. Math. Stat.* **18**, 413-426.

[36] Herget, Paul and Clemence, G. M. (1944). Optimum-interval punched-card tables. *Math Tables and Aids to Comp 1* (1943-1945) 173-176.

[37] Hotelling, Harold (1931). The generalization of Student's ratio, *Ann. Math. Stat.* **2**, 360-378.

[38] Kempthorne, Oscar (1952). *The design and analysis of experiments.* xix and 631pp, especially Chapter 8ff. John Wiley, New York; Chapman and Hall, London.

[39] Keuls, M. (1952). The use of the "studentized range" in connection with an analysis of variance, *Euphytica* **1**, 112-122.

[40] Kurtz, T. E., Link, R. F., Tukey, J. W. and Wallace, D. L., (195). Short analysis for balanced single and double classifications, to be submitted for publication. (See *Biometrika* **52** and *Technometrics* **7**, 1965, Author's note.)

[41] Link, R. F. and Wallace, D. L. (1952p). Short cuts to allowances. Paper before the Institute of Mathematical Statistics and the Biometric Society, Blacksburg, 19 March 1952.

[42] Lord, E. (1947). The use of range in place of standard deviation in the t-test, *Biometrika* **34**, 41-67.

[43] Lord, E. (1950). Power of the modified t-test (u-test) based on range, *Biometrika* **37**, 64-77.

[44] May, Joyce M. (1952). Extended and corrected tables of the upper percentage points of the 'studentized' range, *Biometrika* **39**, 192-193.

[45] Mood, A. M. (1951). Erratum: On the determination of sample sizes in designing experiments, *J. Am. Stat. Assoc.* **46**, 515.

[46] Moshman, Jack (1952). Testing a straggler mean in a two-way table using the range, *Ann. Math. Stat.* **23**, 126-132.

[47] Nair, K. R. — Cannot determine appropriate Nair reference from published works.

[48] Nair, K. R. — Cannot determine appropriate Nair reference from published works.

[49] Nandi, H. K. (1951). On the analysis of variance test, *Calcutta Statist. Assoc. Bull.* **3**, 103-114.

[50] Nemenyi, Peter (1953p). Percentage points of the studentized maximum modulus. *Memorandum Report* 53, Statistical Research Group, Princeton University.

[51] Newman, D. (1939). The distribution of range in samples from a normal population, expressed in terms of an independent estimate of standard deviation, *Biometrika* **31**, 20-30.

[52] Neyman, J. (1934). On the two different aspects of the representative method: the method of stratified sampling and the method of purposive selection, *J. Roy. Stat. Soc.* **97**, 558-625.

[53] Neyman, J. (1935). On the problem of confidence intervals, *Ann. Math. Stat.* **6**, 111-116.

[54] Neyman, J. (1937). Outline of a theory of statistical estimation based on the classical theory of probability, *Phil. Trans. (Roy. Soc. London)*

A236, 333-380 (No. 767).

[55] Olekiewicz, M. (1950). Tables of significance limits for the largest critical ratio out of k ratios, *Ann Univ Mariae Curie-Skoduka* **A4**, 115-121 (No. 11).

[56] Patnaik, P. B. (1950). The use of mean range as an estimator of variance in statistical tests, *Biometrika* **37**, 78-87.

[57] Paulson, Edward (1949). A multiple decision procedure for certain problems in the analysis of variance, *Ann. Math. Stat.* **20**, 95-98.

[58] Paulson, Edward (1952). On the comparison of several experimental categories with a control, *Ann. Math. Stat.* **23**, 239-246.

[59] Pearson, Egon S. (1926). A further note on the distribution of range in samples taken from a normal population, *Biometrika* **18**, 173-194.

[60] Pearson, Egon S. (1942). The probability integral of the range in samples of n observations from a normal population, *Biometrika* **32**, (1941-42) 302-308.

[61] Pearson, E. S. and Sukhatme, A. V. (1935). An illustration of the use of fiducial limits in determining the characteristics of a sampled batch, *Sankhya* **2**, (1935-36) 13-32.

[62] (1931). *Tables for Statisticians and Biometricians Part II.* (K. Pearson, ed.) ool and 262 pages, especially Table XXI bis, page 164, prepared by E. S. Pearson.

[63] Peisakoff, M. P. (1951p). Transformation parameters. Princeton University Thesis. Abstracted in (1951). *Ann. Math. Stat.* **22**, 136-137.

[64] Pitman, E. J. G. (1937a). Significance test which may be applied to samples from any populations, *J Roy. Statist. Soc. Suppl.* **4**, 119-130.

[65] Pitman, E. J. G. (1937b). Significance tests which may be applied to samples from any populations. II. The correlation coefficient test, *J. Roy Statist. Soc. Suppl.* **4**, 225-232.

[66] Pitman, E. J. G. (1938). Significance tests which may be applied to samples from any populations. III. The analysis of variance test, *Biometrika* **29**, (1937-38) 322-335.

[67] Roy, S. N. and Bose, R. C. (1952p). On a relevant set of simultaneous confidence interval statements in discriminant analysis. Abstracted in (1935). *Ann. Math. Stat. 4*, 145.

[68] Scheffé, Henry (1952p). On judging all contrasts in the analysis of variance (Preliminary report). Abstracted in *Ann. Math. Stat. 23*, 477.

[69] Scheffé, Henry, (1952q). Multiple Comparison procedures in the analysis of variance. Address before the Institute of Mathematical Statistics, East Lansing, September 1952.

[70] Scheffé, Henry, (1953). A method for judging all contrasts in the analysis of variance, *Biometrika* **40**.

[71] "STUDENT," (1908). The probable error of a mean, *Biometrika* **6**, (1908-09) 1-25.

[72] Thompson, W. R., (1935). On a criterion for the rejection of observations and the distribution of the ratio of deviation to sample standard deviation, *Ann. Math. Stat.* **6**, 214-219.

[73] Tukey, John W. (1949f). Comparing individual means in the analysis of variance, *Biometrics* **5**, 99-114. [Also in this Volume, 315-333.]

[74] Tukey, John W. (1950). Notes on the use of the range in industrial statistics. Paper before the Baltimore and Washington Sects, Am. Soc. Qual. Cont. on 17 November 1950.

[75] Tukey, John W. (1951d). Quick and dirty methods in statistics. Part II — Simple analyses for standard designs, *Quality Control Conf. Papers 1951* (Am. Soc. Qual. Control) 189-197.

[76] Tukey, John W. (1951). Quick and easy comparisons. Paper before the Philadelphia Section, Am. Soc. Qual. Cont on 16 November 1951 and, in revised form, under title "New methods in the analysis of variance. Range in the numerator and range in the denominator" before the Princeton Conference, ASQC, on 8 December 1951 and the Toronto Sect, ASQC on 9 January 1952.

[77] Tukey, John W. (1952). Allowances for various types of error rates. Paper before the Institute of Mathematical Statistics and the Biometric Society, Blacksburg, 19 March 1952.

[78] Tukey, John W. (1952). Multiple Comparisons. Paper before *Amer. Statist. Assn.* and *Biom. Soc.*, Chicago, 28 December 1952.

[79] Tukey, John W. (1952). Various methods from a unified point of view. Paper before *Inst. Math. Stat.*, Chicago, 29 December 1952.

[80] Wald, A. (1939). Contributions to the theory of statistical estimation and testing hypotheses, *Ann. Math. Stat.* **10**, 299-326.

[80A] Walker, Sir Gilbert (1914). *Indian Meterological Memoirs* **21**.

[81] Welch, B. L. (1937). On the z-test in randomized blocks and latin squares, *Biometrika* **29**, (1937-38) 21-52.

[82] Welch, B. L. (1939). On confidence limits and sufficiency, with particular reference to parameters of location, *Ann. Math. Stat.* **10**, 58-69.

[83] Wilks, S. S. (1938). Shortest average confidence intervals from large samples, *Ann. Math. Stat.* **9**, 166-175.

[84] Wilson, E. B. (1927). Probable inference, the law of succession, and statistical inference, *J. Am. Stat. Assoc.* **22**, 209-212.

[85] Yates, F. (1950). Experimental techniques in plant improvement, *Biometrics* **6**, 200-207.

[15A] Davies, Owen L. (1949). *Statistical Methods in Research and Production.* xi + 292. Oliver and Boyd, Edinburgh.

[67A] Roy, S. N. (1953p). Applications to Multivariate Situations. Address before *Inst. Math. Stat.* and *Biom. Soc.*, Washington, April, 1953.

227. REFERENCES TO EXAMPLES

The references to sources and or discussions, of examples are given in two series. Numbers [101] to [115] apply to the incidental examples of Part A, and are as follows:

[101] Anderson, R. L. and Bancroft, T. A. (1952). *Statistical Theory in Research.* McGraw-Hill, New York.

[102] Beall, Geoffrey, (1942). The transformation of data from entomological field experiments so that the analysis of variance becomes applicable, *Biometrika* **32**, (1941-42) 243-262.

[103] Brownlee, K. A. (1946-1949). *Industrial Experimentation.* 1st edition 1946, 4th edition 1949. His Majesty's Stationery Office, London.

[104] Chapanis, A., Rouse, R. O. and Schaefer, Stanley (1949). The effect of intersensory stimulation on dark adaptation and night vision, *J. Exp. Psych.* **39**, 425-427.

[105] Clark, Charles E. (1953). *An introduction to Statistics.* John Wiley, New York.

[106] Curtis, J. M. (1949). A study of the business organization and operating methods of farmer retail produce markets in North Carolina. Unpub. thesis, North Carolina State College, Raleigh.

[107] Davies, Owen L. (1949). *Statistical Methods in Research and Production*. (2nd ed.) London: Oliver and Boyd.

[108] Dockray, George H., Meadows, Barkley and Smith, Leonard (1950). Cotton Waste, *Textile Research Journal* **20**, 684-718.

[109] Freeman, H. A. (1942). *Industrial Statistics*. John Wiley, New York.

[110] Johnson, Palmer O. (1949). *Statistical Methods in Research*. Prentice-Hall, New York.

[111] Roberts, Donald E. and Jessup, Ralph S. (1951). Heat of polymerization of alpha-methyl styrene from heats of combustion of monomer and four polymer fractions, *J. Res. Natl. Bur. Stds.* **46**, 11-17.

[112] Tippett, L. H. C. (1939). *The Statistical Principles of Experimentation, with Particular Reference to Experimentation in Textile Factories*. Proceedings of the Industrial Statistics Conference held at M.I.T. New York: Pitman Publishing Corp.

[113] Warren, E. (1909). Some Statistical Observation on Termites, *Biometrika* **6**, 329-347.

[114] Wilson, G. B. (1949). "Cytological studies in the *Musae*. IV. Tetrad shape, *Trans. Amer. Microscopical. Soc.* **68**, 217-221.

[115] Yates, F. (1937). The design and analysis of factorial experiments, *Imp. Bur. Soil Sci. Tech. Comm. No. 35*.

2

Equalizing Unequal Accuracies of Means

J. W. Tukey

1. SUMMARY

There are many manipulations and statistical tests which can be conveniently performed on a set of means whose accuracy may be presumed equal and for whose variance there is frequently an independent estimate. The purpose of the present memorandum is to outline a computationally feasible method of extending all such tests to the case of accuracies which may be presumed to be in a known ratio, as is frequently the case when the means are means of differing numbers of observations.

This is done by finding a linear transformation from the observed means to certain dummy means. The dummy means have the variances and covariances of independent means of equal accuracy, and each one corresponds definitely to one of the observed means.

Such procedures as that outlined in MR-6* can, with the aid of this transformation, be easily applied to the case of means of samples of unequal size.

The Collected Works of John W. Tukey (1993) Chapman and Hall, New York, NY.

Prepared in connection with research sponsored by the Office of Naval Research.

*Reference is to the technical report form of Paper 3, following. (JWT)

Previously unpublished manuscript (1948).

2. PLAN

The interesting manipulations and tests are concerned only in the deviations of the observations $y_1, y_2, \ldots, y_n$ from their mean $\bar{y}$. We shall begin then with n independent observations $x_1, x_2, \ldots, x_n$ of weights $w_1, w_2, \ldots, w_n$, by which is meant

$$\text{variance}\,(x_i) = \frac{\sigma^2}{w_i},$$

$$\text{covariance}\,(x_i, x_j) = 0, \quad i \neq j.$$

We denote by $\bar{x}$ the *weighted* mean

$$\bar{x} = \frac{\Sigma\, w_i x_i}{\Sigma\, w_i}, \quad \text{variance}\,(\bar{x}) = \frac{\sigma^2}{w},$$

where $w = \Sigma\, w_i$, and proceed to try to transform the differences

$$x_1 - \bar{x},\ x_2 - \bar{x},\ x_3 - \bar{x}\,, \ldots,\ x_n - \bar{x}$$

into dummy differences

$$y_1 - \bar{y},\ y_2 - \bar{y},\ y_3 - \bar{y}\,, \ldots,\ y_n - \bar{y},$$

whose variances and covariances are those of the differences from the mean of independent observations of equal accuracy.

Since

$$\text{covariance}\,(x_i, \bar{x}) = \frac{\sigma^2}{w},$$

we have,

$$\text{variance}\,(x_i, \bar{x}) = \frac{\sigma^2}{w_i} - \frac{\sigma^2}{w},$$

$$\text{covariance}\,(x_i - \bar{x}, x_j - \bar{x}) = -\frac{\sigma^2}{w}$$

and we want to attain, choosing a convenient normalization,

$$\text{variance}\,(y_i - \bar{y}) = \sigma^2 - \frac{\sigma^2}{n},$$

$$\text{covariance}\,(y_i - \bar{y}, y_j - \bar{y}) = -\frac{\sigma^2}{n}.$$

We shall do this in three steps. First, we convert the $x_i - \bar{x}$ by a linear transformation into uncorrelated quantities, the ratios of whose variance we can calculate. Second, we bring them to suitable variances. Third, we equate the result to transforms of the $y_i - \bar{y}$.

3. CALCULATION OF TRANSFORMATION

We consider a special case as an illustration. Let the weights of x_1, x_2, x_3, x_4 be 5, 3, 1 and 1 respectively. We form the following uncorrelated differences.

$$5x_1 - 5x_2,$$

$$5x_1 + 3x_2 - 8x_3,$$

$$5x_1 + 3x_2 + x_3 - 9x_4,$$

where the coefficients of the positive terms are the corresponding weights and the coefficient of the negative term is the sum of the positive coefficients. (As an example, consider the covariance of the last two differences, which is

$$5 \cdot \frac{5\sigma^2}{5} + 3\,\frac{3\sigma^2}{3} - 8\,\frac{\sigma^2}{1} = 0.$$

All the other covariances vanish for the same reason.) The variances of these uncorrelated differences are

$$\frac{5^2}{5}\,\sigma^2 + \frac{5^2}{3}\,\sigma^2 = \frac{40}{3}\,\sigma^2,$$

$$\frac{5^2}{5}\sigma^2 + \frac{3^2}{3}\sigma^2 + \frac{8^2}{1}\sigma^2 = 72\sigma^2,$$

$$\frac{5^2}{5}\sigma^2 + \frac{3^2}{3}\sigma^2 + \frac{1^2}{1}\sigma^2 + \frac{9^2}{1}\sigma^2 = 90\sigma^2,$$

respectively. The analogous differences of the y's are

$$y_1 - y_2,$$

$$y_1 + y_2 - 2y_3,$$

$$y_1 + y_2 + y_3 - 3y_3,$$

whose variances are $2\sigma^2$, $6\sigma^2$, $12\sigma^2$. The required transformation is thus given by

$$5x_1 - 5x_2 = \sqrt{20/3}\,(y_1 - y_2),$$

$$5x_1 + 3x_2 - 8x_3 = \sqrt{12}\,(y_1 + y_2 - 2y_3),$$

$$5x_1 + 3x_2 + x_3 - 9x_4 = \sqrt{15/2}\,(y_1 + y_2 + y_3 - 3y_4),$$

where

$$\frac{20}{3} = \frac{\frac{40}{3}\sigma^2}{2\sigma^2},\quad 12 = \frac{72\sigma^2}{6\sigma^2},\quad \frac{15}{2} = \frac{90\sigma^2}{12\sigma^2},$$

are the necessary ratios of the variances. An alternative form is, obviously,

$$y_1 - y = \sqrt{3/20}\,(5x_1 - 5x_2),$$

$$y_1 + y_2 - 2y_3 = \sqrt{1/12}\,(5x_1 + 3x_2 - 8x_3),$$

$$y_1 + y_2 + y_3 - 3y_3 = \sqrt{2/15}\,(5x_1 + 3x_2 + x_3 - 9x_4).$$

We next want the values of $y_1 - \bar{y}$, $y_2 - \bar{y}$, $y_3 - \bar{y}$, $y_4 - \bar{y}$ in terms of these differences. Consider

$$\alpha(y_1 - y_2) + \beta(y_1 + y_2 - 2y_3) + \gamma(y_1 + y_2 + y_3 - 3y_4)$$
$$= (\alpha+\beta+\gamma)y_1 + (-\alpha+\beta+\gamma)y_2 + (-2\beta+\gamma)y_3 + (-3\gamma)y_4\,.$$

The desired differences arise when α, β, γ take the following values:

	α	β	γ
$y_1 - \bar{y}$	$^6/_{12}$	$^2/_{12}$	$^1/_{12}$
$y_2 - \bar{y}$	$-^6/_{12}$	$^2/_{12}$	$^1/_{12}$
$y_3 - \bar{y}$		$-^4/_{12}$	$^1/_{12}$
$y_4 - \bar{y}$			$-^3/_{12}$

If we modify the differences first by the simple factors $^1/_2 = {}^6/_{12}$, $^1/_6 = {}^2/_{12}$, $^1/_{12}$, where we note the reappearance of $i(i+1)$ in the numerators, and denote the results by z_1, z_2, z_3 for convenient reference, we have

$$z_1 = {}^1/_2\,(y_1 - y_2) = {}^1/_4\sqrt{^3/_5}\,(5x_1 - 5x_2),$$
$$z_2 = {}^1/_6\,(y_1 + y_2 - 2y_3) = {}^1/_{12}\sqrt{3}\,(5x_1 + 3x_2 - 8x_3),$$
$$z_3 = {}^1/_{12}\,(y_1 + y_2 + y_3 - 3y_4) = {}^1/_{12}\sqrt{^2/_{15}}\,(5x_1 + 3x_2 + x_3 - 9x_4),$$

and

$$y_1 - \bar{y} = z_1 + z_2 + z_3\,,$$
$$y_2 - \bar{y} = -z_1 + z_2 + z_3\,,$$
$$y_3 - \bar{y} = -2z_2 + z_3\,,$$
$$y_4 - \bar{y} = -3z_3\,.$$

Converting to decimals, we have

$$z_1 = .96825x_1 - .96825x_2,$$

$$z_2 = .24056x_1 + .14433x_2 - .38498x_3,$$

$$z_3 = .15215x_1 + .09129x_2 + .03043x_3 - .27387x_4,$$

and

$$y_1 - \bar{y} = 1.36096x_1 - .73263x_2 - .35446x_3 - .27387x_4,$$

$$y_2 - \bar{y} = -.57554x_1 + 1.20387x_2 - .35446x_3 - .27387x_4,$$

$$y_3 - \bar{y} = -.32897x_1 - .19737x_2 + .80021x_3 - .27387x_4,$$

$$y_4 - \bar{y} = -.45645x_1 - .27387x_2 - .09129x_3 + .82161x_4.$$

Let us consider the general case, where we shall put

$$\begin{aligned} W_1 &= w_1, \\ W_2 &= w_1 + w_2, \\ W_3 &= w_1 + w_2 + w_3, \\ \dots &\quad \dots \end{aligned} \tag{1}$$

and

$$\begin{aligned} z_1 &= \sqrt{\frac{w_2}{1 \cdot 2 \cdot W_1 \cdot W_2}}\,(w_1 x_1 - W_1 x_2), \\ z_2 &= \frac{w_2}{2 \cdot 3 \cdot W_2 \cdot W_3}\,(w_1 x_1 + w_2 x_2 - W_2 x_3), \\ \dots &\quad \dots \end{aligned} \tag{2}$$

$$z_i = \sqrt{\frac{w_{i+1}}{i(i+1)\,W_i\,W_{i+1}}}\,(w_1 x_1 + w_2 x_2 + w_3 x_3 + \ldots + w_i x_i - W_i x_{i+1}),$$

$$\ldots \qquad \ldots$$

The variance of the i-th quantity in parenthesis is

$$\frac{w_1^2}{w_1}\sigma^2 + \frac{w_2^2}{w_2}\sigma^2 + \ldots + \frac{w_i^2}{w_i}\sigma^2 + \frac{W_i^2}{w_{i+1}}\sigma^2 = W_i\,\sigma^2 + \frac{W_i^2}{w_{i+1}}\sigma^2$$

$$= W_i\left(\frac{W_i + W_{i+1}}{w_{i+1}}\right)\sigma_2$$

$$= \frac{W_i W_{i+1}}{w_{i+1}}\sigma^2$$

so that the variance of z_i is

$$\frac{\sigma^2}{i(i+1)}$$

and it is appropriate to set

$$z_1 = \frac{1}{1.2}(y_1 - y_2),$$

$$z_2 = \frac{1}{2.3}(y_1 + y_2 - 2y_3),$$

$$\ldots \qquad \ldots$$

$$z_i = \frac{1}{i(i+1)}(y_1 + y_2 + \ldots + y_i - iy_{i+1}).$$

$$\ldots \qquad \ldots$$

Since

$$\frac{1}{i\,(i+1)} = \frac{1}{i} - \frac{1}{i+1}\,,$$

we have

$$\sum_{i-j}^{i=n} z_i = -\frac{1}{n}y_1 - \frac{1}{n}y_2 - \ldots - \frac{1}{n}y_{i-1} + \left(1 - \frac{1}{n}\right)y_i - \frac{1}{n}y_{i+1} - \ldots - \frac{1}{n}y_n$$

$$= y_i - \bar{y}\,.$$

Thus the $y_i - \bar{y}$ are given in terms of the z_i by

$$\begin{aligned} y_1 - \bar{y} &= z_1 + z_2 + z_3 + \ldots + z_n \\ y_2 - \bar{y} &= -z_1 + z_2 + z_3 + \ldots + z_n \\ y_3 - \bar{y} &= -2z_2 + z_3 + \ldots + z_n \\ y_4 - \bar{y} &= \qquad -3z_3 + \ldots + z_n \\ \ldots & \qquad \ldots \end{aligned} \tag{3}$$

Taken together, formulas (1), (2) and (3) define the transformation.

The transformation *will* depend on the order in which the x_i are arranged. It is recommended that an order of decreasing weights (as in the example above) be adopted as a standard.

4. EXAMPLE OF APPLICATION OF THE PROCEDURE OF MR-6

The following data on androsterone excretion in clinically distinct group was obtained by Dr. Konrad Dobriner of the Sloan-Kettering Institute. The variate analyzed, x, is the common logarithm of the milligrams excreted per 2400 hours.

Table 1

Serial Number i	Identifying Symbol	Number of cases W_i	Observed Mean x_i	Transformed "Dummy" deviation $y_i - \bar{y}$
1	NM	10	2.059	1.848
2	NF	9	1.794	1.030
3	CBD	7	1.448	0.093
4	CS	7	1.284	− .419
5	CBL	6	1.552	.281
6	CP	5	0.580	−1.958
7	CAREU	5	1.812	0.750
8	HT	5	1.012	−1.065
9	LL	4	0.865	−1.268
10	CP(CA)	1	0.350	−1.282
11	CL	1	0.350	−1.339
12	TT	1	2.827	1.089
13	CC	1	1.477	−0.182
14	HG	1	3.267	1.654
15	PA	1	2.344	0.859
16	AA	1	1.380	−0.114
Grand Mean				−0.004

The transformation is computed in the Appendix.

Rearranging in order of size, we have the dummy deviations and related quantities shown in Table 2. The straggler test has thus indicated that NM and HG are high, while CP is low, compared to the whole collection of cases. Application of an F-test to the remaining 13 deviates leads to $F = 2.42$ on 12 and 49 degrees of freedom. This is nearly at the 1% level, and we conclude that there are real differences between the remaining groups, although we cannot designate a specific group or groups as definitely different from the others.

Table 2

Application of Straggler Test

Symbol	Dummy deviation	Grand Mean	$\frac{m_1 - m}{s}^{+}$	$6/5 \log_{10} k$	Approx. Normal[++] Deviative	Significance at 5%
NM	1.845	0.130*	2.96	1.42*	1.89	yes
HG	1.645	0.008**	2.83	1.37**	1.80	yes
TT	1.089					
NF	1.030					
PA	0.859					
CAREU	0.750					
CBL	0.281					
CBD	0.093					
AA	– .114					
CC	– .182					
CS	– .419					
HT	–1.065					
CP(CA)	–1.262					
LL	–1.268					
CL	–1.339	–0.118***	2.11	1.34***	0.95	no
CP	–1.956	0.000	3.38	1.46	2.36	yes

+ $s = 0.578$ on 49 degrees of freedom

++ Denominator $= 3\sqrt{1/4 + 1/49} = .814$

* After removing CP

** After removing CP and NM

*** After removing CP, NM and HG

5. UNSETTLED QUESTIONS

The transformation found above does produce dummy deviations, but is this the "best" set of dummy deviations? How is "best" to be defined?

Some would question the combination of the transformation with the procedure of MR-6, particularly if the gap test were used. The writer feels that the use is both legitimate and wise, although the reasons for its wisdom are hard to write down.

APPENDIX
CALCULATION OF TRANSFORMATION

The following tables contain the necessary computations.

Table 3

First steps in finding the transformation

(1) i	(2) w_i	(3) W_i	(4) iW_i	(5)*	(6) $\frac{W_{i+1}}{(5)}$	(7) $(6)^{1/2}$
1	10	10	10	380	.02368421	.1539
2	9	19	38	2964	.00236167	.0486
3	7	26	78	10296	.00067987	.0261
4	7	33	132	25740	.00023310	.0153
5	6	39	195	51480	.00009712	.0099
6	5	44	264	90552	.00005522	.0074
7	5	49	343	148176	.00003374	.0059
8	5	54	432	225504	.00001774	.0042
9	4	58	522	307980	.00000325	.0018
10	1	59	590	389400	.00000257	.0016
11	1	60	660	483120	.00000207	.0014
12	1	61	732	589992	.00000169	.0013
13	1	62	806	710892	.00000141	.0012
14	1	63	882	84672	.00000118	.0011
15	1	64	960	998400	.00000100	.0010
16	1	65	1040			

* Product of entry in column (4) by next lower entry in that column $= i(i+1)W_i W_{i+1}$

Table 4

Second step in finding transformation

		i	1	2	3	4	5	6	7
		w_i	10	9	7	7	6	5	5
		w_{i-1}		10	19	26	33	39	44

	(7) from Table 3	Coefficients of x_i in z_5						
1	.1539	1.539	−1.539	0	0	0	0	0
2	.0486	.486	.437	−.923	0	0	0	0
3	.0261	.261	.235	.183	−.679	0	0	0
4	.0153	.153	.138	.107	.107	−.505	0	0
5	.0099	.099	.089	.069	.069	.059	−.385	0
6	.0074	.074	.067	.052	.052	.044	.037	−.326
7	.0059	.059	.053	.041	.041	.035	.030	.030
8	.0042	.042	.038	.029	.029	.025	.021	.021
9	.0081	.081	.016	.013	.013	.011	.009	.009
10	.0016	.016	.014	.011	.011	.010	.008	.008
11	.0014	.014	.013	.010	.010	.008	.007	.007
12	.0013	.013	.012	.009	.009	.008	.006	.006
13	.0012	.012	.011	.008	.008	.007	.006	.006
14	.0011	.011	.010	.008	.008	.007	.006	.006
15	.0010	.009	.007	.007	.006	.005	.005	.005

Coefficients of x_i in $y_j - \bar{y}$

1	2.807	−.397	−.376	−.315	−.285	−.250	−.228	−.221
2	−.271	2.681	−.376	−.315	−.285	−.250	−.228	−.221
3	−.190	−.169	2.393	−.315	−.285	−.250	−.228	−.221
4	−.262	−.235	−.185	2.401	−.285	−.250	−.228	−.221
5	−.244	−.220	−.171	−.171	2.240	−.250	−.228	−.221
6	−.226	−.202	−.157	−.157	−.134	2.060	−.228	−.221
7	−.249	−.226	−.176	−.176	−.147	−.124	2.054	−.221
8	−.277	−.248	−.192	−.192	−.163	−.142	−.142	2.091
9	−.242	−.219	−.166	−.166	−.143	−.121	−.121	−.121
10	−.086	−.075	−.064	−.064	−.053	−.043	−.043	−.043
11	−.100	−.085	−.068	−.068	−.064	−.050	−.050	−.050
12	−.108	−.101	−.078	−.078	−.060	−.054	−.054	−.054
13	−.123	−.114	−.085	−.085	−.076	−.055	−.055	−.055
14	−.135	−.124	−.089	−.089	−.078	−.067	−.067	−.067
15	−.144	−.131	−.105	−.105	−.092	−.079	−.079	−.079
16	−.150	−.135	−.105	−.105	−.090	−.075	−.075	−.075

Table 4 (Cont'd)

i	8	9	10	11	12	13	14	15	16
w_i	5	4	1	1	1	1	1	1	1
w_{i-1}	49	54	58	59	60	61	62	63	64
(7) from Table 3			Coefficients of x_i in z_5						
1	0	0	0	0	0	0	0	0	0
2	0	0	0	0	0	0	0	0	0
3	0	0	0	0	0	0	0	0	0
4	0	0	0	0	0	0	0	0	0
5	0	0	0	0	0	0	0	0	0
6	0	0	0	0	0	0	0	0	0
7	−.289	0	0	0	0	0	0	0	0
8	.021	−.226	0	0	0	0	0	0	0
9	.009	.007	−.105	0	0	0	0	0	0
10	.008	.006	.002	−.094	0	0	0	0	0
11	.007	.006	.001	.001	−.084	0	0	0	0
12	.006	.005	.001	.001	.001	−.077	0	0	0
13	.006	.005	.001	.001	.001	.001	−.073	0	0
14	.006	.004	.001	.001	.001	.001	.001	−.071	0
15	.004	.001	.001	.001	.001	.001	.001	.001	−.064
Coefficients of x_i in $y_j - \bar{y}$									
1	−.189	−.098	−.089	−.080	−.074	−.071	−.070	−.064	
2	−.189	−.098	−.089	−.080	−.074	−.071	−.070	−.064	
3	−.189	−.098	−.089	−.080	−.074	−.071	−.070	−.064	
4	−.189	−.098	−.089	−.080	−.074	−.071	−.070	−.064	
5	−.189	−.098	−.089	−.080	−.074	−.071	−.070	−.064	
6	−.189	−.098	−.089	−.080	−.074	−.071	−.070	−.064	
7	−.189	−.098	−.089	−.080	−.074	−.071	−.070	−.064	
8	−.189	−.098	−.089	−.080	−.074	−.071	−.070	−.064	
9	1.845	−.098	−.089	−.080	−.074	−.071	−.070	−.064	
10	−.033	.952	−.089	−.080	−.074	−.071	−.070	−.064	
11	−.036	−.015	.945	−.080	−.074	−.071	−.070	−.064	
12	−.048	−.007	−.007	.928	−.074	−.071	−.070	−.064	
13	−.047	−.009	−.009	−.009	.927	−.071	−.070	−.064	
14	−.067	−.057	−.011	−.011	−.011	.951	−.070	−.064	
15	−.052	−.013	−.013	−.013	−.013	−.013	.995	−.064	
16	−.060	−.015	−.015	−.015	−.015	−.015	−.015	.960	

3

Comparing Individual Means in the Analysis of Variance

John W. Tukey

The practitioner of the analysis of variance often wants to draw as many conclusions as are reasonable about the relation of the true means for individual "treatments," and a statement by the *F*-test (or the *z*-test) that they are not all alike leaves him thoroughly unsatisfied. The problem of breaking up the treatment means into distinguishable groups has not been discussed at much length, the solutions given in the various textbooks differ and, what is more important, seem solely based on intuition.

After discussing the problem on a basis combining intuition with some hard, cold facts about the distributions of certain test quantities (or "statistics") a simple and definite procedure is proposed for dividing treatments into distinguishable groups, and for determining that the treatments within some of these groups are different, although there is not enough evidence to say "which is which." The procedure is illustrated on examples.

2. DISCUSSION OF THE PROBLEM

Let us begin by considering how the latest and most advanced statistical theory would approach this problem and then explain why such a solution

The Collected Works of John W. Tukey (1993) Chapman and Hall, New York, NY.

Prepared in connection with research sponsored by the Office of Naval Research.

Reprinted with the permission of the American Statistical Association, from (1949) *Biometrics* **5**, 99-114.

seems impractical. To make things more precise, let us suppose as a fictitious example that seven varieties of buckwheat; *A*, *B*, *C*, *D*, *E*, *F*, and *G* have been tested for yield in each of 12 locations, and that our interest is in the average yield of the buckwheat varieties in a region of which the 12 locations are a respectable sample, and in years exactly like the one in which the experiment was made. We will then have a simple and straightforward analysis of variance into varieties, locations, and interaction. We shall be concerned with the seven observed variety means and with an unbiased estimate of their variance, which will be given by $^1/_{12}$th of the interaction mean square, which is itself on 66 degrees of freedom. What can we say about the varieties under these conditions?

We will wish to say, for example, that *B* and *F* yield better than *A*, *C* and *G*, which yield better than *D* and *E*. Perhaps we might wish to add that *A*, *C* and *G* are not alike, although we do not know which one yields better. The most modern approach would require us to proceed as follows: Write down all the possible conclusions to which we might come — the one illustrated above is one of the 120,904 similar possibilities for seven "treatments." Then for each combination of seven true mean yields we should decide how much it would "cost" us to make each of these 120,904 decisions. Making the usual assumptions about the distribution of fluctuations in yield, we would have begun to state a mathematically well-posed problem. We are unlikely to get this far in a practical problem in my lifetime! Then we find, to our horror, that there are many competing methods of decision, and that which one risks the least will depend on the true variety yields, which we will never know. The problem is not as hopeless as it sounds, for Wald has taken a large step forward, and shown that any decision method can be replaced by one derived from *a priori* probability considerations without increasing the risk under any set of true variety yields. This is a great simplification — but the mathematical complications of dealing with 120,904 functions of seven variables are still awe-inspiring. If we were able to carry through this program — to set the risks intelligently, to carry out the mathematics, and to choose wisely among the admissible decision functions — we would surely do much better than we can hope to do now, but for the present we need to adopt a simpler procedure. (Note. The case of 3 or 4 means has been attacked within the scope of Wald's theory by Duncan [7] using a different philosophy which emphasizes conclusions about pairs of means.)

At a low and practical level, what do we wish to do? We wish to separate the varieties into distinguishable groups, as often as we can without too frequently separating varieties which should stay together. Our criterion of "not too frequently" is a rough one, and may frequently be expressed by saying "at the 5% level" or "at the 1% level." The meaning of these words deserves a little discussion. To the writer they do not mean, "so that an entirely nonexistent effect will be called real once in twenty times, or once in a hundred times", but rather that "with the same sort of protection against false positives that I usually have when I make tests of significance on hypotheses suggested by the results tested, successive tests of hypotheses, tests of regression on selected variables, etc." For these reasons, working "at the 5% level" may involve the successive use of tests, each of which yields false positives five times in a hundred, but, when used together, will yield seven, eight or nine false positives in a hundred. It is such a primitive and rough standard that we wish to combine with a primitively and roughly outlined desire to detect effects which are really there. From these primitive desires we are to seek a method.

When the real differences between variety means are large, how do we realize this fact? Three vague criteria come naturally to mind:

(1) There is an unduly wide gap between adjacent variety means when arranged in order of size;

(2) One variety means straggles too much from the grand mean;

(3) The variety means taken together are too variable.

It is these three criteria we are going to apply in order to break up an observed set of means. We need, then quantitative tests for detecting (1) excessive gaps, (2) stragglers, (3) excess variability. These must be used when the variance of an individual observed mean is not known exactly, but rather when it is estimated from some other line of an analysis of variance table. The tests which we use must therefore be Studentized tests. Exact tests for (2) and (3) are available, but for the present we shall confine ourselves to an approximate and conservative test for (1).

If there are only two variety means, the largest gap between adjacent means is the same as the absolute value of the difference of the means. If $m_1 > m_2$, and s_m^2 is the estimated variance of a single mean, then

$$\frac{m_1 - m_2}{s_m 2^{1/2}}$$

has one-half of a t-distribution and assuming normality, exceeds 2.447 only 5% of the time when the two true means are equal and s_m is based on 6 degrees of freedom. There are good reasons based on experimental sampling (Section 9) and numerical integration (Section 8) to believe that the one-sided 5%, 2%, 1% points of

$$\frac{\text{largest gap between adjacent means}}{s_m 2^{1/2}}$$

are smaller than the corresponding two-sided percentage points of t. If this is true we will be conservative to use this ratio and the two-sided percentage points of t as a test of excessive gapping. The reasons are discussed in a later section.

The exact test of

$$\frac{m_1 - \overline{m}}{s_m}$$

where m_1 is the largest mean and $\overline{m}$ is the grand mean has been discussed for the case of normality by K. R. Nair [4] in a very recent number of *Biometrika*. Simple and satisfactory *empirical* approximation to the *upper* percentage points (between 10% and 0.1%) can be obtained by treating

$$\frac{\left(\frac{m_1 - \overline{m}}{s_m}\right) - \frac{6}{5}\log_{10} k}{3\left(\frac{1}{4} + \frac{1}{n}\right)} \qquad (k > 3 \text{ means})$$

or

$$\frac{\left(\frac{m_1 - \overline{m}}{s_m}\right) - \frac{1}{2}}{3\left(\frac{1}{4} + \frac{1}{n}\right)} \qquad (3 \text{ means})$$

as unit normal deviates, where s_m is based on n degrees of freedom. The adequacy of this approximation — which avoids the use of multiple entry tables — is also discussed in Section 6.

The exact test of excessive spread in general will of course be the familiar F-test (or z-test).

We propose to use these tests successively, and in the following order and manner. First, apply the gap test to break up the means into one or more broad groups. Second, apply the straggler test *within* these groups to further break off stragglers within groups. Third, apply the F-test to these new subgroups to detect excess variability. It is hard to see how to find the frequency of false positives with the whole system analytically, but the writer conjectures that, if the same level, such as 5%, is used in all three tests, the frequency of false positives will be between 1.2 and 1.6 times the level used (i.e., between 6% and 8% when a 5% level is used). This is about where the frequency of false positives stands for many repeated and result-guided tests of significance now in actual practice.

4. DETAILED PROCEDURE ILLUSTRATED BY EXAMPLES

The two examples we are going to use are those discussed by Newman [5] in connection with the use of the Studentized range. The advantages of continuing with the same examples may compensate for disadvantages of lack of simplicity, and in the case of the first example, lack of appropriateness. This first example is a 6×6 Latin square with potatoes, cited by Fisher [1] in Article 36 of *The Design of Experiments*. As first presented this example is stated to be six fertilizer treatments in a Latin Square, and Newman seems to have based his example on this discussion. Later on in the book (Article 64), Fisher points out that these treatments were a 2×3 factorial design in nitrogen and phosphorus, so that there were specific individual degrees of freedom whose analysis was planned when the experiment was designed. These were *not* 6 treatments all on an equal footing, and overall analysis is not appropriate, *but* we shall proceed to analyze them as if they were six treatments about which there is no advance information. The six means were (*A*) 345.0, (*B*) 426.5, (*C*) 477.8, (*D*) 405.2, (*E*) 520.2, (*F*) 601.8, and the estimated standard deviation of a mean was $s_m = 15.95$.

Step 1. Choose a level of significance. For this example we shall choose 5%.

Step 2. Calculate the difference which would have been significant if there were but two varieties.

The two-sided 5% point of t on 20 degrees of freedom is 2.086. For this example, then, this least significant difference is $2.086\,(2^{1/2})\,15.95 = 47.0$.

Step 3. Arrange the means in order and consider any gap longer than the value found in Step 2 as a group boundary.

Arranged in order, the means are 345.0, 405.2, 426.5, 477.8, 520.2, 601.8 and the differences $405.2 - 345.0 = 60.2$, $477.8 - 426.5 = 51.3$, and $601.8 - 520.2 = 81.6$ exceed 45.7, so that we have divided the varieties into four groups: 345.0 (*A*) by itself, 405.2 (*D*) and 426.5 (*B*) together, 477.8 (*C*) and 520.2 (*E*) together, and 601.8 (*F*) by itself.

If no group contained more than two means, the process terminates.

The first example having terminated, we must pass to another to illustrate the continuance of the process. Snedecor [6] gives as Example 11.28 on

p. 274 (of the 4th edition) the results of a 7×7 Latin Square with potatoes. The means were (*A*) 341.9, (*B*) 363.1, (*C*) 360.5, (*D*) 360.4, (*E*) 379.9, (*F*) 386.3, (*G*) 387.1 and s_m on 30 degrees of freedom was 9.52. Choosing the 5% level, for which t on 30 degrees of freedom is 2.042, we find $t\,(2^{1/2})s_m = 27.5$. In order, the means are 341.9, 360.4, 360.6, 363.1, 379.9, 386.3, and 387.1. No difference between adjacent means exceed 27.5, so that there is only one group at the end of Step 3.

Step 4. In each group of 3 or more means find the grand mean, the most straggling mean and the difference of these two divided by s_m. Convert these ratios into approximate unit normal deviates by finding

$$\frac{\dfrac{m-\overline{m}}{s_m} - \dfrac{6}{5}\log_{10} k}{3\left(\dfrac{1}{4}+\dfrac{1}{n}\right)}, \qquad (k > 3 \textit{ means in the group}),$$

$$\frac{\dfrac{m-\overline{m}}{s_m} - \dfrac{1}{2}}{3\left(\dfrac{1}{4}+\dfrac{1}{n}\right)}, \qquad (3 \textit{ means in the group}).$$

Separate off any straggling mean for which this is significant at the chosen two-sided significance level for the normal.

For the Snedecor example we find $\overline{m} = 368.5$, and the most straggling mean is $m = 341.9$. The ratio is $26.6/9.51 = 2.80$. Further $\log_{10} 7 = .845$ and we are to consider

$$\frac{2.80 - \dfrac{6}{5}.845}{3\left(\dfrac{1}{4}+\dfrac{1}{30}\right)} = \frac{60}{51}(2.80 - 1.01) = 2.10.$$

Since the two-sided 5% level for the unit normal is well known to be 1.96, we must separate 341.9 (*A*).

Step 5. If Step 4 changed any group, repeat the process until no further means are separated in the old groups. The means separated off from one side of a group form a subgroup. If there are any subgroups of three or more when no more means are being separated from groups, apply the same process (Steps 4 and 5) *to the subgroups.*

The old group in the Snedecor example now contains 6 means, and its grand mean has increased to $\overline{m} = 372.9$. The most straggling mean is 387.1 for which $(387.1 - 372.9)/9.51 = 1.49$. The approximate unit normal deviate is $^{60}/_{51}$ $(1.49 - 0.93) = 0.66$, which is far from significance. Step 5 has produced no further effect.

Step 6. Calculate the sum of squares of deviations from the group mean and the corresponding mean square for each group of or subgroup 3 or more resulting from Step 5. Using s_m^2 as the denominator, calculate the variance ratios and apply the F-test.

In the Snedecor example, we have one group of six, for which the sum of squares of deviations is 829 and the mean square 166. The denominator is $(9.51)^2 = 90.4$ and the F-ratio 1.83 on 4 and 30 degrees of freedom, which is near the 12% point. Thus there is no overall evidence of difference in yield for these six varieties.

If varieties (B) 363.1, (C) 360.6, and (D) 360.4 had been *known in advance* to be different *as a class* from varieties (E) 379.9, (F) 386.3, and (G) 387.1, it would be fair to introduce a single degree of freedom for this comparison, giving an analysis of variance (in terms of means) like this.

	Degrees of Freedom	Mean Square
BCD vs *EFG*	1	794
Varieties within classes	4	35
Error	30	90.4

From this we could conclude that *BCD* and *EFG* were different, even at the 1% level. There is no valid basis for this particular conclusion ***unless*** the classes are ***uniquely*** known in advance of the experiment. (There are 20 ways to split six varieties into two classes of three varieties each, so that the apparent significance of the most significant split would be expected to be at a percentage level near $^1/_{20}$th of the percentage level of the whole group. The actual figures are, approximately, 0.6% and 12% and their agreement with the 1-to-20 ratio is unusually close.)

In the Fisher example, the proposed procedure gave the following result: Variety A (345.0) *is significantly lower than varieties D (405.2) and B (426.5), these in turn are significantly lower than C (477.8) and E (520.2), and in turn these are significantly lower than F (601.8). All significance statements are statistical, and are at the 5% level or better.*

In the Snedecor example, the proposed procedure gave the following result: *Variety A (341.9) was significantly lower than some of the varieties C (360.1), D (360.4), B (363.1), E (379.9), F (386.3), and G (387.1) at the 5% level or better, the group of 6 varieties showed no overall evidence of internal differences at the 5% level.*

These conclusions should be compared with those of Newman, who used the Studentized range to conclude in the first case that even taking *ADB* and *CEF* as two groups, neither was homogeneous. This is consistent with the result of the present analysis, but far less detailed. For the Snedecor example, Newman found that if either *A* or *F* and *G* together were made a separate group, the remainder seemed homogeneous. This is again consistent, but less detailed, since the present process finds definite reason to suppose that it is *A* which is inhomogeneous. (How *much* stronger is the evidence we have against *A* than against *F* and *G* is another matter.)

The writer feels that the proposed procedure is direct, reasonably simple, involves no new tables, and is ready to be used in practice and thereby put to the ultimate test.

5. THE DISTRIBUTION OF THE MAXIMUM GAP

We are interested in the following problem:

"Let a sample of k values (in our case means) be drawn from a normal distribution, of which we know only an independent estimate s of its standard deviation, based on n degrees of freedom. What is the distribution of

$$\frac{\text{largest gap between ordered observed values}}{s}$$

?" The methods of Hartley, reviewed in detail by Nair [4], would allow us to solve this problem for finite n if we knew the answer for infinite n, that is for the case where we know σ, the standard deviation of the normal population.

The problem of the distribution of the largest gap in a sample of k values from a unit normal distribution can easily be attacked by experimental sampling (see Section 9). The fact that the random normal deviates of Mahalanobis [3] are printed in blocks of five leads one to study $k = 5$ and $k = 10$ first. The first 1000 blocks of five in that table were used (skipping block 768, which was marked as an error in the copy available to the author).

The results are shown below:

Table 1

Upper Percentage Points of the Largest Gap in an Ordered Sample of k From a Unit Normal

%	$k = 2$ theory	$k = 5$ sample of 1000 cases	$k = 10$ sample of 500 cases
10	2.33	1.86	<1.50
5	2.77	2.13	1.68
2	3.29	2.49	1.95
1	3.64	2.77	2.42

The theoretical values for $k = 2$ are values of $(2^{1/2})t$ and are accurate, the others are as found by experimental sampling and may deviate from accuracy by perhaps 1 or 2 in the first decimal. They are sufficiently accurate, however, to indicate that the upper percentage point decreases as k increases. Thus if we use the values for $k = 2$ we will make a conservative test. This is true for $n = \infty$, and by the nature of Hartley's expansion it will continue to hold for all reasonable values of n.

The discussion in Section 2 suggests, of course, that it would be correct and wise to find accurately the percentage points of the largest gap for various values of k and then use the appropriate values of k. This is not being suggested for the present, because:

(1) the necessary table does not exist;

(2) it would complicate the procedure;

(3) there are problems in choosing the appropriate value of k;

(4) the simpler proposed procedure has not yet been used enough to show its characteristics.

6. THE STUDENTIZED EXTREME DEVIATE

In his recent paper, Nair [4] has given the following upper percentage points for 3 to 9 samples: (*A*) the 10%, 5%, 2.5%, 1%, 0.5% points for $n = \infty$, (*B*) the 5% points for n from 10 to 20 and 24, 30, 40, 60, 120, ∞, (*C*) the 1% points for the same values of n. The accuracy of our rough approximation is most easily considered by transforming them into percentage points for the approximate unit normal deviates — these are what should be used for accuracy, — and comparing these with the percentage points of the normal — these are what we propose to use. Such a comparison has the following results, (Table 2).

Table 2

Quality of Approximation of Percentage Points for the Straggler Test

Normal percentage point minus accurate percentage point	Occurs for	Cases
0.15 to 0.20	3 means, $n \leq 15$	6
0.10 to 0.15	5%, 3 or 4 means, $n \leq 24$ 1%, 4 means, $n \leq 11$ 1%, 3 means, $n \leq 30$	33
0.05 to 0.10	5%, 5 means, $n \leq 24$ 5%, 3 or 4 means, $n \leq 60$ 1%, 4 means, $n \leq 11$ 1%, 3 means, $n \leq 120$	21
−0.05 to + 0.05	otherwise	154
−0.10 to − 0.05	10%, all cases 5%, 9 means, $n = 10, 11$ 1%, 8 or 9 means, $n = 20$	20

Thus for about two-thirds of the cases tabulated by Nair, the error is less than 0.05, and is surely negligible in practice.

In doubtful cases, a more precise approximate test may be made as follows. Let

$$w = \frac{|\, m - \overline{m} \,|}{s_m} \quad (m \text{ an extreme mean}).$$

Then treat

$$\left(\frac{k}{k-1}\right)^{1/2} \left(w - \frac{10(w-1.2)}{3n}\right)$$

as a unit normal deviate and multiply the tail area by k if only one kind of straggler (high or low) could be considered, and by $2k$ otherwise. Thus if $\overline{m} = 52, m = 43, s = 4, k = 13, n = 28$

$$w = \frac{143-521}{2} = \frac{9}{4} = 2.25,$$

$$\left(\frac{13}{12}\right)^{1/2} \left(2.25 - \frac{10(1.05)}{3.28}\right) = 1.041(2.25-0.13) = 2.20.$$

Now the probability of a unit normal deviate = 2.14 is 0.01390 (from any normal table, e.g., Fisher and Yates [2] Table IX where 98.610% corresponds to a probit of 7.1200). Multiplying by 11 gives 15.3% as the approximate significance, if only low means are of interest, while the level is 30.6% when either high or low means are involved.

This approximation is discussed by Nair [4] for the case $n = \infty$, where it is due to McKay. Nair shows that it is very good indeed. The effectiveness of the term in n^{-1} may be tested by calculating the true percentage points for $w - 3n^{-1}(w-1.2)$ from Nair's tables. The errors involved in the use of the values at the bottom of the columns of Table 3 instead of those above them can hardly ever be of practical importance.

The previous approximation is recommended for routine work since it involves less computation and no changing of significance levels. Both approximations are only good for upper percentage points in the significance test range. The latter approximation should meet all practical needs.

The writer would rarely bother with the more precise approximation except possibly for the cases where the error of the rough test is between

−0.10 and −0.05. The original experimental values are likely to be somewhat non-normal with large tails. An accurate allowance for this would be hard to compute, but it would increase the accurate percentage point slightly, more for smaller n. The rough approximation tends to compensate for this fact in most cases.

Table 3

Upper Percentage Points for $w - 10/3n\ (w - 1.2)$

	5%				1% points			
n	$k = 3$	5	7	9	$k = 3$	5	7	9
10	1.75	2.06	2.24	2.35	2.24	2.57	2.73	2.85
15	1.76	2.08	2.26	2.39	2.27	2.62	2.81	2.93
20	1.76	2.08	2.27	2.39	2.25	2.62	2.82	2.92
30	1.75	2.09	2.27	2.40	2.25	2.61	2.82	2.93
∞	1.74	2.08	2.27	2.39	2.22	2.57	2.76	2.88

7. THE DISTRIBUTION OF LONG GAPS IN A SAMPLE OF k FROM ANY POPULATION

While we could concern ourselves with the distribution of the longest gap, the next longest gap, and so on, it seems theoretically better and practically simpler to do something somewhat different. We are going to calculate the expected number of gaps longer than a length G, which we denote by p_1. For the sort of test considered above, there is much reason to use p_1. For p_1 is the fraction of gaps per sample which will be falsely judged significant. If it is as bad to find two false gaps in a sample as to find one false gap in each of two samples, then we should consider p_1.

Now we shall take the definition of a gap starting at y to be that y is the left hand of the gap. If y is the left-hand end of a gap of length at least G, we have the following table of elementary probabilities:

Event	Probability
One observation must fall between y and $y + dy$	$k\,dF(y)$
$k - 1$ observations must fall between $-\infty$ and y or between $y + G$ and $+\infty$	$\{F(y) + 1 - F(y+G)\}^{k-1}$
Not all $k - 1$ observations can fall between $-\infty$ and y	$-(F(y))^{k-1}$

hence

$$p_1 = k \int_{-\infty}^{+\infty} \{(F(y) + 1 - F(y+G))^{k-1} - (F(y))^{k-1}\}\, dF(y).$$

8. THE SYMMETRICAL CASE

If the distribution of x is symmetrical about zero, we may count only the gaps with centers to the left of the origin and then double. The expression for p_1 follows from:

Event	Probability
One observation must fall between y and $y + dy$	$k\,dF(y)$
$k - 1$ observations must fall between $-\infty$ and y or $y + G$ and $+\infty$	$(F(y) + 1 - F(y+G))^{k-1}$
Not all $k - 1$ observations can fall between $-\infty$ and y or $-y$ and $+\infty$	$-(2F(y))^{k-1}$

Since $y \leq -{}^1/_2 G$, and since the result is to be doubled, we have

$$p_1 = 2k \int_{-\infty}^{-{}^1/_2 G} \{(F(y) + 1 - F(y+G)^{k-1} - (2F(y))^{k-1}\}\, dF(y).$$

Making the substitutions $u = F(y)$, $h(u) = F(y) + 1 - F(y+G)$, this becomes

$$p_1 = 2k \int_{-\infty}^{F(-\frac{1}{2}G)} h^{k-1} du - \left\{ 2F\left(-\frac{G}{2} \right) \right\}^k .$$

For reasonably large G, the second term is fairly small and we can get an accurate value of p_1 with a reasonable amount of labor.

As an example, let us take the unit normal distribution and $G = 2$. Since $h(u)$ is non-analytic near 1 and has a minimum at $F(-1) = .1587$, it is natural to break the integral up into parts as follows:

$$p_1 = 2k \int_0^{.0004} h^{k-1} du + 2k \int_{.0004}^{.004} h^{k-1} du + 2k \int_{.004}^{.04} h^{k-1} du$$
$$+ 2k \int_{.04}^{.16} h^{k-1} du - 0.0013(2k)(h(.1587))^{k-1} - (.3174)^k .$$

Calculating h to four decimals, applying Simpson's rule to the range from 0 to .004, and the corresponding six-panel rule to the other three ranges yields the following results, where the terms are given in the order of the formula above:

k	+	+	+	+	−	−	p_1
2	.00151	.01168	.07875	.16781	.00165	.10074	.15736
3	.00214	.01426	.06603	.08885	.00079	.03206	.13843
4	.00270	.01553	.04227	.04224	.00032	.01014	.09228
5	.00320	.01590	.03035	.01908	.00013	.00322	.06518
6	.00363	.01567	.02635	.00835	.00005	.00102	.05293
7	.00402	.01508	.01880	.00353	.00002	.00032	.04109
8	.00436	.01426	.01342	.00154	.00001	.00010	.03347
9	.00468	.01330	.00959	.00065	.00000	.00003	.02819
10	.00493	.01230	.00691	.00029	.00000	.00000	.02443

The value for $k = 2$ can of course be calculated directly as

$$2(1 - F(2^{1/2})) = 2(.0787) = .1574.$$

The results are probably accurate to 1 or 2 in the fourth place. They can be conveniently stated as in the following table:

Table 4

Number of Gaps Longer Than 2.00 Expected Per 100 Samples of k From the Unit Normal

k	2	3	4	5	6	7	8	9	10
gaps / 100 samples	15.74	13.84	9.23	6.52	5.29	4.11	3.35	2.82	2.44

9. RESULTS OF EXPERIMENTAL SAMPLING

The results of the experimental sampling of 1000 sets of 5 from Mahalanobis' approximation to the unit normal are given in the following table, (Table 5).

The approximate normality of (largest gap)$^{1/2}$ in this sample, as indicated by the correspondence of the last two columns between the 2% points is striking. For comparison it seemed worthwhile to examine the normality of (largest gap)$^{1/2}$ for $k = 2$, where the probability of a gap $\geq G$ is $2N(G/2)$, where $N(u)$ is the unit normal cumulative. This gives the following results, (Table 6).

Here the fit is good between the 10% points. This suggests that the (largest gap)$^{1/2}$ may be a convenient interpolation variable.

The number of cases ≥ 2.00 actually found was 68, while the number to be expected according to the last section was 65.2 less an allowance for large double gaps which might amount to one unit. Finding 68 instead of 64 is a deviation of 0.5σ, and is highly reasonable.

For $k = 10$, the count was only made for gaps ≥ 1.5, with the following results, (Table 7).

Table 5

Results of Experimental Sampling.
Distribution of Largest Gaps in 1000 Samples of 5

Cell	Number	Cumu.	Equiv. Norm. Dev.	$\frac{(\text{gap})^{1/2} - 1.07}{.23}$
.185- .199	2	2	−2.88	(−2.70)
.200- .299	9	11	−2.29	(−2.26)
.300- .399	20	31	−1.87	−1.90
.400- .499	28	59	−1.56	−1.57
.500- .699	97	156	−1.01	−1.00
.700- .899	141	297	−0.53	−0.52
.900-1.099	172	469	− .08	− .09
1.100-1.299	149	618	0.30	0.30
1.300-1.499	126	744	0.66	0.68
1.500-1.699	110	854	1.05	1.00
1.700-1.899	56	910	1.34	1.34
1.900-2.099	36	946	1.61	1.64
2.100-2.299	24	970	1.88	1.90
2.300-2.499	11	981	2.07	2.12
2.500-2.699	8	989	2.29	(2.51)
2.700-2.899	4	993	2.46	(2.77)
2.900-3.099	4	997	2.75	(2.99)
3.100-3.299	2	999	3.09	(3.20)
. . .				
4.000-4.099	1	1000	∞	

Table 6

Cumulative for $(\text{Largest Gap})^{1/2}$ in Samples of 2 From the Unit Normal

%	gap	$(\text{gap})^{1/2}$	Equiv. Norm. Deviate	$\frac{(\text{gap})^{1/2} - .98}{.44}$
1	.0177	.134	−2.33	(−1.95)
2	.0357	.189	−2.05	(−1.80)
5	.0891	.299	−1.64	(−1.55)
10	.1781	.423	−1.28	−1.26
20	.360	.600	−0.84	−0.86
50	.960	.980	0.00	0.00
80	1.825	1.353	0.84	0.85
90	2.350	1.536	1.28	1.26
95	2.794	1.672	1.64	(1.57)
98	3.308	1.821	2.05	(1.91)
99	3.650	1.914	2.33	(2.12)

Table 7

Results of Experimental Sampling
Distribution of Largest Gaps in 500 Samples of 10

Cell	Number	Cumul.	Equiv. Norm. Deviate	$\frac{(\text{gap})^{1/2} - 0.85}{.24}$
-1.499	454	454	1.33	1.38
1.500-1.599	15	469	1.54	1.53
1.600-1.699	9	478	1.71	1.68
1.700-1.799	9	487	1.94	1.82
1.800-1.899	2	489	2.01	1.98
1.900-1.999	2	491	2.10	2.08
2.000-2.199	1	492	2.14	(2.39)
2.200-2.399	2	494	2.26	(2.48)
2.400-2.599	3	497	2.51	(2.93)
2.600-2.799	2	499	2.88	(3.13)
...				
3.100-3.199	1	500	∞	

The fit here is reasonably good out to the 5% point. Since theory predicts about 12.2 beyond 2.00 instead of 9 observed, there is no serious disagreement here.

If we want to make real use of this $(\text{gap})^{1/2}$ variable, we may use the known percentages beyond 1.414, found for k between 2 and 10 in the last section to fix lines in the plane of the mean and standard deviation of the approximation. A little bold, dashing, freehand, two-dimensional interpolation produces the following results:

Table 8

Tentative Behavior of $(\text{Largest Gap})^{1/2}$ for Samples of k From the Unit Normal

Parameters			Levels for $(\text{gap})^{1/2}$			Levels for gap		
k	m	s	5%	2.5%	1%	5%	2.5%	1%
2	0.98	0.43	1.69	1.82	1.98	2.8	3.3	3.9
3	1.03	0.36	1.62	1.74	1.87	2.6	3.0	3.5
4	1.06	0.27	1.50	1.59	1.69	2.3	2.5	2.8
5	1.06	0.23	1.43	1.51	1.60	2.0	2.3	2.6
6	1.06	0.22	1.42	1.49	1.57	2.0	2.2	2.5
7	1.04	0.21	1.39	1.45	1.53	1.9	2.1	2.3
8	1.02	0.21	1.37	1.43	1.51	1.9	2.0	2.3
9	1.00	0.21	1.35	1.41	1.49	1.8	2.0	2.2
10	0.99	0.22	1.33	1.40	1.48	1.8	2.0	2.2

By a stroke of luck, the levels for the gap itself might be accurate to one or two tenths. These are, of course, unstudentized levels.

REFERENCES

[1] Fisher, R. A. (1947). *The Design of Experiments.* 4th edition. Oliver and Boyd, Edinburgh.

[2] Fisher, R. A. and Yates, F. (1948). *Statistical Tables.* 4th edition. Oliver and Boyd, Edinburgh.

[3] Mahalanobis, P. C. et al. (1934). Tables of Random Samples from a Normal Population, *Sankhyā* **1**, 289-328.

[4] Nair, K. R. (1948). The distribution of the extreme deviate from the sample mean and its Studentized form, *Biometrika* **35**, 118-144.

[5] Newman, D. (1939). The distribution of range in samples from a normal population, expressed in terms of an independent estimate of standard deviation, *Biometrika* **31**, 20-30.

[6] Snedecor, G. W. (1946). *Statistical methods*. 4th edition. Collegiate Press, Ames, IA.

[7] Duncan, D. B. (1947). Iowa State College Thesis.

4

Reminder Sheets for "Allowances for Various Types of Error Rates"

John W. Tukey

The view presented here should be contrasted with that presented by David B. Duncan in three ways. First, however, the similarity in basic point of view, the agreement that multiple comparisons *must* be treated individually yet jointly should be stressed. This agreement is more important than the differences, which arise in answering these questions (any of the other $6 = 2^3 - 2$ combinations of choices are feasible!):

(1) What form of answer do we seek?

(2) On what basis do we wish to set our error rates?

(3) What balance between various individual comparisons is desired?

Duncan chooses to seek an answer as to which means are shown to be unequal. The writer chooses to seek an allowance which can be applied to the quantitative appearance of any comparison. (When, for example, the treatments are known to be different, this choice has obvious advantages.)

The Collected Works of John W. Tukey (1993) Chapman and Hall, New York, NY.

Prepared in connection with research sponsored by the Office of Naval Research.

Presented before the Institute of Mathematical Statistics and the Eastern North American Region of the Biometric Society, Blacksburg, VA 19 March 1952.

Previously unpublished manuscript (1952).

Duncan chooses to take his error rate in terms of work and expense, as measured by the number of treatments or degrees of freedom. The writer is prepared to accept this in some circumstances, but feels that it is usually better to consider the error rate in terms of experiments (or lines in complex cases). The effects of this choice are two-fold. First, the statements appear more or less conservative. Second, equality of conservatism between experiments of different sizes is differently defined. (Fisher's basic approach to the analysis of variance has been in terms of experiments or lines.)

Duncan chooses to treat all linear contrasts on the same footing. The writer chooses to favor (in the sense of expending on them a greater fraction of the error rate) the comparisons between pairs of treatments. (This is definitely *not* a matter of principle.) Here concentration on paired differences gives a little bit more simplicity in computation, and a little different balance of errors of the first and second kinds; analytically it brings in the studentized range in the place of the square root of F.

Hopefully, the writer would claim greater flexibility with a given amount of tabulation for his method.

GENERAL

The procedure using allowances can be superposed on any conventional analysis of variance (as we shall indicate), or be reduced to a short-cut procedure (as Link and Wallace will indicate later). A summary of formulas and an adequate set of tables for most practice follow:

Parallel determinations [batch of k]

Allowance to fix:

error rate per item

$$LCA = (t_{DDF})\,(\sqrt{\text{estimated variance}}),$$

error rate per batch

$$HCA' = (BF(k))\,(LCA),$$

error rate batchwise

$$HCA = (BW(k))\,(LCA).$$

Multiple comparisons [$^1/_2 k(k-1)$ comparisons of k determinations]

Allowance to fix:

error rate per comparison

$$LSD = (t_{DDF}\sqrt{2})\ (\sqrt{\text{estimated variance}}),$$

error rate per experiment (= per line)

$$HSD' = (BF(\frac{k(k-1)}{2}))\ (LSD),$$

error rate experimentwise (= linewise)

$$HSD = (HSF(k))\ (LSD),$$

error rate intermediatewise

$$MSD = (0.5 + 0.5\ HSF(k))\ (LSD).$$

ANOVA Relations

Estimated variance of mean = DEN/DIV,

Average mean square = terms in other variance components + DIV (own variance component).

Terminology for consideration

LCA = least confident allowance

HCA = honestly confident allowance

DDF = denominator degrees of freedom

BF = batch factor

BW = batchwise factor

LSD = least significant difference

HSD = honestly significant difference

MSD = moderately significant difference

DEN = appropriate denominator for *F*-test (if perchance desired)

5% Table (approximate)

x	$\frac{60}{x}$	t_x	$t_x\sqrt{2}$	$BF(x)$	$BW(x)$	$BF(\frac{1}{2}x(x-1))$	$HSF(x)$
1		12.7	18.0	$1.00 + 0.0/DDF$	-	————	————
2		4.30	6.09	$1.14 + 0.4/DDF$	-	$1.00 + 0.0/DDF$	$1.00 + 0.0/DDF$
3		3.18	4.50	$1.22 + 0.7/DDF$	-	$1.22 + 0.7/DDF$	$1.20 + 0.5/DDF$
4		2.78	3.94	$1.27 + 0.9/DDF$	-	$1.34 + 1.2/DDF$	$1.31 + 0.9/DDF$
5		2.57	3.64	$1.31 + 1.0/DDF$	-	$1.43 + 1.7/DDF$	$1.39 + 1.0/DDF$
6		2.45	3.46	$1.34 + 1.2/DDF$	-	$1.49 + 2.1/DDF$	$1.46 + 1.3/DDF$
7		2.36	3.34	$1.37 + 1.3/DDF$	-	$1.54 + 2.4/DDF$	$1.51 + 1.4/DDF$
8		2.31	3.27	$1.39 + 1.5/DDF$	-	$1.59 + 2.9/DDF$	$1.55 + 1.4/DDF$
9		2.26	3.20	$1.41 + 1.6/DDF$	-	$1.62 + 3.2/DDF$	$1.59 + 1.5/DDF$
10	6	2.23	3.15	$1.43 + 1.7/DDF$	-	$1.65 + 3.6/DDF$	$1.62 + 1.6/DDF$
	5	2.18	3.08	$1.47 + 1.8/DDF$	-	$1.70 + 4.0/DDF$	$1.67 + 1.7/DDF$
	4	2.13	3.01	$1.49 + 2.1/DDF$	-	$1.79 + 4.8/DDF$	$1.73 + 2.0/DDF$
	3	2.09	2.95	$1.54 + 2.4/DDF$	-	$1.86 + 5.5/DDF$	$1.81 + 2.5/DDF$
	2	2.04	2.89	$1.50 + 3.0/DDF$	-	$1.97 + 7.6/DDF$	$(1.9 + 3/DDF)$
	1	2.00	2.83	$1.70 + 4.0/DDF$	-	$2.15 + 10.7/DDF$	$(2.1 + 5/DDF)$
	0	1.96	2.77	————	-	————	————

Examples for Complex Analyses

Balanced crossed design — three way — real interactions only

LINE	A/P (etc.)	DIV	DEN/MS
M = overall	$1/?$	$crsk$	(**)
C = cols.	$c/C = f_c = 1 - e_c$	rsk	$e_rCR + e_sCS - e_re_sCRS + f_rf_sD$
R = rows	$r/R = f_r = 1 - e_r$	csk	$e_cCR + e_sRS - e_ce_sCRS + f_cf_sD$
S = slices	$s/S = f_s = 1 - e_s$	crk	$e_cCS + e_rRS - e_ce_rCRS + f_cf_rD$
CR	cr/CR (*)	sk	$e_sCRS + f_sD$
CS	cs/CS (*)	rk	$e_rCRS + f_rD$
RS	rs/RS (*)	ck	$e_cCRS + f_cD$
CRS	crs/CRS (*)	k	D
D = duplicates	$crsk/\infty$	1	O

(*) Not sampled, but consequences of sampling

(**) $e_cC + e_rR + e_sS - e_ce_rCR - e_ce_sCS - e_re_sRS + e_ce_re_sCRS + f_cf_rf_sD$

Balanced nested design — three layers

LINE	A/P (etc.)	*DIV*	*DEN/MS*
M = overall	$1/?$	$nmpq$	$e_n C + f_n e_m \mathbf{C} + f_n f_m e_p CCC + f_n f_m f_p e_q D$
C = *cols.*	$n/N = f_n = 1 - e_n$	mpq	$e_m \mathbf{C} + f_m e_p CCC + f_m f_p e_q D$
C = subcols.	m/M (n times) $= f_m = 1 - e_m$	pq	$e_p CCC + f_p e_q D$
CCC = cells	p/P (mn times) $= f_p = 1 - e_p$	q	$e_q D$
D = duplicates	q/∞ (mnp times) $= f_q = 1 - e_q$	1	0

Unbalanced nested design — two layers (no cells)

As above, with n/N, m_i/M_i (ith column), P_{ij}/∞ (ij th subcolumn).

Divisors are $(nmp)^*$, $(mp)^*$, p^*, and $1 - f_n - e_n$, $1 - f_m - e_m$, where

$(m_+ - n)p^* = p_{++} - \Sigma_i(\Sigma_j P_{ij}^2 / p_{1+})$,

$(n-1)(mp)^* = p_{++} - \Sigma_i p_{i+}^2 / p_{++}$,

$(nmp)^* = p_{++}$,

$e_n(mp)^* = (\Sigma_i p_{i+}^2 / p_{++}) - p_{++}/N$,

$e_m(n-1)p^* = \Sigma_i(\Sigma_j p_{ij}^2 / p_{i+}) - \Sigma_i(p_{i+}/M_i)$

Notation: A/P = actual number of items/number of items in population

5

Reminder Sheets for "Multiple Comparisons"

John Tukey

1. Error rates may have various bases. Both a numerator and a denominator must be selected. Standard examples are as follows ("number of" may be prefixed to numerators and denominators):

Type	Name	Numerator	Denominator
Parallel	per determination	wrong determinations	determinations
	per batch	wrong determinations	batches
	batchwise	batches with one or more wrong determinations	batches
Comparative	per comparison	wrong comparisons	comparisons
	per family	wrong comparisons	families
	familywise	families with one or more wrong comparisons	families
Contrast	familywise	families with one or more wrong contrasts	families
Universal	familywise	families with one or more wrong linear combinations	families

The Collected Works of John W. Tukey (1993) Chapman and Hall, New York, NY.

Prepared in connection with research sponsored by the Office of Naval Research.

Presented before the American Statistical Association, 28 December 1952.

Previously unpublished manuscript (1952).

2. One standard system of comparative, contrast and universal allowances with error rates familywise operates as follows (alternate systems will be compared at the Institute of Mathematical Statistical session at 4 p.m. on Monday):

$s^2 =$ estimated variance of determination (on *DDF* degrees of freedom)

$LSD = [LSD/s\,]\beta = (t\sqrt{2})\beta$

$WSD = [WSD/LSD\,](LSD)$

allowance $= \pm(WSD)$(norm)

where norms are computed as follows:

Linear combination	Norm
simple comparison (example: $x_5 - x_3$)	1 (1)
contract $\Sigma\, c_i x_i$ (example: $2x_4 - x_2 - x_{12}$)	$\Sigma^+ c_1 = -\Sigma^- c_1$ (2)
General *L.C.* $\Sigma\, c_i x_i$ (example: $3x_2 + 2x_4 - x_{17}$)	max $\{\Sigma^+ c_i, -\Sigma^- c_i\}$ (5)

and the other necessary factors are in Table 1.

3. Some examples may be helpful:

A. Tippett (as in Freeman, *Industrial Statistics* pp. 65ff) has reported a 3×4 textile experiment. The results may be summarized as follows:

Variable measured: strength of yarn

Means by roving set: (1) 1568, (2) 1661, (3) 1798, (4) 1642

Means by roller weighting: (A) 1677, (B) 1645, (C) 1679

Residual mean squares: 227 on 6 *DF*.

The allowance computation runs as follows:

Item	s^2	s	$[LSD/s\,]$	LSD	$[WSD/LSD\,]$	WSD
Roving sets	76	8.7	3.46	30.2	1.42	43
Roller wts.	57	7.6	3.46	26.2	1.27	33

Table 1

Values of [LSD/s] and [WSD/LSD]

DDF	$60/DDF$	[LSD/s]	Family size	[WSD/LSD]
1		18.0	2 (comparison)	1.00 + 0.0/DDF
2		6.10	2 (universal)	1.02 + 0.2/DDF
3		4.50	3	1.20 + 0.4/DDF
4		3.92	4	1.32 + 0.6/DDF
5		3.64	5	1.40 + 0.8/DDF
6		3.46	6	1.46 + 1.0/DDF
7		3.34	7	1.51 + 1.2/DDF
8		3.26	8	1.55 + 1.4/DDF
9		3.20	9	1.59 + 1.5/DDF
10	6	3.15	10	1.62 + 1.6/DDF
	5	3.08	11	1.65 + 1.7/DDF
	4	3.01	12	1.67 + 1.8/DDF
	3	2.95	13	1.70 + 1.9/DDF
	2	2.89	14	1.72 + 2.0/DDF
	1	2.83	15	1.74 + 2.1/DDF
	0	2.77	16	1.75 + 2.2/DDF
			17	1.77 + 2.3/DDF
			18	1.78 + 2.4/DDF
			19	1.80 + 2.4/DDF
			20	1.81 + 2.5/DDF

Sample conclusions are:

Roller weighting C produces stronger yarn than roller weighting A by 35 ± 33 = 2 to 68 units.

Roving set 3 produces stronger yarn than roving set 1 by 330 ± 43 = 287 to 373 units.

B. Lome (as in Snedecor, *Statistical Methods* pp. 217ff) has reported a cooking experiment on fat absorption which may be summarized as follows:

Variable measured: fat absorption of doughnuts

Means by fat used: (1) 172, (2) 178, (3) 182, (4) 185, (5) 165, (6) 176, (7) 161, (8) 162

Residual mean square: 142 on 40 *DF*

Replication: 8 measurements/fat.

The allowance computation runs as follows:

s^2	s	$[LSD/s]$	LSD	$[WSD/LSD]$	WSD
17.8	4.2	2.86	12.0	1.59	19

Sample conclusions are:

Fat 4 is more absorbed than Fat 7 by $24 \pm 19 = 5$ to 43 units.

Fat 7 differs from Fat 8 by $-1 \pm 19 = -20$ to 18 units.

C. Rider (as in Freeman *Industrial Statistics* pp. 72ff) has reported an experiment on insulating material which may be summarized as follows:

Variable measured: impact strength in foot pounds

Means by lot: (1) 0.83, (2) 1.01, (3) 0.65, (4) 0.88, (5) 0.54

Means by strip direction (L = lengthwise) 0.80, (C = crosswise) 0.76

		1	2	3	4	5
Means by double classification	L	.92	1.00	.69	.87	.55
classification (cell means):	C	.74	1.02	.62	.90	.53

Residual mean square 0.0099 on 90 *DF*

Replication: 10 measurements/cell.

The calculation of allowances for cell means proceeds as follows:

s^2	s	$[LSD/s]$	LSD	$[WSD/LSD]$	WSD
0.00099	.0315	2.81	.0885	1.80	.16

Sample conclusions are:

Only in lot 1 has a difference between lengthwise and crosswise been shown.

In the others, any difference between $+.07 - .19 = -.12$ and $-.03 + .19 = +.16$ is not rejected by the observations for "lengthwise - crosswise."

Further analysis should probably treat lot 1 separately from lots 2, 3, 4 and 5.

4. The same techniques can be applied to contrasts and general linear combinations.

5. All error rates above were at 5%, but simple approximate formulas for shifting error rates exist.

6

Souvenir Sheets for "Confidence Procedures are Better"

John W. Tukey

I. GENERAL

(1) The first work in this country on multiple comparisons was the thesis of the first speaker, D. B. Duncan. The next was a procedure, now obsolete, of unknown properties as offered in BIOMETRICS by the present speaker. By now, a complete bibliography would involve several dozen entries. Almost everything except matters of principle are cleared up for the simple situations.

(2) The present speaker recognizes a minimum of three sorts of multiple comparison situations:

(a) those where our aim is to add knowledge (Newman, Duncan, Tukey, Scheffé, Keuls, etc.);

(b) those where the "best" is to be picked after enough experimentation to control the risk of an error by more than so much (Bechhofer, Dunnett, Sobel);

The Collected Works of John W. Tukey (1993) Chapman and Hall, New York, NY.

Prepared in connection with research sponsored by the Office of Naval Research.

For presentation before ENAR, The Biometric Society, 19 March 1954, Gainesville.

Previously unpublished manuscript (1954).

(c) those where an economic amount (costing both experimentation and errors) of experimentation is to be used for selection (Sommerville, Grundy, Healy, Yates).

The second and third sorts are inherently sequential, and, the speaker believed at time of writing these words, are outside the scope of the present discussion.

(3) Differences in opinion between Duncan and the speaker have, in the main, fallen in the following areas:

(a) whether tests of significance, confidence procedures are appropriate;

(b) along what lines error rates should be chosen;

(c) whether ranges in the numerator are as completely legitimate and moral as F-tests.

Of these, the first is extremely crucial, the second is crucial, while the third matters but is not of the same importance.

II. CONFIDENCE PROCEDURES

(4) The basic issues are more primitive than the problem of multiple comparisons — they are the closely related issues of the sacredness of particular null hypotheses and the interpretation of nonsignificant significance tests.

(5) The foundations of modern statistics (by Student and Fisher) came in fields of science and technology where *it was felt* that an experiment had to show a difference to be meaningful — to be significant in the popular sense. This is far from universally true, thus (i) a demonstration of the agreement in speeds of light and radio waves to a closer tolerance has won a prize from the Franklin Institute, (ii) a demonstration that a battery additive, such as AD-X2 for example, has no appreciable effect on battery performance is surely meaningful, (iii) a demonstration that fertilizer A, which costs half as much as fertilizer B, gives substantially the same results would certainly interest a farmer. Results of the form "the difference between A and B is no more than $\pm e$" can be very definite and very meaningful!

(6) For such a result to be meaningful, it is essential to know the magnitude of $\pm e$, and insufficient to know only that the particular null hypothesis $A = B$ was not rejected. (If rejections of such null hypotheses were the only definite and meaningful results of science and technology, the world would be a far different place.)

(7) It is only possible to focus attention on the null null hypotheses if we forget to say or think "this experiment was insufficient to reject the null hypothesis" and make the serious error of saying or thinking "as a result of this experiment we should accept the null hypotheses." Today this may not be egregious, but it is a serious blunder, since in almost every case we know (by analogy) that the null hypothesis is not *exactly* true.

(8) It is only by bringing in the other null hypothesis — by considering not only $A = B$ but also $A = B + 3$, $A = B + 1$, $A = B + 0.732$, and so on — that we force ourselves to keep the right attitude. Now, many are the null hypotheses which are not rejected. Clearly we cannot "accept" them all, and no one is distinctly preferred above all others. The resulting confidence procedure keeps our attitude from falling into error.

(9) If we do not exalt one particular null hypothesis above all others we see that our results will always be of about the same definiteness. (Whether we find $-2 \le A - B \le +5$ or $98 \le A - B \le 105$, there remain seven units of null hypotheses not rejected.) Some results will be more interesting than others, but this is the fault of the world (and the experimenter's choice) not of the experiment.

(10) Confidence procedures have the advantage that they have the same error rate no matter what the true situation. (Sensible significance tests cannot make serious errors when the true values are widely separated — they attain this state by never saying anything interesting when the true values are widely separated.) This constancy of error rate makes it easier to choose an appropriate error rate.

III. ERROR RATES

(11) The Duncan philosophy on error rates is based on two foundations:

(a) Given m measurements, a particular advance choice of $m - 1$ orthogonal degrees of freedom, and a hankering for the 5% level, it is appropriate to test each of the single degrees of freedom at 5%.

(b) Given that (a) holds, but that we have not prechosen $m-1$ orthogonal degrees of freedom, we should be willing, if we wish to work at 5%, to make one or more errors at the same rate as in (a).

With neither of these do I agree!

(12) Given the advance choice of $m-1$ orthogonal degrees of freedom, where do we stand? Perhaps they were chosen at random — if so all we have done is to raise our error rate from 5% for an F-test to $100\ (1-(1-.05)^{m-1})\%$ for $m-1$ t-tests (utterly without excuse!). Perhaps they were independently meaningful comparisons, but then they would be most unlikely to come out orthogonally — and somehow orthogonality is a sacred essential — it is all right to replace one 5% test by $m-1$ 5% tests so long as they are orthogonal, but not otherwise (Generally speaking, in the nonorthogonal case the error rate is not increased as much as in the orthogonal case! Why exclude the less dangerous case? How can orthogonality matter if we are to consider $m-1$ separate things? How can we justify the increased error rate if we are really considering them together?). I would propose that (a) is only valid in some subset of those cases where the $m-1$ prechosen degrees of freedom are such that the experimenter would be willing, if necessary, to run a separate experiment — a really *separate* experiment — on each one.

(13) Duncan proposes to support (b), and to an extent (a) on the basis of a principle which the speaker likes to call (or miscall) "equal errors for equal work." This seems entirely unacceptable! If we admit that the natural basis for error rates is what the experimenter would be willing to make a separate experiment or experimental program about — if we prefer "equal errors for equal experiments" — then we cannot accept "equal errors for equal work."

(14) One basis of reconciling these two views *might* be by the identification of two different views on nonsignificant results. If one believes that a nonsignificant result of a test of significance means that the identical null hypothesis should be accepted, one may be ready to do almost anything, even ready to raise the error rate per experiment far above 5%, in order to avoid accepting too many identical null hypotheses. If one believes that a nonsignificant result of a test of significance means only that the

interval of non-rejected null hypothesis happens to include the identical null hypothesis, and that asymmetric placement of the interval is a legitimate source of hints, even if they are not proved, then one may be willing and eager to hold to 5% error rate per experiment (just as we have always done with the *F*-tests!). In other words if we can agree on an approach, perhaps we can agree on error rates.

IV. NUMERATORS

(15) After a careful investigation, the speaker is convinced that range-like numerators are preferable to root-mean-square numerators, because they expend a given error rate in a more useful way, by reducing the length of confidence intervals for simple comparisons and, in compensation, increasing the lengths of those for complex contrasts. This seems very desirable to the speaker.

Duncan, on the other hand, seems to feel that principles of abstract justice imply that we should treat all contrasts alike. The logical extension of this view to two or more variable factorial designs would seem to imply that we should treat all contrasts whether main effects, interactions, or mixtures alike. (It seems to the speaker that this would be acceptable to no one.)

V. TECHNIQUES

(16) It seems to the speaker that knowledge of constructional, computational, and Monte Carlo techniques has advanced to the point where if we know what we want to do in multiple comparisons, we can go ahead and do it. (Witness the new significance procedure discussed at the Monte Carlo meeting.) Our problems are problems of principles.

7

Short-Cut Multiple Comparisons for Balanced Single and Double Classifications: Part 1, Results

T. E. Kurtz
R. F. Link
J. W. Tukey
D. L. Wallace

Methods are provided for the rapid analysis of data in balanced single and double classifications. Sums of ranges provide short-cut measures of variability with only slight loss of efficiency. Tables of factors are provided to convert these sums directly into widths of simultaneous confidence intervals for simple comparisons of group totals. Two illustrative examples are treated in detail. The proposed procedures are recommended for routine use in initial analyses, though not to the exclusion of more refined procedures.

The Collected Works of John W. Tukey (1993) Chapman and Hall, New York, NY.

Prepared in connection with research sponsored by the Office of Naval Research. Completion of the work has been facilitated by a grant from the National Science Foundation NSF-G-21058. Some of the computations were performed under Public Health Service facilities grant USPHS-FR-00013. Final revisions were aided by a grant (DA-ARO(D)- 31-214-G130) from the Army Research Office (Durham). Reproduction for any purpose of the United States Government is encouraged.

Now at Dartmouth College [TEK]

Now at The University of Chicago [DLW]

Reprinted from (1965) *Technometrics* **7**, 95-161 by permission of the American Statistical Association.

The paper includes a general discussion of multiple comparison procedures: when they should and should not be used, the importance of confidence procedures and their advantages over significance procedures, choice among multiple comparisons confidence procedures, choice and description of error rates. It does *not* attempt to compare multiple comparison and multiple decision procedures.

1. SUMMARY

Possible choices of measures of variability in balanced single and double classifications are discussed (Sections 3 to 5). Tables of factors which convert suitable sums of ranges (measuring variability in such patterns) into widths of simultaneous confidence intervals for all simple comparisons (expressed as differences of pairs of group totals) are presented, and illustrated by application to examples (Section 6 to 8). These factors can also be used to fix widths of simultaneous confidence intervals for all contrasts. [Preliminary versions of these tables have been presented elsewhere (Tukey, 1953a). The techniques of approximation used in calculating these tables are given in part 2 of this paper, to appear in another journal (Kurtz, et al. 1965).] The nature of the conclusions to be drawn from these specific illustrative examples is then further considered (Sections 9 to 11). The various questions which arise in the choice of any multiple comparison procedure are outlined, first as to broad approach (Sections 12 to 14), then as to relative emphasis (Sections 15 and 16), and finally as to error rates (Sections 17 and 18). The loss of efficiency in using ranges to estimate variability is discussed (Section 19). Recommendations are summarized (Section 20). The nature of the approximations used to compute the table is indicated (Section 21) and approximate formulas are given that are useful (i) for approximately representing the content of the tables on stored-program computers, and (ii) for extending the scope of the tables (Section 22).

The appropriate tables for the various situations treated as follows (two alternative procedures are provided for double classifications):

Confidence Coefficient (Two-Sided)	Level of Significance (Two-Sided)	Factors for Single Classifications	Factors for Double Classifications "Easy"	Factors for Double Classifications "Intermediate"
95%	5%	Table 4 (page 366)	Table 6 (page 372)	Table 8 (page 384)
99%	1%	Table 5 (page 369)	Table 7 (page 378)	Table 9 (page 392)

2. INTRODUCTION

Many sets of observations fall into, or can be analyzed as if they fell into, the simple pattern conventionally called "balanced single classifications," or "balanced double classifications." Data are in a balanced single classification when each of a number of treatments, sets of conditions, varieties of wheat or raw material, etc. is represented by an equal number of observations, and when there is no natural correspondence between individual observations across "treatments."

A set of data is in a balanced (unreplicated) double classification when it is appropriately regarded as arrayed in rows and columns, one observation for each combination of a row and a column, and when it is reasonable to contemplate that additive shifts describe row-to-row, or column-to-column, differences in behavior with adequate precision. Sometimes, as in conventional randomized-block experiments, one classification (say rows) is into the "treatments" being studied, while the other classification is into "blocks," i.e. into sets of related "experimental units." Alternatively, as in simple factorial experiments, the individual observations may be defined by combinations of "treatments," one from each of two classifications according to treatments, varieties, etc. In either case, the "individual observations" may in fact be (equally variable) means of several results each falling into the same cell.

When only two "treatments" are being compared the situation is as simple as possible. Not only is this the situation in which Student's t was first illustrated ('Student', 1908) but it is the situation in which Student's t is least affected by deviations from normality of distribution (Gayen, 1949). In many practical circumstances, ease of analysis is more important than a slight difference in stringency of conclusion. The use of ranges to estimate underlying variability in the two-treatment situation was proposed by various authors and made effective by Lord (1947), who showed how very slight the loss in stringency really was (Lord, 1950).

When more than two "treatments" are to be compared, the situation is more complex. Fisher's original solution (Fisher, 1925, Section 40), making an overall F-test among all "treatments" of a given classification, evolved through the years in two quite distinct directions. The first direction involved sharpening the possible inferences by concentrating action on the particular groups of degrees of freedom, usually a very few single degrees of freedom, corresponding to separate and specific purposes of the investigator. Used in moderation, this procedure has often been helpful. It was followed by a tendency to make undue use of the apparent freedom of choice thus made available, by "spending" a separate 5% error rate on each of many single degrees of freedom, without regard to whether these degrees of freedom were meaningfully related to the investigator's purposes. Partially in reaction to such spendthrift actions, the second direction involved the development of a variety of "multiple-comparison procedures" in which comparisons between all pairs of treatments, at least, could be made while the total probabilities of false significance or false confidence were controlled at well-understood levels. Experiments may often require two sorts of examination, one as to what is indicated or suggested, the other as to what is reasonably established.

Today, many experiments are best analyzed by methods involving specified single degrees of freedom (including regression). Others, particularly experiments intended to discover "what is going on," are naturally analyzed by looking at all comparisons. (Even in experiments with clearly defined, limited objectives, it is difficult, and often unwise, to resist searching the data for surprises.) This latter analysis is well carried out by one sort (Tukey, 1953b) of multiple-comparisons procedure that obtains its numerical factors from the distribution of the studentized range, and that produces simultaneous confidence statements about all linear contrasts among "treatments," focusing its primary attention on simple comparisons between pairs of "treatments."

Again ease of calculation is often more important than slight differences in stringency of conclusion. Again ranges can be used to assess the underlying variability effectively. The two parts of the present paper discuss the development and use of multiple-comparison procedures using ranges. (The use of these procedures has been illustrated (Link and Wallace, 1952; Tukey, 1953a) with preliminary versions of the present tables, which have been reproduced in part by several authors (Duncan, 1952; Federer, 1955; Mosteller and Bush, 1954; Natrella, 1963). Neither tables of the present quality nor a description of their construction have previously been available.) Part 1 discusses the problem, presents the necessary tables, illustrates their use, and sets up the mathematical problems to which the tables are solutions. Part 2 (to appear elsewhere, Kurtz et al. 1965) treats the problems of approximation underlying the computation of the tables.

CHOICE IN MEASURING VARIABILITY

3. CHOICES IN A BALANCED SINGLE CLASSIFICATION

In a single-classification analysis-of-variance situation it is usual to assume that (observation) = (column contribution) + (individual contribution)

where the "treatments" are specified in terms of columns, and where the individual contributions may be thought of as all coming randomly from some population (perhaps a population of errors of measurements, perhaps a population of variations in behavior of the actual items measured). Two sorts of quantities need to be calculated from such data:

(1) those which indicate the underlying column contributions; and

(2) a measure of the typical size of the individual contributions.

The latter is needed so that the apparent behavior of the column contributions can be wisely interpreted, making due allowance for the perturbing effects of the individual contributions.

We are free, at costs of labor in computation and of possible losses in accuracy or flexibility, to adopt any of a number of measures for either individual column behavior or residual variability. A few possibilities are illustrated in Table 1, using warp breakage rates for 9 replications on each of 6 yarns, as reported by Tippett (1950, p. 106).

As indicators of column behavior we have shown totals, means, and medians. As indicators of variability we have shown, both separately for individual columns and in total, the sums of squares of deviations (*SSD*), the usual quadratic estimate (s^2) of the variance of the individual contributions, and the range of each column.

If the reader will redo the arithmetic for any one column, he will conclude, we believe, that the total range is much easier to compute than the usual estimate s^2, and that there is little difference between the computations of totals, means, and medians.

Ease of computation is, of course, not the whole story. When the individual contributions are sampled from a normal population, we can calculate the relative qualities of the various choices. If we have 9 replications per column, as in the example, medians are approximately as useful as the means of only 6 entries would have been. This waste exceeds any gain in computation and it seems reasonable in many instances to use either totals or means.

At first glance, it might seem that we would face the same situation with measures of variability. While in samples of 9 the range is not too bad an estimator of variability, it is widely known that "the range has very low efficiency for large samples." This sounds as if using the range could be dangerous! If we were estimating the variability coming from individual contributions for its own sake, this could be true. But we are not doing this. We are estimating variability as a guide to judgment about indicated column behavior. For this purpose, as an observant reader of the *t*-table

Table 1

Various Possible Ways of Approaching Tippett's (1950, p. 106) Warp-Breakage-Rate Data for Apparent Behavior and Variability

	Yarns							
	AL	AM	AH	BL	BM	BH		
	26	18	36	27	42	20		
	30	21	21	14	26	21		
	54	29	24	29	19	24		
	25	17	18	19	16	17		
	70	12	10	29	39	13		
	52	18	43	31	28	15		
	51	35	28	41	21	15		
	26	30	15	20	39	16		
	67	36	26	44	29	28		
Totals	401	216	221	254	259	169		
Means	44.6	24.0	24.6	28.2	28.8	18.8		
Medians	51	21	24	29	28	17		
*SSD**	2620.2	600.00	844.2	777.6	711.6	191.6	5745.2	= total
s^2	327.5	75.0	105.5	97.2	88.9	23.9	119.7	= mean
Range	45	24	33	30	26	15	173	= total

* SSD = Sum of squares of deviations
$= \Sigma\,(\text{entries})^2 - (\text{total})^2/(\text{number of entries})$
$s^2 = (SSD)/(\text{one less than number of entries})$.

would conjecture, we do not lose much from a rather poor estimate of variability.

How are we to judge the importance of precision in the assessment of variability? Clearly, in terms of the relative tightness of the statements about *column contributions* which result. By this standard, the total range is only slightly poorer than the conventional s^2, and the difference becomes smaller as the number of observations increase. (For the ranges of larger groups do estimate variability more precisely per group, even if less precisely per observation.) Numerical details are given in Section 19. Thus, as far as efficiency is concerned, we can afford to use the total range as our measure of residual variability in almost any situation. For further discussion of related topics, see Lord (1950), Pearson (1950) and David (1953).

Before completing our choice by specifying the procedure by which the uncertainties of the comparisons are to be described, we shall turn to the choice of measures of variability in a double classification.

4. CHOICES IN A BALANCED DOUBLE CLASSIFICATION

The simple models for a double classification have all followed the pattern

(observation) = (column contribution) +

(row contribution) + (individual contribution)

where varying assumptions have been made about the distribution of the individual contributions and their relationship to the row and column contributions. These differences between assumptions are of particular importance once methods are even slightly more complex, but the simplicity of a balanced double classification with one entry per cell allows us to pass them by here without detailed attention. We shall act as if the individual contributions were a random sample from a single population, and shall treat this population as normally distributed whenever necessary.

We need to choose three sorts of summary quantities now:

(1) indicators of column-contribution behavior;

(2) indicators of row-contribution behavior;

(3) as indicator of residual variability.

We then want to use the last to aid us in making wise judgments about differences between column contributions, or between row contributions. We should like a certain sort of sterility for our indicators. If we raise all the observed values in one row by ten, a hundred, a thousand, or even a million units, we should like the indicated *differences* between column contributions to be unchanged, since this change in values is exactly that corresponding to a change in the corresponding *row* contribution.

This need of keeping column indicator differences free of contamination by row contributions, and vice versa, immediately forces us to give up medians. Only some sort of weighted total or mean will give the proper sterility in a simple way. In some circumstances unequal weights will be desirable, but the simplest, and most widely useful choice is the unweighted (i.e. equally weighted) total or mean. We make this choice.

There remains a choice of the indicator of residual variability. The classical procedure is to calculate the sum of squares, and thence the mean square, for interaction (also called for discrepance, for individuals, for residuals, for error, etc.) whose square root is again called s. While quite manageable, this is the hardest part of the arithmetic, and in many situations there is again a place for an easy substitute.

At this point we have to make up our mind about the need for "objectivity" in our procedure. The calculation of total range in the single-classification method is not affected by rearranging columns, or by rearranging entries in a column. The simplest calculation of a total range of differences for the two-way situation *does depend upon* the order of rows and columns among themselves. The result of a somewhat less simple calculation does not depend on order, but does depend on which is called "rows" and which "columns." General principles do not seem to give an unequivocal answer as to which of these prefer, so we present methods of analysis based on both.

The "easy" method involves taking differences between successive entries in a row (*including* between the last and first entries), and then taking ranges down the columns of differences. It is illustrated in Table 2 for data given by Snedecor (1948) on the amount of lactic acid (milligrams per gram) found, during five successive periods, in alfalfa silage treated in four ways. This method gives a result depending on the order in which the columns are arranged. Two orders are treated in Table 2, and lead to total ranges of differences of 76.8 and 101.7, respectively. The only other essentially different arrangement of the columns is in the order I-II-IV-III-I, which gives a total range of differences of 80.7. The variation of total range of differences with the order of the columns is rather large in this example. We shall return to a discussion of this point later. (cf. Section 11.)

The "easy" method might have been defined to use differences between fewer or more pairs of columns than those actually used. Tables could be prepared for any choice. Our choice aims at a balance of efficiency of estimation and convenience of calculation.

In the "intermediate" method, we calculate the mean of each row, and then form a new table giving

$$\text{(original entry)} - \text{(row mean)}.$$

The ranges of these differences are then taken in each column, and the results totaled. This is done for the same example in Table 3. The total range of these differences, which is not directly comparable with the total

Table 2

The "Easy" Calculation of an Indicator of Residual Variability in Snedecor's (1948) Data on Lactic Acid in Alfalfa Silage

Two alternative calculations, for two alternative orderings of columns are shown in detail. Differences between values in adjacent columns are shown in parentheses.

Alternative 1 (Order I-II-III-IV-I)
Treatment

Period	I	(II-I)	II	(III-II)	III	(IV-III)	IV	(I-IV)	Row Totals
1	13.4	(+2.6)	16.0	(−1.6)	14.4	(+5.6)	20.0	(−6.6)	63.8
2	37.5	(+5.2)	42.7	(−13.4)	29.3	(+5.2)	34.5	(+3.0)	144.0
3	65.2	(−10.3)	54.9	(−18.5)	36.4	(+3.3)	39.7	(+25.5)	196.2
4	60.8	(−3.7)	57.1	(−18.0)	39.1	(−0.4)	38.7	(+22.1)	195.7
5	37.7	(+11.5)	49.2	(−9.8)	39.4	(+0.3)	39.7	(−2.0)	166.0
Column Totals	214.6		219.9		158.6		172.6		
Range of Differences		21.8		16.9		6.0		32.1	76.8 = Total range

Alternative 2 (Order I-III-II-IV-I)
Treatment

	I	(III-I)	III	(II-III)	II	(IV-II)	IV	(I-IV)	Row Totals
1	13.4	(+1.0)	14.4	(+1.6)	16.0	(+4.0)	20.0	(−6.6)	63.8
2	37.5	(−8.2)	29.3	(+13.4)	42.7	(−8.2)	34.5	(+3.0)	144.0
3	65.2	(−28.8)	36.4	(+18.5)	54.9	(−15.2)	39.7	(25.5)	196.2
4	60.8	(−21.7)	39.1	(+18.0)	57.1	(−18.4)	38.7	(22.1)	195.7
5	37.7	(+1.7)	39.4	(+0.8)	49.2	(−9.5)	39.7	(−2.0)	166.0
Column Totals	214.6		158.6		219.9		172.6		
Range of Differences		30.5		16.9		22.4		32.1	101.9 = Total range

range of column-to-column differences, comes out to be 53.4. This value is independent of rearrangements of either rows or columns.

An interchange of rows with columns will change the result of either method. If we desire our results to be independent of the choice, it is always possible to do the essential arithmetic of a two-way analysis of variance, calculating the interaction sum of squares and thence a value of s. Thus three sorts of multiple-comparison procedures may be considered for

Table 3

The "Intermediate" Calculation of an Indicator of Residual Variability for Data on Lactic Acid in Alfalfa Silage

Period	(Treatment) I	II	III	IV	Totals	Means
	(original values)					
1	13.4	16.0	14.4	20.0	63.8	16.0
2	37.5	42.7	29.3	34.5	144.0	36.0
3	65.2	54.9	36.4	39.7	196.2	49.0
4	60.8	57.1	39.1	38.7	195.7	48.9
5	37.7	49.2	39.4	39.7	166.0	41.5
Totals	214.6	219.9	158.6	172.6		
	(Differences from row means)				(Sum of differences)	
1	−2.6	+0.0	−1.6	+4.0	(−0.2)	
2	+1.5	+6.7	−6.7	−1.5	(+0.0)	
3	+16.2	+5.9	−12.6	−9.3	(+0.2)	
4	+11.9	+8.2	−9.8	−10.2	(+0.1)	
5	−3.8	+7.7	−2.1	−1.8	(+0.0)	
Column ranges of differences	20.0 +	8.2 +	11.0 +	14.2 =	53.4	= Total range

the double-classification situation: one based upon s, one based upon the "intermediate" indicator of variability, one based upon the "easy" indicator of variability.

In Section 20 we set forth recommendations for choosing among procedures. Here we note one — that the "easy" procedure for double classifications is worthy of routine application as an initial analysis, though not to the exclusion of more refined procedures.

5. THE NEXT STEP

There remain choices of nature and form for the procedures by which these indicators are used. We postpone discussion of these choices to Sections 12-18, and describe here only the final choice.

If $q\ (c, v\ ;5\%)$ designates the upper 5% point of the studentized range, where the range of c items is studentized by an estimate of variability using v degrees of freedom, then, in the single-classification situation with r observations in each of c columns, the statement, made simultaneously for *all* pairs (i, j):

$$\mu_i - \mu_j \text{ differs from } m_i - m_j \text{ by at most } (s/\sqrt{r}) \cdot q\ (c, c\ (r-1); 5\%) \qquad (*)$$

where

$$\mu_i = \text{actual contribution from column i}$$

$$m_i = \text{observed mean for column i}$$

will hold in 19 out of 20 single-classification experiments. This statement about simple comparisons (differences of pairs) among column means has a logical consequence for every contrast among means, i.e. for every linear combination of means, the sum of whose coefficients is zero. In the 19 experiments out of 20 in which *all* statements (*) hold, *all* logical consequences of these statements must also hold. Thus we may have 95% confidence in the *simultaneous* holding of many more statements, one about each contrast. Provided that $\Sigma\, b_i = 0$, so that $1/2\ \Sigma\ |\ b_i\ | = \Sigma$ (positive b_i's), these logically consequent statements take the form

$$\Sigma\, b_i \mu_i \text{ differs from } \Sigma\, b_i m_i \text{ by no more than} \qquad (**)$$

$$\left(1/2\ \Sigma\ |\ b_i\ |\right) \cdot \frac{s}{\sqrt{r}}\, q\ (c, c\ (r-1); 5\%).$$

When (i) the total range is used as an indicator of variability in place of s and (ii) statements are made about differences of column totals instead of about differences of column means (which has some arithmetic convenience) the factor $(1/\sqrt{r})q\ (c, c\ (r-1); 5\%)$ has to be replaced by a new factor, whose values may be obtained from Table 4 or Table 5.

In the case of a double classification with c columns and r rows, the whole discussion proceeds similarly, so far as either column contributions alone or row contributions alone are concerned. In 19 out of 20 experiments all statements about row contributions will be correct. Thus in somewhat more than 18 experiments out of 20, all of both sets of statements will be correct.

As factors multiplying $\left(\frac{1}{2}\Sigma \mid b_i \mid\right) \cdot s_i$ we should use

$$\frac{1}{\sqrt{r}}\, q\,(c,\,(c-1)(r-1);\,5\%)$$

and

$$\frac{1}{\sqrt{c}}\, q\,(r,\,(c-1)(r-1);\,5\%)$$

for column means and row means, respectively. To find the corresponding factors multiplying either sort of total range of differences to yield allowances for contrasts among totals we shall need to use pairs of values from an appropriate new table (Tables 6 through 9).

THE TABLES AND THEIR APPLICATION

6. SINGLE CLASSIFICATIONS

The calculation of the total range for the various groups has already been illustrated, in Table 1, for the warp-breakage-rate data. Most of the computations in that table are not needed; only those leading to the following summary are relevant to the use of Table 4:

Treatment	=	*AL*	*AM*	*AH*	*BL*	*BM*	*BH*
Total breakage	=	401	216	221	254	259	169

Number of groups = Number of ranges = 6,
Number per range = Number of replications = 9,
Total range = 173.

Reference to Table 4 shows that the factor by which the *total* range must be multiplied to reach an allowance for *totals* is .71. Hence

allowance for totals = (.71)(173) = 123.

Thus we are to believe every simple difference ± 123. For example:

$$
\begin{aligned}
(AL) - (BH) &= 401 - 169 \pm 123 = 232 \pm 123 \\
&= \text{between 109 and 355 in total,} \\
(AL) - (BL) &= 401 - 254 \pm 123 = 147 \pm 123 \\
&= \text{between 24 and 270 in total,} \\
(AM) - (BM) &= 216 - 259 \pm 123 = -43 \pm 123 \\
&= \text{between } -166 \text{ and } +80 \text{ in total.}
\end{aligned}
$$

These statements are about *totals*. If we want to change to statements about means we have only to divide through by the number of replications. In this example, dividing by 9 yields

$$
\begin{aligned}
(AL) - (BH) &= \text{between 12 and 38 in mean,} \\
(AL) - (BL) &= \text{between 3 and 30 in mean,} \\
(AM) - (BM) &= \text{between } -18 \text{ and } +9 \text{ in mean,}
\end{aligned}
$$

three of the $^1/_2\,(6)(5) = 15$ possible statements about possible differences in *means*.

Table 4 was set up to give allowances for totals rather than means for two reasons:

(1) in many real examples we can interpret statements about totals as well as those about means — when this is so we can save a minute amount of arithmetic by not calculating the means;

(2) in the most used portion of the table, the values are more nearly constant and nearer unity in terms of totals — this tends to avoid possible difficulties with decimal points in computation and allows a more compact table.

In addition to the statements about simple differences, we can of course make statements about such contrasts as

$$
\begin{aligned}
2(AL) - (BH) - (BL) = 2(401) - 169 - 254 \pm 2(123) = \\
379 = 246 = \text{between 133 and 625 in total.}
\end{aligned}
$$

This statement is easily seen to be a logical consequence of the first two statements about simple differences, or it may be obtained formally by

Table 4

Factors from Total Range to Totals in a One-Way Classification*.
(Entries are to be Multiplied by the Sum of the Ranges Within Groups to Obtain Allowances for Group Totals.)
5% ERROR RATE**
Number of groups = Number of ranges

Number in group = Number per range	2	3	4	5	6	7	8	9	10	11	12	13	14	15
2	3.43	2.35	1.74	1.39	1.15	.99	.87	.77	.70	.63	.58	.54	.50	.47
3	1.90	1.44	1.14	.94	.80	.70	.62	.56	.51	.47	.43	.40	.38	.35
4	1.62	1.25	1.01	.84	.72	.63	.57	.51	.47	.43	.40	.37	.35	.33
5	1.53	.1.19	.96	.81	.70	.61	.55	.50	.45	.42	.39	.36	.34	.32
6	1.50	1.17	.95	.80	.69	.61	.55	.49	.45	.42	.39	.36	.34	.32
7	1.49	1.17	.95	.80	.69	.61	.55	.50	.45	.42	.39	.36	.34	.32
8	1.49	1.18	.96	.81	.70	.62	.55	.50	.46	.42	.39	.37	.35	.33
9	1.50	1.19	.97	.82	.71	.62	.56	.51	.47	.43	.40	.37	.35	.33
10	1.52	1.20	.98	.83	.72	.63	.57	.52	.47	.44	.41	.38	.36	.34
11	1.54	1.22	.99	.84	.73	.64	.58	.52	.48	.44	.41	.38	.36	.34
12	1.56	1.23	1.01	.85	.74	.65	.58	.53	.49	.45	.42	.39	.37	.35
13	1.58	1.25	1.02	.86	.75	.66	.59	.54	.49	.46	.42	.40	.37	.35
14	1.60	1.26	1.03	.87	.76	.67	.60	.55	.50	.46	.43	.40	.38	.36
15	1.62	1.28	1.05	.89	.77	.68	.61	.55	.51	.47	.44	.41	.38	.36
16	1.64	1.30	1.06	.90	.78	.69	.62	.56	.52	.48	.44	.41	.39	.37
17	1.66	1.32	1.08	.91	.79	.70	.63	.57	.52	.48	.45	.42	.39	.37
18	1.68	1.33	1.09	.92	.80	.71	.64	.58	.53	.49	.46	.43	.40	.38
19	1.70	1.35	1.10	.93	.81	.72	.64	.59	.54	.50	.46	.43	.41	.38
20	1.72	1.36	1.12	.95	.82	.73	.65	.59	.54	.50	.47	.44	.41	.39
30	1.92	1.52	1.24	1.05	.91	.81	.73	.66	.60	.56	.52	.49	.46	.43
40	2.08	1.66	1.35	1.14	.99	.88	.79	.72	.66	.61	.57	.53	.50	.47
50	2.23	1.77	1.45	1.22	1.06	.94	.85	.77	.71	.65	.61	.57	.53	.50
100	2.81	2.23	1.83	1.55	1.34	1.19	1.07	.97	.89	.83	.77	.72	.67	.64
200	3.61	2.88	2.35	1.99	1.73	1.53	1.38	1.25	1.15	1.06	.99	.93	.87	.82
500	5.15	4.10	3.35	2.84	2.47	2.19	1.97	1.79	1.64	1.52	.1.42	1.32	1.24	1.17
1000	6.81	5.43	4.44	3.77	3.28	2.90	2.61	2.37	2.18	2.02	1.88	1.76	1.65	1.56

* Also the factors from [(total range)/(number in group)] to allowances for means in a one-way classification.

** All confidence statements will be correct in each of 95% of one-way classifications to which applied.

Table 4 (Cont'd)

Factors from Total Range to Totals in a One-Way Classification*.
(Entries are to be Multiplied by the Sum of the Ranges Within Groups to Obtain Allowances for Group Totals.)
5% ERROR RATE**
Number of groups = Number of ranges

Number in group = Number per range	16	17	18	19	20	30	40	50	100	200	500	1000
2	.443	.418	.396	.376	.358	.245	.187	.151	.079	.042	.0177	.0093
3	.335	.317	.301	.287	.274	.189	.146	.119	.063	.033	.0143	.0075
4	.310	.294	.279	.266	.254	.177	.136	.112	.060	.0316	.0136	.0071
5	.303	.287	.273	.260	.249	.173	.134	.110	.059	.0312	.0134	.0070
6	.302	.287	.273	.260	.249	.174	.135	.110	.059	.0314	.0135	.0071
7	.304	.289	.275	.262	.251	.175	.136	.111	.060	.032	.0137	.0072
8	.308	.292	.278	.265	.254	.178	.138	.113	.061	.032	.0139	.0073
9	.312	.297	.282	.269	.258	.180	.140	.115	.062	.033	.0141	.0074
10	.317	.301	.287	.274	.262	.183	.142	.117	.063	.033	.0143	.0075
11	.322	.306	.291	.278	.266	.186	.145	.119	.064	.034	.0416	.0077
12	.327	.311	.296	.282	.270	.189	.147	.121	.065	.034	.0148	.0078
13	.332	.316	.300	.287	.274	.192	.149	.122	.066	.035	.0151	.0079
14	.337	.320	.305	.291	.279	.195	.152	.124	.067	.036	.0153	.0080
15	.342	.325	.310	.295	.283	.198	.154	.126	.068	.036	.0156	.0082
16	.348	.330	.314	.300	.287	.201	.156	.128	.069	.037	.0158	.0083
17	.352	.335	.319	.304	.291	.204	.158	.130	.070	.037	.0160	.0084
18	.357	.339	.323	.308	.295	.207	.161	.132	.071	.038	.0162	.0085
19	.362	.344	.327	.312	.299	.210	.163	.134	.072	.038	.0165	.0087
20	.367	.348	.332	.317	.303	.212	.165	.135	.073	.039	.0167	.0088
30	.408	.387	.369	.352	.337	.237	.184	.151	.081	.043	.0187	.0098
40	.444	.422	.402	.384	.367	.258	.201	.165	.089	.047	.0204	.0107
50	.476	.453	.431	.412	.394	.277	.216	.177	.095	.051	.0219	.0115
100	.60	.573	.546	.521	.499	.351	.273	.224	.121	.064	.0277	.0146
200	.78	.74	.70	.67	.64	.454	.353	.290	.156	.083	.0358	.0188
500	1.11	1.06	1.01	.96	.92	.65	.504	.414	.223	.119	.0513	.0269
1000	1.47	1.40	1.33	1.27	1.22	.86	.669	.549	.296	.158	.068	.0357

* Also the factors from [(total range)/(number in group)] to allowances for means in a one-way classification.

** All confidence statements will be correct in each of 95% of one-way classifications to which applied.

Table 4a

Factors from Mean Range to Column Totals in a One-Way Classification.* (Entries are to be Multiplied by the Mean of the Ranges Within Groups to Obtain Allowances for Group Totals.)
WARNING: These instructions for using this table differ from those for using Table 4.
5% ERROR RATE**
Number of groups = Number of ranges

Number in group = Number per range	10	15	20	30	40	50	100	200	500	1000
2	7.0	7.1	7.2	7.3	7.5	7.6	7.9	8.3	8.8	9.3
3	5.1	5.3	5.5	5.7	5.8	6.0	6.3	6.7	7.2	7.5
4	4.7	4.9	5.1	5.3	5.5	5.6	6.0	6.3	6.8	7.1
5	4.5	4.8	5.0	5.2	5.4	5.5	5.9	6.2	6.7	7.0
6	4.5	4.8	5.0	5.2	5.4	5.5	5.9	6.3	6.7	7.1
7	4.5	4.8	5.0	5.3	5.4	5.6	6.0	6.4	6.8	7.2
8	4.6	4.9	5.1	5.3	5.5	5.6	6.1	6.4	6.9	7.3
9	4.6	4.9	5.1	5.4	5.6	5.7	6.2	6.6	7.1	7.4
10	4.7	5.0	5.2	5.5	5.7	5.8	6.3	6.7	7.2	7.5
11	4.8	5.1	5.3	5.6	5.8	5.9	6.4	6.8	7.3	7.7
12	4.9	5.2	5.4	5.7	5.9	6.0	6.5	6.9	7.4	7.8
13	4.9	5.3	5.5	5.8	6.0	6.1	6.6	7.0	7.5	7.9
14	5.0	5.3	5.6	5.9	6.1	6.2	6.7	7.1	7.7	8.0
15	5.1	5.4	5.6	5.9	6.2	6.3	6.8	7.2	7.8	8.2
16	5.1	5.5	5.7	6.0	6.2	6.4	6.9	7.3	7.9	8.3
17	5.2	5.6	5.8	6.1	6.3	6.5	7.0	7.4	8.0	8.4
18	5.3	5.6	5.9	6.2	6.4	6.6	7.1	7.5	8.1	8.5
19	5.4	5.7	6.0	6.3	6.5	6.7	7.2	7.7	8.2	8.7
20	5.4	5.8	6.0	6.4	6.6	6.8	7.3	7.8	8.3	8.8
30	6.0	6.5	6.7	7.1	7.4	7.6	8.1	8.7	9.3	9.8
40	6.6	7.0	7.3	7.8	8.0	8.2	8.9	9.5	10.2	10.7
50	7.1	7.5	7.9	8.3	8.6	8.9	9.5	10.2	10.9	11.5
100	8.9	9.6	10.0	10.5	10.9	11.2	12.1	12.9	13.9	14.6
200	11.5	12.3	12.9	13.6	14.1	14.5	15.6	16.6	17.9	18.8
500	16.4	17.6	18.4	19.5	20.2	20.7	22.3	23.8	25.6	26.9
1000	21.8	23.4	24.4	25.8	26.8	27.5	29.6	31.6	34.0	35.7

* Also the factors from [(mean range)/(number in group)] to allowances for means in a one-way classification.

** All confidence statements will be correct in each of 95% of one-way classifications to which applied.

Table 5

Factors from Total Range to Totals in a One-Way Classification*.
(Entries are to be Multiplied by the Sum of the Ranges Within Groups to Obtain Allowances for Group Totals.)
1% ERROR RATE**
Number of groups = Number of ranges

Number in group = Number per range	2	3	4	5	6	7	8	9	10	11	12	13	14	15
2	7.92	4.32	2.84	2.10	1.66	1.38	1.17	1.02	.91	.82	.74	.68	.63	.58
3	3.14	2.12	1.57	1.25	1.04	.89	.78	.69	.62	.57	.52	.48	.45	.42
4	2.48	1.74	1.33	1.08	.91	.78	.69	.62	.56	.51	.47	.44	.41	.38
5	2.24	1.60	1.24	1.02	.86	.75	.66	.59	.54	.49	.46	.42	.40	.37
6	2.14	1.55	1.21	.99	.85	.74	.65	.59	.53	.49	.45	.42	.39	.37
7	2.10	1.53	1.20	.99	.84	.73	.65	.59	.53	.49	.45	.42	.39	.37
8	2.09	1.53	1.20	.99	.85	.74	.66	.59	.54	.49	.46	.43	.40	.37
9	2.09	1.54	1.21	1.00	.85	.75	.66	.60	.54	.50	.46	.43	.40	.38
10	2.10	1.55	1.22	1.01	.86	.76	.67	.61	.55	.51	.47	.44	.41	.38
11	2.11	1.56	1.23	1.02	.87	.76	.68	.61	.56	.51	.48	.44	.42	.39
12	2.13	1.58	1.25	1.04	.89	.78	.69	.62	.57	.52	.48	.45	.42	.40
13	2.15	1.60	1.26	1.05	.90	.79	.70	.63	.58	.53	.49	.46	.43	.40
14	2.18	1.62	1.28	1.06	.91	.80	.71	.64	.58	.54	.50	.46	.43	.41
15	2.20	1.63	1.30	1.08	.92	.81	.72	.65	.59	.54	.50	.47	.44	.41
16	2.22	1.65	1.31	1.09	.93	.82	.73	.66	.60	.55	.51	.48	.45	.42
17	2.25	1.67	1.33	1.10	.95	.83	.74	.67	.61	.56	.52	.48	.45	.43
18	2.27	1.69	1.34	1.12	.96	.84	.75	.68	.62	.57	.53	.49	.46	.43
19	2.30	1.71	1.36	1.13	.97	.85	.76	.68	.62	.57	.53	.50	.46	.44
20	2.32	1.73	1.38	1.14	.98	.86	.77	.69	.63	.58	.54	.50	.47	.44
30	2.59	1.95	1.54	1.27	1.09	.96	.85	.77	.70	.65	.60	.56	.52	.49
40	2.80	2.11	1.66	1.38	1.18	1.04	.93	.84	.76	.70	.65	.61	.57	.54
50	2.99	2.25	1.78	1.48	1.27	1.11	.99	.90	.82	.75	.70	.65	.61	.57
100	3.74	2.83	2.24	1.86	1.60	1.40	1.25	1.13	1.03	.95	.88	.82	.77	.73
200	4.79	3.63	2.88	2.39	2.06	1.81	1.61	1.46	1.33	1.23	1.14	1.06	.99	.94
500	6.81	5.16	4.10	3.41	2.93	2.58	2.30	2.08	1.90	1.75	1.62	1.52	1.42	1.34
1000	9.01	6.83	5.42	4.52	3.88	3.41	3.05	2.76	2.52	2.32	2.15	2.01	1.88	1.77

* Also the factors from [(total range)/(number in group)] to allowances for means in a one-way classification.

** All confidence statements will be correct in each of 99% of one-way classifications to which applied.

Table 5 (Cont'd)

Factors from Total Range to Totals in a One-Way Classification*.
(Entries are to be Multiplied by the Sum of the Ranges Within
Groups to Obtain Allowances for Group Totals.)
1% ERROR RATE**
Number of groups = Number or ranges

Number in group = Number per range	16	17	18	19	20	30	40	50	100	200	500	1000
2	.54	.51	.480	.454	.430	.285	.214	.172	.088	.045	.0191	.0099
3	.39	.37	.352	.334	.318	.217	.165	.134	.070	.036	.0155	.0081
4	.36	.34	.323	.307	.293	.200	.153	.125	.066	.034	.0416	.0076
5	.35	.33	.314	.299	.285	.196	.151	.123	.065	.034	.0145	.0076
6	.35	.33	.313	.298	.284	.196	.151	.123	.065	.034	.0145	.0076
7	.35	.33	.314	.299	.286	.198	.152	.124	.066	.035	.0147	.0077
8	.35	.33	.318	.303	.289	.200	.154	.126	.066	.035	.0149	.0078
9	.36	.34	.322	.307	.293	.203	.156	.127	.068	.036	.0152	.0079
10	.36	.34	.327	.311	.297	.206	.159	.129	.069	.036	.0155	.0081
11	.37	.35	.332	.316	.302	.209	.161	.132	.070	.037	.0157	.0082
12	.37	.35	.337	.321	.306	.213	.164	.134	.071	.037	.0160	.0084
13	.38	.36	.342	.326	.311	.216	.166	.136	.072	.038	.0162	.0085
14	.39	.36	.347	.330	.316	.219	.169	.138	.073	.039	.0165	.0086
15	.39	.37	.352	.335	.320	.222	.171	.140	.074	.039	.0168	.0088
16	.40	.38	.357	.340	.325	.226	.174	.142	.075	.040	.0170	.0089
17	.40	.38	.362	.345	.329	.229	.176	.144	.077	.040	.0173	.0090
18	.41	.39	.367	.350	.334	.232	.179	.146	.078	.041	.0175	.0091
19	.41	.39	.372	.354	.338	.235	.181	.148	.079	.042	.0177	.0093
20	.42	.40	.376	.359	.343	.238	.184	.150	.080	.042	.0180	.0094
30	.46	.44	.419	.399	.381	.266	.205	.168	.089	.047	.0201	.0005
40	.51	.48	.456	.435	.415	.289	.223	.183	.097	.051	.0219	.0115
50	.54	.51	.489	.466	.446	.310	.240	.196	.104	.055	.0236	.0123
100	.69	.65	.62	.590	.564	.393	.304	.248	.132	.070	.0299	.0156
200	.88	.84	.80	.76	.73	.507	.392	.320	.171	.090	.0386	.0202
500	1.26	1.20	1.14	1.09	1.04	.73	.560	.458	.244	.129	.0552	.0289
1000	1.68	1.59	1.51	1.44	1.38	.96	.743	.608	.324	.171	.0732	.0383

* Also the factors from [(total range)/(number in group)] to allowances for means in a one-way classification.

** All confidence statements will be correct in each of 99% of one-way classifications to which applied.

Table 5a

Factors from Mean Range to Column Totals in a One-Way Classification*.
(Entries are to be Multiplied by the Mean of the Ranges Within
Groups to Obtain Allowances for Group Totals.)
WARNING: These instructions for using this table differ
from those for using Table 5.
1% ERROR RATE**
Number of groups = Number of ranges

Number in group = Number per range	10	15	20	30	40	50	100	200	500	1000
2	8.9	8.7	8.6	8.6	8.6	8.6	8.8	9.1	9.5	9.9
3	6.2	6.3	6.4	6.5	6.6	6.7	7.0	7.3	7.7	8.1
4	5.6	5.7	5.9	6.0	6.1	6.2	6.6	6.9	7.3	7.6
5	5.4	5.6	5.7	5.9	6.0	6.1	6.5	6.8	7.2	7.6
6	5.3	5.5	5.7	5.9	6.0	6.1	6.5	6.8	7.3	7.6
7	5.4	5.6	5.7	5.9	6.1	6.2	6.6	6.9	7.4	7.7
8	5.4	5.6	5.8	6.0	6.2	6.3	6.6	7.0	7.5	7.8
9	5.5	5.7	5.9	6.1	6.2	6.4	6.8	7.1	7.6	7.9
10	5.5	5.8	5.9	6.2	6.3	6.5	6.9	7.2	7.7	8.1
11	5.6	5.9	6.0	6.3	6.4	6.6	7.0	7.4	7.9	8.2
12	5.7	5.9	6.1	6.4	6.6	6.7	7.1	7.5	8.0	8.4
13	5.8	6.0	6.2	6.5	6.7	6.8	7.2	7.6	8.1	8.5
14	5.8	6.1	6.3	6.6	6.8	6.9	7.3	7.7	8.3	8.6
15	5.9	6.2	6.4	6.7	6.9	7.0	7.4	7.9	8.4	8.8
16	6.0	6.3	6.5	6.8	7.0	7.1	7.5	8.0	8.5	8.9
17	6.1	6.4	6.6	6.9	7.1	7.2	7.7	8.1	8.6	9.0
18	6.2	6.5	6.7	7.0	7.2	7.3	7.8	8.2	8.8	9.1
19	6.2	6.5	6.8	7.0	7.2	7.4	7.9	8.3	8.9	9.3
20	6.3	6.6	6.8	7.1	7.3	7.5	8.0	8.4	9.0	9.4
30	7.0	7.4	7.6	8.0	8.2	8.4	8.9	9.4	10.1	10.5
40	7.6	8.0	8.3	8.7	8.9	9.1	9.7	10.3	11.0	11.5
50	8.2	8.6	8.9	9.3	9.6	9.8	10.4	11.0	11.8	12.3
100	10.3	10.9	11.3	11.8	12.1	12.4	13.2	14.0	14.9	15.6
200	13.3	14.0	14.5	15.2	15.7	16.0	17.1	18.1	19.3	20.2
500	19.0	20.1	20.8	21.8	22.4	22.9	24.4	25.8	27.6	28.9
1000	25.2	26.6	27.5	28.8	29.7	30.4	32.4	34.3	36.6	38.3

* Also the factors from [(mean range)/(number in group)] to allowances for means in a one-way classification.

** All confidence statements will be correct in each of 99% of one-way classifications to which applied.

NW Quadrant of Table 6

Factors from Total Range of Differences in a Two-Way Classification to Allowances for Totals* (According to the "Easy" Method.)

5% ERROR RATE**

Entries are to be Multiplied by the Sum of Ranges of Differences between Adjacent Observations to Obtain Allowances for {Column Totals (Use Upper Entry). Row Totals (Use Lower Entry).

(Differences Taken Horizontally, Their Ranges Taken Vertically.)

Number of ranges = Number of columns

Number of differences per range = Number of rows	2	3	4	5	6	7	8	9	10	11	12	13	14	15
2	6.35	2.19	1.52	1.16	.94	.79	.69	.60	.54	.490	.448	.412	.382	.356
	6.35	1.96	1.39	1.12	.95	.84	.76	.70	.65	.609	.576	.547	.522	.500
3	1.96	1.14	.88	.72	.61	.53	.47	.42	.38	.346	.319	.297	.277	.260
	2.19	1.14	.90	.76	.67	.61	.56	.52	.49	.458	.435	.415	.398	.383
4	1.43	.96	.76	.63	.54	.47	.42	.38	.34	.313	.290	.270	.253	.238
	1.54	.93	.76	.65	.58	.53	.49	.45	.43	.404	.384	.367	.352	.339
5	1.27	.89	.71	.60	.51	.45	.40	.36	.33	.303	.281	.262	.245	.231
	1.28	.84	.69	.60	.53	.49	.45	.42	.40	.375	.357	.342	.328	.316
6	1.19	.87	.70	.58	.50	.44	.39	.36	.33	.300	.279	.260	.244	.229
	1.14	.78	.64	.56	.50	.46	.43	.40	.38	.356	.340	.325	.313	.301
7	1.16	.86	.69	.58	.50	.44	.40	.36	.33	.301	.280	.261	.245	.230
	1.06	.74	.62	.54	.48	.44	.41	.38	.36	.344	.328	.314	.302	.291
8	1.15	.86	.69	.58	.50	.44	.40	.36	.33	.304	.282	.263	.247	.233
	1.01	.71	.59	.52	.47	.43	.40	.37	.35	.334	.319	.306	.294	.283
9	1.15	.86	.70	.59	.51	.45	.40	.36	.33	.307	.286	.267	.250	.236
	.97	.69	.58	.51	.46	.42	.39	.36	.34	.327	.312	.299	.287	.277
10	1.15	.87	.71	.60	.51	.45	.41	.37	.34	.312	.289	.270	.254	.239
	.93	.67	.56	.50	.45	.41	.38	.36	.34	.321	.306	.294	.282	.272
11	1.16	.88	.71	.60	.52	.46	.41	.37	.34	.316	.294	.274	.257	.243
	.91	.66	.55	.49	.44	.40	.38	.35	.33	.316	.302	.289	.278	.268
12	1.16	.89	.72	.61	.53	.47	.42	.38	.35	.321	.298	.278	.261	.246
	.89	.65	.55	.48	.43	.40	.37	.35	.33	.312	.298	.286	.275	.265
13	1.17	.90	.73	.62	.54	.47	.42	.38	.35	.325	.302	.282	.265	.250
	.87	.64	.54	.47	.43	.39	.37	.34	.32	.308	.295	.282	.272	.262
14	1.19	.91	.74	.63	.54	.48	.43	.39	.36	.330	.307	.287	.269	.254
	.85	.63	.53	.47	.42	.39	.36	.34	.32	.305	.292	.280	.269	.260

* Lower entries are also factors from mean range of differences to allowances for *row* means; upper entries are also factors from [(total range)/(number of rows)] to allowances for *column* means.

** All confidence statements about ***column*** totals will be correct in each of 95% of two-way classifications to which applied: similar result holds for *row* totals, hence all statements about ***all*** totals will be correct in each of over 90% of all two-way classifications.

NE Quadrant of Table 6

Factors from Total Range of Differences in a Two-Way Classification to Allowances for Totals*. (According to the "Easy" Method.)

5% ERROR RATE**

Entries are to be Multiplied by the Sum of Ranges of Differences between Adjacent Observations to Obtain Allowances for {Column Totals (Use Upper Entry). / Row Totals (Use Lower Entry).}

(Differences Taken Horizontally, Their Ranges Taken Vertically.)

Number of ranges = Number of columns

Number of differences per range = Number of rows	16	17	18	19	20	30	40	50	100	200	500	1000
2	.334	.314	.297	.281	.267	.181	.137	.111	.057	.0297	.0126	.0066
	.481	.464	.448	.434	.421	.338	.288	.255	.177	.124	.078	.0550
3	.245	.232	.220	.209	.199	.137	.105	.086	.045	.0238	.0102	.0053
	.369	.356	.345	.335	.326	.263	.226	.201	.140	.099	.062	.0439
4	.224	.212	.202	.192	.184	.127	.098	.080	.042	.0224	.0096	.0050
	.327	.317	.307	.298	.290	.234	.202	.179	.126	.089	.056	.0395
5	.218	.207	.196	.187	.179	.124	.096	.078	.042	.0221	.0095	.0050
	.305	.295	.286	.278	.271	.219	.189	.168	.118	.083	.052	.0371
6	.217	.206	.196	.186	.178	.124	.096	.078	.042	.0222	.0096	.0050
	.291	.282	.273	.266	.258	.209	.180	.161	.113	.080	.050	.0355
7	.218	.207	.197	.187	.179	.125	.097	.079	.042	.0225	.0097	.0051
	.281	.272	.264	.257	.250	.203	.175	.156	.110	.077	.049	.0340
8	.220	.209	.199	.189	.181	.127	.098	.080	.043	.0228	.0098	.0052
	.274	.265	.257	.250	.243	.198	.170	.152	.107	.075	.048	.0037
9	.223	.212	.201	.192	.184	.128	.099	.082	.044	.0232	.0100	.0052
	.268	.260	.252	.245	.238	.194	.167	.149	.105	.074	.047	.0330
10	.226	.215	.204	.195	.186	.130	.101	.083	.044	.0236	.0101	.0053
	.263	.255	.248	.241	.234	.190	.164	.147	.103	.073	.046	.0325
11	.230	.218	.207	.198	.189	.132	.103	.084	.045	.0240	.0103	.0054
	.259	.251	.244	.237	.231	.188	.162	.145	.102	.072	.045	.0321
12	.233	.221	.211	.201	.192	.135	.104	.086	.046	.0244	.0105	.0055
	.256	.248	.241	.234	.228	.186	.160	.143	.101	.071	.045	.0317
13	.237	.224	.214	.204	.195	.137	.106	.087	.047	.0248	.0107	.0056
	.253	.246	.238	.232	.226	.184	.159	.142	.100	.070	.044	.0314
14	.240	.228	.217	.207	.198	.139	.108	.088	.047	.0252	.0108	.0057
	.251	.243	.236	.230	.224	.182	.157	.140	.099	.070	.044	.0312

* Lower entries are also factors from mean range of differences to allowances for *row* means; upper entries are also factors from [(total range)/(number of rows)] to allowances for *column* means.

** All confidence statements about *column* totals will be correct in each of 95% of two-way classifications to which applied: similar result holds for *row* totals, hence all statements about *all* totals will be correct in each of over 90% of *all* two-way classifications.

SW Quadrant of Table 6

Factors from Total Range of Differences in a Two-Way Classification to Allowances for Totals*. (According to the "Easy" Method.)

5% ERROR RATE**

Entries are to be Multiplied by the Sum of Ranges of Differences between Adjacent Observations to Obtain Allowances for {Column Totals (Use Upper Entry). Row Totals (Use Lower Entry).

(Differences Taken Horizontally, Their Ranges Taken Vertically.)

Number of ranges = Number of columns

Number of differences per range = Number of rows	2	3	4	5	6	7	8	9	10	11	12	13	14	15
15	1.20	.92	.75	.63	.55	.49	.44	.40	.362	.335	.311	.291	.273	.257
	.84	.62	.53	.46	.42	.39	.36	.34	.318	.303	.289	.277	.267	.257
16	1.21	.93	.76	.64	.56	.49	.44	.40	.367	.339	.315	.295	.277	.261
	.83	.61	.52	.46	.42	.38	.36	.33	.316	.300	.287	.275	.265	.255
17	1.22	.94	.77	.65	.56	.50	.45	.41	.372	.344	.320	.299	.281	.265
	.82	.61	.52	.46	.41	.38	.35	.33	.313	.298	.285	.273	.263	.254
18	1.24	.95	.78	.66	.57	.51	.45	.41	.377	.348	.324	.303	.284	.268
	.81	.60	.51	.45	.41	.38	.35	.33	.311	.296	.283	.271	.261	.252
19	1.25	.97	.79	.67	.58	.51	.46	.42	.382	.353	.328	.307	.288	.272
	.80	.60	.51	.45	.41	.37	.35	.33	.309	.294	.281	.270	.260	.250
20	1.26	.98	.80	.67	.59	.52	.46	.42	.387	.357	.332	.310	.292	.275
	.79	.59	.50	.45	.40	.37	.35	.33	.308	.293	.280	.268	.258	.249
30	1.40	1.09	.89	.75	.65	.58	.52	.47	.430	.397	.369	.350	.324	.306
	.75	.57	.49	.43	.39	.36	.33	.314	.297	.283	.270	.259	.250	.241
40	1.51	1.18	.96	.81	.71	.63	.56	.51	.468	.432	.402	.380	.353	.333
	.73	.56	.47	.42	.38	.35	.33	.307	.291	.277	.265	.254	.244	.236
50	1.61	1.27	1.03	.87	.76	.67	.60	.54	.501	.463	.431	.403	.378	.357
	.71	.55	.47	.41	.37	.35	.32	.302	.286	.273	.261	.250	.241	.232
100	2.02	1.59	1.30	1.10	.95	.85	.76	.69	.63	.585	.544	.509	.478	.451
	.67	.52	.45	.40	.36	.33	.309	.291	.275	.262	.251	.241	.232	.224
200	.258	2.04	1.67	1.41	1.23	1.09	.98	.89	.82	.75	.70	.66	.62	.582
	.64	.30	.43	.38	.35	.32	.300	.282	.267	.255	.244	.234	.225	.217
500	3.67	2.91	2.38	2.02	1.75	1.55	1.40	1.27	1.16	1.08	1.00	.94	.88	.83
	.61	.49	.42	.37	.34	.312	.291	.274	.260	.247	.237	.227	.219	.211
1000	4.85	3.85	3.15	2.67	2.32	2.06	1.85	1.68	1.54	1.43	1.33	1.24	1.17	1.10
	.60	.48	.41	.36	.33	.306	.286	.269	.255	.243	.233	.224	.215	.208

* Lower entries are also factors from mean range of differences to allowances for *row* means; upper entries are also factors from [(total range)/number of rows)] to allowances for *column* means.

** All confidence statements about *columns* totals will be correct in each of 95% of two-way classifications to which applied: similar result holds for *row* totals, hence all statements about *all* totals will be correct in each of over 90% of *all* two-way classifications.

SE Quadrant of Table 6

Factors from Total Range of Differences in a Two-Way Classification
to Allowances for Totals*. (According to the "Easy" Method.)
5% ERROR RATE**

Entries are to be Multiplied by the Sum of Ranges of Differences between Adjacent Observations to Obtain Allowances for {Column Totals (Use Upper Entry). / Row Totals (Use Lower Entry).

(Differences Taken Horizontally, Their Ranges Taken Vertically.)
Number of ranges = Number of columns

Number of differences per range = Number of rows

	16	17	18	19	20	30	40	50	100	200	500	1000
15	.243	.231	.220	.210	.201	.141	.109	.090	.048	.0256	.0100	.0058
	.249	.241	.234	.228	.222	.180	.156	.139	.098	.069	.044	.0309
16	.247	.234	.223	.213	.204	.143	.111	.091	.049	.0260	.0112	.0059
	.247	.239	.232	.226	.220	.179	.155	.138	.097	.069	.043	.0307
17	.250	.238	.226	.216	.207	.145	.112	.092	.049	.0263	.0113	.0060
	.245	.238	.231	.224	.219	.178	.154	.137	.097	.068	.043	.0305
18	.254	.241	.229	.219	.209	.147	.114	.093	.050	.0267	.0115	.0060
	.244	.236	.229	.223	.217	.177	.153	.136	.096	.068	.043	.0303
19	.257	.244	.232	.222	.212	.149	.115	.095	.051	.0271	.0117	.0061
	.242	.235	.228	.222	.216	.176	.152	.136	.096	.068	.043	.0302
20	.260	.247	.235	.225	.215	.151	.117	.096	.052	.0274	.0118	.0062
	.241	.234	.227	.221	.215	.175	.151	.135	.095	.067	.042	.0300
30	.289	.275	.262	.250	.239	.168	.131	.107	.058	.0307	.0132	.0069
	.233	.226	.219	.213	.208	.169	.146	.130	.092	.065	.041	.0290
40	.315	.299	.285	.272	.260	.183	.142	.117	.063	.033	.0144	.0076
	.228	.221	.215	.209	.203	.166	.143	.128	.090	.064	.040	.0285
50	.338	.321	.306	.292	.279	.197	.153	.125	.067	.036	.0155	.0081
	.225	.218	.212	.206	.201	.163	.141	.126	.089	.063	.0400	.0281
100	.427	.406	.387	.369	.353	.249	.193	.159	.085	.046	.0196	.0103
	.217	.210	.204	.198	.193	.157	.136	.122	.086	.061	.038	.0271
200	.551	.523	.499	.476	.456	.321	.250	.205	.110	.059	.0253	.0133
	.210	.204	.198	.193	.188	.153	.132	.118	.084	.059	.037	.0264
500	.79	.75	.71	.68	.65	.459	.357	.293	.158	.084	.0362	.0190
	.205	.198	.193	.188	.183	.149	.129	.115	.081	.058	.036	.0257
1000	1.04	.99	.94	.90	.86	.608	.473	.389	.209	.112	.0481	.0253
	.201	.195	.190	.185	.180	.147	.127	.113	.080	.057	.036	.0253

* Lower entries are also factors from mean range of differences to allowances for *row* means; upper entries are also factors from [(total range)/(number of rows)] to allowances for *column* means.

** All confidence statements about *column* totals will be correct in each of 95% of two-way classifications to which applied: similar result holds for *row* totals, hence all statements about *all* totals will be correct in each of over 90% of all two-way classifications.

Table 6a

Modified Version of Table 6 to Facilitate Interpolation When the Number of Ranges Exceeds 20. The Upper Entries are to be Divided by the Number of Ranges, the Lower Entries are to be Divided by $10\sqrt{\text{Number of Ranges}}$, then Used Exactly as the Corresponding Entries of Table 6. (See Table 6 for Further Instructions.)

5% ERROR RATE

Number of ranges = Number of columns

Number of differences per range = Number of rows

	10	15	20	30	40	50	100	200	500	1000
2	5.4	5.4	5.4	5.4	5.5	5.5	5.7	5.9	6.3	6.6
	20.8	19.6	19.1	18.5	18.2	18.0	17.7	17.5	17.4	17.4
3	3.8	3.9	4.0	4.1	4.2	4.3	4.5	4.8	5.1	5.3
	15.5	15.0	14.7	14.4	14.3	14.2	14.0	13.9	13.9	13.9
4	3.4	3.6	3.7	3.8	3.9	4.0	4.2	4.5	4.8	5.0
	13.6	13.2	13.0	12.8	12.7	12.7	12.6	12.5	12.5	12.5
5	3.3	3.5	3.6	3.7	3.8	3.9	4.2	4.4	4.7	5.0
	12.6	12.3	12.1	12.0	11.9	11.9	11.8	11.8	11.7	11.7
6	3.3	3.4	3.6	3.7	3.8	3.9	4.2	4.4	4.8	5.0
	11.9	11.7	11.6	11.5	11.4	11.4	11.3	11.3	11.2	11.2
7	3.3	3.5	3.6	3.8	3.9	4.0	4.2	4.5	4.8	5.1
	11.5	11.3	11.2	11.1	11.0	11.0	11.0	10.9	10.9	10.9
8	3.3	3.5	3.6	3.8	3.9	4.0	4.3	4.6	4.9	5.2
	11.2	11.0	10.9	10.8	10.8	10.7	10.7	10.7	10.7	10.6
9	3.3	3.5	3.7	3.9	4.0	4.1	4.4	4.6	5.0	5.2
	10.9	10.8	10.7	10.6	10.6	10.5	10.5	10.5	10.5	10.4
10	3.4	3.6	3.7	3.9	4.0	4.1	4.4	4.7	5.1	5.3
	10.7	10.6	10.5	10.4	10.4	10.4	10.3	10.3	10.3	10.3
11	3.4	3.6	3.8	4.0	4.1	4.2	4.5	4.8	5.2	5.4
	10.6	10.4	10.4	10.3	10.2	10.2	10.2	10.2	10.2	10.2
12	3.5	3.7	3.8	4.0	4.2	4.3	4.6	4.9	5.2	5.5
	10.4	10.3	10.2	10.2	10.1	10.1	10.1	10.1	10.0	10.0
13	3.5	3.7	3.9	4.1	4.2	4.3	4.7	5.0	5.3	5.6
	10.3	10.2	10.1	10.1	10.0	10.0	10.0	10.0	9.9	9.9
14	3.6	3.8	4.0	4.2	4.3	4.4	4.7	5.0	5.4	5.7
	10.2	10.1	10.0	10.0	9.9	9.9	9.9	9.9	9.9	9.9

Table 6a (Cont'd)

Modified Version of Table 6 to Facilitate Interpolation When the Number of Ranges Exceeds 20. The Upper Entries are to be Divided by the Number of Ranges, the Lower Entries are to be Divided by $10\sqrt{\text{Number of Ranges}}$, then Used Exactly as the Corresponding Entries of Table 6. (See Table 6 for Further Instructions.)

5% ERROR RATE

Number of ranges = Number of columns

Number of differences per range = Number of rows	10	15	20	30	40	50	100	200	500	1000
15	3.6	3.9	4.0	4.2	4.4	4.5	4.8	5.1	5.5	5.8
	10.1	10.0	9.9	9.9	9.9	9.8	9.8	9.8	9.8	9.8
16	3.7	3.9	4.1	4.3	4.4	4.5	4.9	5.2	5.6	5.9
	10.0	9.9	9.9	9.8	9.8	9.8	9.7	9.7	9.7	9.7
17	3.7	4.0	4.1	4.3	4.5	4.6	4.9	5.3	5.7	6.0
	10.0	9.8	9.8	9.7	9.7	9.7	9.7	9.7	9.7	9.6
18	3.8	4.0	4.2	4.4	4.6	4.7	5.0	5.3	5.7	6.0
	9.9	9.8	9.7	9.7	9.7	9.7	9.6	9.6	9.6	9.6
19	3.8	4.1	4.2	4.5	4.6	4.7	5.1	5.4	5.8	6.1
	9.8	9.7	9.7	9.6	9.6	9.6	9.6	9.6	9.5	9.5
20	3.9	4.1	4.3	4.5	4.7	4.8	5.2	5.5	5.9	6.2
	9.8	9.7	9.6	9.6	9.6	9.6	9.5	9.5	9.5	9.5
30	4.3	4.6	4.8	5.0	5.2	5.4	5.8	6.1	6.6	6.9
	9.4	9.3	9.3	9.3	9.2	9.2	9.2	9.2	9.2	9.2
40	4.7	5.0	5.2	5.5	5.7	5.8	6.3	6.7	7.2	7.6
	9.2	9.1	9.1	9.1	9.1	9.0	9.0	9.0	9.0	9.0
50	5.0	5.4	5.6	5.9	6.1	6.3	6.7	7.2	7.7	8.1
	9.1	9.0	9.0	8.9	8.9	8.9	8.9	8.9	8.9	8.9
100	6.3	6.8	7.1	7.5	7.7	7.9	8.	9.1	9.8	10.3
	8.7	8.7	8.6	8.6	8.6	8.6	8.6	8.6	8.6	8.6
200	8.2	8.7	9.1	9.6	10.0	10.2	11.0	11.8	12.7	13.3
	8.5	8.4	8.4	8.4	8.4	8.4	8.4	8.4	8.4	8.4
500	11.6	12.5	13.0	13.8	14.3	14.7	15.8	16.8	18.1	19.0
	8.2	8.2	8.2	8.2	8.2	8.1	8.1	8.1	8.1	8.1
1000	15.4	16.5	17.3	18.3	18.9	19.4	20.9	22.3	24.0	25.3
	8.1	8.1	8.0	8.0	8.0	8.0	8.0	8.0	8.0	8.0

NW Quadrant of Table 7

Factors from Total Range of Differences in a Two-Way Classification
to Allowances for Totals*. (According to the "Easy" Method.)
1% ERROR RATE**

Entries are to be Multiplied by the Sum of Ranges of Differences between Adjacent Observations to Obtain Allowances for {Column Totals (Use Upper Entry). Row Totals (Use Lower Entry).}

(Differences Taken Horizontally, Their Ranges Taken Vertically.)
Number of ranges = Number of columns

Number of differences per range = Number of rows	2	3	4	5	6	7	8	9	10	11	12	13	14	15
2	31.83	5.00	2.91	2.00	1.51	1.20	1.00	.86	.75	.66	.60	.543	.497	.459
	31.83	4.51	2.72	1.99	1.59	1.35	1.19	1.07	.97	.90	.84	.793	.751	.715
3	4.51	1.84	1.31	1.01	.82	.70	.60	.53	.48	.43	.39	.363	.337	.314
	5.00	1.84	1.35	1.10	.94	.83	.76	.69	.65	.61	.57	.544	.519	.497
4	2.63	1.40	1.04	.83	.69	.59	.52	.46	.42	.38	.35	.322	.300	.281
	2.75	1.35	1.04	.87	.76	.68	.63	.58	.54	.51	.48	.461	.442	.424
5	2.11	1.25	.95	.77	.64	.56	.49	.44	.40	.36	.33	.309	.288	.270
	2.05	1.14	.90	.77	.68	.61	.56	.52	.49	.46	.44	.419	.402	.386
6	1.88	1.18	.91	.74	.63	.54	.48	.43	.39	.36	.33	.305	.284	.267
	1.71	1.02	.82	.71	.63	.57	.52	.49	.46	.43	.41	.393	.377	.363
7	1.78	1.15	.89	.73	.62	.54	.48	.43	.39	.36	.33	.305	.285	.267
	1.52	.95	.77	.66	.59	.54	.50	.46	.44	.41	.39	.375	.360	.347
8	1.72	1.14	.89	.73	.62	.54	.48	.43	.39	.36	.33	.307	.287	.269
	1.40	.90	.73	.63	.57	.52	.48	.45	.42	.40	.38	.362	.348	.335
9	1.69	1.14	.89	.73	.62	.54	.48	.43	.39	.36	.33	.310	.290	.272
	1.31	.86	.71	.61	.55	.50	.46	.43	.41	.39	.37	.352	.338	.325
10	1.67	1.14	.89	.74	.63	.55	.48	.44	.40	.36	.34	.313	.293	.275
	1.24	.83	.68	.59	.53	.49	.45	.42	.40	.38	.36	.344	.330	.318
11	1.67	1.15	.90	.74	.63	.55	.49	.44	.40	.37	.34	.318	.297	.279
	1.19	.80	.67	.58	.52	.48	.44	.41	.39	.37	.35	.337	.324	.312
12	1.67	1.15	.91	.75	.64	.56	.50	.45	.41	.37	.35	.322	.301	.283
	1.15	.78	.65	.57	.51	.47	.43	.41	.38	.36	.35	.331	.318	.307
13	1.67	1.16	.92	.76	.65	.57	.50	.45	.41	.38	.35	.326	.305	.287
	1.11	.77	.64	.56	.50	.46	.43	.40	.38	.36	.34	.326	.314	.302
14	1.68	1.17	.93	.77	.65	.57	.51	.46	.42	.38	.36	.331	.310	.291
	1.08	.75	.63	.55	.49	.45	.42	.39	.37	.35	.34	.322	.310	.298

* Lower entries are also factors from mean range of differences to allowances for ***row*** means; upper entries are also factors from [(total range)/(number of rows)] to allowances for ***column*** means.

** All confidence statements about ***column*** totals will be correct in each of 99% of two-way classifications to which applied: similar result holds for ***row*** totals, hence all statements about ***all*** totals will be correct in each of over 98% of all two-way classifications.

NE Quadrant of Table 7

Factors from Total Range of Differences in a Two-Way Classification
to Allowances for Totals*. (According to the "Easy" Method.)
1% ERROR RATE**
Entries are to be Multiplied by the Sum of Ranges of Differences between Adjacent
Observations to Obtain Allowances for {Column Totals (Use Upper Entry). Row Totals (Use Lower Entry).
(Differences Taken Horizontally, Their Ranges Taken Vertically.)
Number of ranges = Number of columns

Number of differences per range = Number of rows	16	17	18	19	20	30	40	50	100	200	500	1000
2	.426	.397	.372	.350	.331	.215	.160	.127	.064	.0325	.0136	.0070
	.683	.655	.631	.608	.588	.466	.393	.346	.236	.164	.103	.072
3	.294	.277	.261	.248	.235	.159	.120	.097	.050	.0260	.0110	.0057
	.478	.461	.445	.431	.419	.338	.288	.255	.176	.123	.078	.0548
4	.264	.249	.236	.224	.213	.145	.111	.090	.047	.0244	.0104	.0054
	.408	.394	.382	.370	.360	.290	.248	.220	.154	.108	.068	.0479
5	.254	.240	.227	.216	.206	.141	.108	.088	.046	.0241	.0102	.0053
	.372	.360	.349	.338	.329	.265	.228	.202	.141	.099	.063	.0443
6	.251	.238	.225	.214	.204	.141	.108	.088	.046	.0242	.0103	.0054
	.350	.338	.328	.318	.310	.250	.215	.191	.134	.094	.059	.0420
7	.252	.238	.226	.215	.205	.142	.109	.088	.047	.0245	.0104	.0054
	.334	.324	.314	.305	.296	.240	.206	.184	.129	.091	.057	.0404
8	.254	.240	.228	.217	.207	.143	.110	.089	.047	.0248	.0106	.0055
	.323	.313	.303	.295	.287	.232	.200	.178	.125	.088	.055	.0392
9	.256	.243	.230	.219	.209	.145	.111	.091	.048	.0252	.0108	.0056
	.314	.304	.295	.287	.279	.226	.194	.173	.122	.086	.054	.0382
10	.260	.246	.233	.222	.212	.147	.113	.092	.049	.0257	.0109	.0057
	.307	.297	.288	.280	.273	.221	.190	.170	.119	.084	.053	.0375
11	.263	.249	.237	.225	.215	.149	.115	.094	.050	.0261	.0111	.0058
	.301	.292	.283	.275	.268	.217	.187	.167	.117	.083	.052	.0368
12	.267	.253	.240	.229	.218	.151	.117	.095	.050	.0265	.0113	.0059
	.296	.287	.278	.271	.263	.214	.184	.164	.115	.081	.051	.0363
13	.271	.256	.244	.232	.222	.154	.118	.097	.051	.0270	.0115	.0060
	.292	.283	.274	.267	.260	.211	.182	.162	.114	.080	.051	.0359
14	.275	.260	.247	.235	.225	.156	.120	.098	.052	.0274	.0117	.0061
	.288	.279	.271	.263	.257	.208	.180	.160	.113	.079	.050	.0355

* Lower entries are also factors from mean range of differences to allowances for *row* means; upper entries are also factors from [(total range)/(number of rows)] to allowances for *column* means.

** All confidence statements about *column* totals will be correct in each of 99% of two-way classifications to which applied: similar result holds for *row* totals, hence all statements about *all* totals will be correct in each of over 98% of all two-way classifications.

SW Quadrant of Table 7

Factors from Total Range of Differences in a Two-Way Classification
to Allowances for Totals*. (According to the "Easy" Method.)
1% ERROR RATE**
Entries are to be Multiplied by the Sum of Ranges of Differences between Adjacent
Observations to Obtain Allowances for {Column Totals (Use Upper Entry). / Row Totals (Use Lower Entry).}
(Differences Taken Horizontally, Their Ranges Taken Vertically.)
Number of ranges = Number of columns

Number of differences per range = Number of rows	2	3	4	5	6	7	8	9	10	11	12	13	14	15
15	1.69	1.19	.94	.78	.66	.58	.52	.46	.42	.39	.36	.335	.314	.295
	1.06	.74	.62	.54	.49	.45	.41	.39	.37	.35	.33	.318	.306	.295
16	1.70	1.20	.95	.78	.67	.59	.52	.47	.43	.39	.36	.340	.318	.299
	1.04	.73	.61	.53	.48	.44	.41	.38	.36	.34	.33	.315	.303	.292
17	1.71	1.21	.96	.79	.68	.59	.53	.48	.43	.40	.37	.344	.322	.303
	1.02	.72	.60	.53	.48	.44	.41	.38	.36	.34	.33	.312	.300	.289
18	1.73	1.22	.97	.80	.69	.60	.53	.48	.44	.40	.37	.349	.327	.307
	1.00	.71	.60	.52	.47	.43	.40	.38	.36	.34	.32	.310	.298	.287
19	1.74	1.24	.98	.81	.69	.61	.54	.49	.45	.41	.38	.353	.331	.311
	.99	.70	.59	.52	.47	.43	.40	.37	.35	.34	.32	.307	.295	.285
20	1.75	1.25	.99	.82	.70	.61	.55	.49	.45	.41	.38	.357	.335	.315
	.97	.70	.59	.51	.46	.43	.40	.37	.35	.33	.32	.305	.293	.283
30	1.93	1.41	1.11	.92	.78	.68	.61	.55	.50	.46	.43	.398	.373	.350
	.90	.66	.56	.49	.44	.41	.38	.35	.34	.32	.304	.292	.281	.271
40	2.07	1.52	1.20	.99	.85	.74	.66	.60	.54	.50	.46	.433	.405	.381
	.85	.64	.54	.47	.43	.39	.37	.34	.33	.309	.296	.283	.273	.263
50	2.20	1.62	1.28	1.06	.91	.79	.71	.64	.58	.54	.50	.463	.434	.408
	.83	.62	.53	.46	.42	.39	.36	.34	.32	.303	.290	.278	.267	.258
100	2.72	2.02	1.60	1.33	1.14	1.00	.89	.80	.73	.68	.63	.585	.548	.516
	.76	.58	.49	.44	.40	.37	.34	.32	.303	.288	.275	.264	.254	.245
500	4.90	3.67	2.92	2.43	2.08	1.83	1.63	1.48	1.35	1.24	1.15	1.07	1.01	.95
	.68	.53	.45	.40	.36	.34	.314	.295	.280	.266	.255	.244	.235	.227
1000	6.45	4.85	.3.85	3.21	2.76	2.42	2.16	1.95	1.79	1.64	1.53	1.42	1.33	1.26
	.65	.51	.44	.39	.36	.33	.306	.288	.273	.260	.249	.239	.230	.222

* Lower entries are also factors from mean range of differences to allowances for *row* means; upper entries are also factors from [(total range)/(number of rows)] to allowances for ***column*** means.

** All confidence statements about ***column*** totals will be correct in each of 99% of two-way classifications to which applied: similar result holds for *row* totals, hence all statements about ***all*** totals will be correct in each of over 98% of all two-way classifications.

SE Quadrant of Table 7

Factors from Total Range of Differences in a Two-Way Classification
to Allowances for Totals.* (According to the "Easy" Method.)
1% ERROR RATE**
Entries are to be Multiplied by the Sum of Ranges of Differences between Adjacent
Observations to Obtain Allowances for { Column Totals (Use Upper Entry). / Row Totals (Use Lower Entry).
(Differences Taken Horizontally, Their Ranges Taken Vertically.)
Number of ranges = Number of columns

Number of differences per range = Number of rows	16	17	18	19	20	30	40	50	100	200	500	1000
15	.279	.264	.251	.239	.228	.158	.122	.099	.053	.0278	.0119	.0062
	.285	.276	.268	.261	.254	.206	.178	.159	.112	.079	.050	.0351
16	.282	.267	.254	.242	.231	.161	.124	.101	.054	.0282	.0120	.0063
	.282	.273	.265	.258	.251	.204	.176	.157	.111	.078	.049	.0348
17	.286	.271	.258	.245	.234	.163	.124	.102	.054	.0286	.0122	.0064
	.280	.271	.263	.256	.249	.202	.175	.156	.110	.077	.049	.0345
18	.290	.275	.261	.249	.237	.165	.127	.104	.055	.0290	.0124	.0065
	.277	.269	.261	.254	.247	.201	.173	.155	.109	.077	.048	.0343
19	.294	.278	.264	.252	.240	.167	.129	.105	.056	.0294	.0126	.0066
	.275	.267	.259	.252	.245	.199	.172	.154	.108	.076	.048	.0340
20	.297	.282	.268	.255	.243	.169	.130	.106	.056	.0298	.0127	.0066
	.273	.265	.257	.250	.243	.198	.171	.153	.107	.076	.048	.0338
30	.331	.314	.298	.284	.271	.189	.145	.119	.063	.033	.0142	.0074
	.262	.254	.246	.239	.233	.189	.163	.146	.103	.073	.046	.0324
40	.360	.341	.324	.309	.295	.205	.158	.129	.069	.036	.0155	.0081
	.254	.246	.239	.233	.227	.184	.159	.142	.100	.071	.045	.0316
50	.386	.366	.348	.331	.317	.220	.170	.139	.074	.039	.0167	.0087
	.249	.242	.235	.228	.222	.181	.156	.139	.098	.069	.044	.0310
100	.487	.462	.439	.418	.400	.279	.215	.176	.094	.049	.0211	.0110
	.237	.230	.223	.217	.212	.172	.149	.133	.094	.066	.042	.0296
200	.63	.595	.566	.539	.515	.359	.278	.227	.121	.064	.0273	.0143
	.228	.221	.215	.209	.204	.166	.143	.128	.090	.064	.040	.0286
500	.90	.85	.81	.77	.74	.513	.397	.324	.173	.091	.0390	.0204
	.220	.213	.207	.201	.196	.160	.138	.124	.087	.062	.039	.0276
1000	1.19	1.13	1.07	1.02	.98	.680	.526	.430	.229	.121	.0518	.0271
	.215	.208	.202	.197	.192	.156	.135	.121	.085	.060	.038	.0270

* Lower entries are also factors from mean range of differences to allowances for *row* means; upper entries are also factors from [(total range)/(number of rows)] to allowances for *column* means.

** All confidence statements about *column* totals will be correct in each of 99% of two-way classifications to which applied: similar result holds for *row* totals, hence all statements about *all* totals will be correct in each of over 98% of all two-way classifications.

Table 7a

Modified Version of Table 7 to Facilitate Interpolation When the Number of Ranges Exceeds 20. The Upper Entries are to be Divided by the Number of Ranges, the Lower Entries are to be Divided by $10\sqrt{\text{Number of Ranges}}$, then Used Exactly as the Corresponding Entries of Table 7. (See Table 7 for Further Instructions.)

1% ERROR RATE

Number of ranges = Number of columns

Number of differences per range = Number of rows	10	15	20	30	40	50	100	200	500	1000
2	7.2	6.8	6.6	6.5	6.4	6.4	6.4	6.5	6.8	7.0
	30.8	28.2	26.8	25.5	24.8	24.4	23.6	23.2	23.0	22.9
3	4.8	4.7	4.7	4.8	4.8	4.8	5.0	5.2	5.5	5.7
	20.9	19.7	19.1	18.5	18.2	18.0	17.6	17.5	17.4	17.3
4	4.2	4.2	4.3	4.4	4.4	4.5	4.7	4.9	5.2	5.4
	17.4	16.6	16.3	15.9	15.7	15.6	15.4	15.2	15.2	15.1
5	4.0	4.1	4.1	4.2	4.3	4.4	4.6	4.8	5.1	5.3
	15.7	15.1	14.8	14.5	14.4	14.3	14.1	14.1	14.0	14.0
6	3.9	4.0	4.1	4.2	4.3	4.4	4.6	4.8	5.1	5.4
	14.6	14.2	13.9	13.7	13.6	13.5	13.4	13.3	13.3	13.3
7	3.9	4.0	4.1	4.2	4.3	4.4	4.7	4.9	5.2	5.4
	13.9	13.5	13.3	13.1	13.0	13.0	12.9	12.8	12.8	12.8
8	3.9	4.1	4.2	4.3	4.4	4.5	4.7	5.0	5.3	5.5
	13.4	13.0	12.9	12.7	12.6	12.6	12.5	12.4	12.4	12.4
9	4.0	4.1	4.2	4.3	4.5	4.5	4.8	5.0	5.4	5.6
	13.0	12.7	12.5	12.4	12.3	12.3	12.2	12.1	12.1	12.1
10	4.0	4.1	4.3	4.4	4.5	4.6	4.9	5.1	5.5	5.7
	12.6	12.4	12.2	12.1	12.0	12.0	11.9	11.9	11.9	11.8
11	4.0	4.2	4.3	4.5	4.6	4.7	5.0	5.2	5.6	5.8
	12.4	12.1	12.0	11.9	11.8	11.8	11.7	11.7	11.7	11.7
12	4.1	4.3	4.4	4.5	4.7	4.8	5.0	5.3	5.7	5.9
	12.2	11.9	11.8	11.7	11.6	11.6	11.5	11.5	11.5	11.5
13	4.1	4.3	4.4	4.6	4.7	4.8	5.1	5.4	5.7	6.0
	12.0	11.8	11.7	11.5	11.5	11.5	11.4	11.4	11.3	11.3
14	4.2	4.4	4.5	4.7	4.8	4.9	5.2	5.5	5.8	6.1
	11.8	11.6	11.5	11.4	11.4	11.3	11.3	11.2	11.2	11.2

Table 7a (Cont'd)

Modified Version of Table 7 to Facilitate Interpolation When the Number of Ranges Exceeds 20. The Upper Entries are to be Divided by the Number of Ranges, the Lower Entries are to be Divided by $10\sqrt{\text{Number of Ranges}}$, then Used Exactly as the Corresponding Entries of Table 7. (See Table 7 for Further Instructions.)

1% ERROR RATE

Number of ranges = Number of columns

Number of differences per range = Number of rows	10	15	20	30	40	50	100	200	500	1000
15	4.3	4.4	4.6	4.7	4.9	5.0	5.3	5.6	5.9	6.2
	11.7	11.5	11.4	11.3	11.2	11.2	11.2	11.1	11.1	11.1
16	4.3	4.5	4.6	4.8	4.9	5.0	5.4	5.6	6.0	6.3
	11.6	11.4	11.3	11.2	11.1	11.1	11.1	11.0	11.0	11.0
17	4.4	4.6	4.7	4.9	5.0	5.1	5.4	5.7	6.1	6.4
	11.4	11.3	11.2	11.1	11.0	11.0	11.0	10.9	10.9	10.9
18	4.4	4.6	4.8	4.9	5.1	5.2	5.5	5.8	6.2	6.5
	11.3	11.2	11.1	11.0	11.0	10.9	10.9	10.9	10.8	10.8
19	4.5	4.7	4.8	5.0	5.1	5.3	5.6	5.9	6.3	6.6
	11.2	11.1	11.0	10.9	10.9	10.9	10.8	10.8	10.8	10.8
20	4.5	4.7	4.9	5.1	5.2	5.3	5.6	6.0	6.4	6.6
	11.2	11.0	10.9	10.8	10.8	10.8	10.7	10.7	10.7	10.7
30	5.0	5.3	5.4	5.7	5.8	5.9	6.3	6.7	7.1	7.4
	10.6	10.5	10.4	10.4	10.3	10.3	10.3	10.3	10.2	10.2
40	5.4	5.7	5.9	6.2	6.3	6.5	6.9	7.3	7.8	8.1
	10.3	10.2	10.1	10.1	10.1	10.0	10.0	10.0	10.0	10.0
50	5.8	6.1	6.3	6.6	6.8	6.9	7.4	7.8	8.3	8.7
	10.1	10.0	9.9	9.9	9.9	9.9	9.8	9.8	9.8	9.8
100	7.3	7.7	8.0	8.4	8.6	8.8	9.4	9.9	10.6	11.0
	9.6	9.5	9.5	9.4	9.4	9.4	9.4	9.4	9.4	9.4
200	9.5	10.0	10.3	10.8	11.1	11.3	12.1	12.8	13.6	14.3
	9.2	9.1	9.1	9.1	9.1	9.1	9.0	9.0	9.0	9.0
500	13.5	14.2	14.7	15.4	15.9	16.2	17.3	18.3	19.5	20.4
	8.8	8.8	8.8	8.8	8.7	8.7	8.7	8.7	8.7	8.7
1000	17.9	18.8	19.5	20.4	21.0	21.5	22.9	24.2	25.9	27.1
	8.6	8.6	8.6	8.6	8.6	8.6	8.5	8.5	8.5	8.5

NW Sixth of Table 8

Factors from Total Range of Differences of Adjusted Values in a Two-Way Classification to Allowances for Column Totals and for Row Totals*.
(According to the Intermediate Method.)
5% ERROR RATE**
Entries to be Multipled by the Sum of Ranges between Adjusted Observations to Obtain
Allowances for {Column Totals (Use Upper Entry). Row Totals (Use Lower Entry).
(Observations Adjusted by Subtraction of Row Means, Ranges then Taken Within each Column.)
Number of ranges = Number of columns

Number of differences per range = Number of rows	2	3	4	5	6	7	8	9	10
2	12.71	3.79	2.31	1.69	1.34	1.11	.96	.84	.75
	12.71	3.39	2.15	1.66	1.39	1.22	1.09	1.00	.93
3	3.91	1.98	1.40	1.09	.90	.77	.68	.60	.54
	4.37	1.98	1.43	1.17	1.01	.90	.82	.75	.70
4	2.87	1.66	1.21	.97	.81	.70	.61	.55	.50
	3.07	1.62	1.21	1.00	.88	.79	.72	.67	.62
5	2.53	1.54	1.15	.92	.78	.67	.59	.53	.48
	2.57	1.45	1.10	.92	.81	.73	.67	.62	.58
6	2.38	1.50	1.13	.91	.77	.66	.59	.53	.48
	2.29	1.35	1.04	.87	.77	.69	.63	.59	.55
7	2.32	1.49	1.12	.91	.77	.67	.59	.53	.48
	2.13	1.28	.99	.84	.74	.67	.61	.57	.53
8	2.30	1.49	1.12	.91	.77	.67	.59	.53	.49
	2.01	1.23	.96	.81	.71	.65	.59	.55	.52
9	2.29	1.49	1.13	.92	.78	.68	.60	.54	.49
	1.93	1.19	.93	.79	.70	.63	.58	.54	.51
10	2.30	1.51	1.15	.93	.79	.69	.61	.55	.50
	1.87	1.16	.91	.78	.68	.62	.57	.53	.50
11	2.31	1.52	1.16	.94	.80	.70	.62	.56	.51
	1.81	1.14	.90	.76	.67	.61	.56	.52	.49
12	2.33	1.54	1.17	.96	.81	.71	.63	.56	.51
	1.77	1.12	.88	.75	.66	.60	.55	.52	.49
13	2.35	1.56	1.19	.97	.82	.72	.64	.57	.52
	1.74	1.10	.87	.74	.66	.60	.55	.51	.48
14	2.37	1.57	1.20	.98	.83	.73	.65	.58	.53
	1.71	1.09	.86	.73	.65	.59	.54	.51	.48

* Lower entries are also factors from mean range of differences to allowances for *row* means; upper entries are also factors from [(total range)/(number of rows)] to allowances for *column* means.

** All confidence statements about *column* totals will be correct in each of 95% of two-way classifications to which applied: similar result holds for *row* totals, hence all statements about *all* totals will be correct in each of over 90% of all two-way classifications.

N Cent. Sixth of Table 8

Factors from Total Range of Differences of Adjusted Values in a Two-Way Classification to Allowances for Column Totals and for Row Totals*.

(According to the Intermediate Method.)

5% ERROR RATE**

Entries to be Multipled by the Sum of Ranges between Adjusted Observations to Obtain

Allowances for { Column Totals (Use Upper Entry). / Row Totals (Use Lower Entry).

(Observations Adjusted by Subtraction of Row Means, Ranges then Taken Within each Column.)

Number of ranges = Number of columns

Number of differences per range = Number of rows	11	12	13	14	15	16	17	18	19	20
2	.68	.62	.57	.53	.49	.461	.434	.410	.388	.369
	.87	.82	.78	.74	.71	.68	.66	.64	.62	.597
3	.50	.46	.42	.39	.37	.347	.328	.311	.296	.282
	.66	.63	.60	.57	.55	.53	.51	.49	.48	.466
4	.45	.42	.39	.36	.34	.321	.304	.288	.274	.262
	.59	.56	.53	.51	.49	.47	.46	.44	.43	.416
5	.44	.41	.38	.35	.33	.314	.297	.282	.268	.256
	.55	.52	.50	.48	.46	.44	.43	.41	.40	.389
6	.44	.41	.38	.35	.33	.313	.296	.281	.268	.256
	.52	.50	.47	.45	.44	.42	.41	.39	.38	.372
7	.44	.41	.38	.36	.33	.315	.298	.283	.270	.258
	.50	.48	.46	.44	.42	.41	.39	.38	.37	.360
8	.45	.41	.38	.36	.34	.319	.302	.287	.273	.261
	.49	.47	.45	.43	.41	.40	.38	.37	.36	.351
9	.45	.42	.39	.36	.34	.323	.306	.291	.277	.265
	.48	.46	.44	.42	.40	.39	.38	.36	.35	.344
10	.46	.42	.39	.37	.35	.328	.311	.295	.281	.269
	.47	.45	.43	.41	.40	.38	.37	.36	.35	.338
11	.47	.43	.40	.38	.35	.333	.316	.300	.286	.273
	.47	.44	.42	.41	.39	.38	.36	.35	.34	.334
12	.47	.44	.41	.38	.36	.338	.321	.305	.290	.277
	.46	.44	.42	.40	.39	.37	.36	.35	.34	.330
13	.48	.44	.41	.39	.36	.344	.326	.309	.295	.282
	.45	.43	.41	.40	.38	.37	.36	.35	.34	.326
14	.49	.45	.42	.39	.37	.349	.330	.314	.299	.286
	.45	.43	.41	.39	.38	.36	.35	.34	.33	.323

* Lower entries are also factors from mean range of differences to allowances for *row* means; upper entries are also factors from [(total range)/(number of rows)] to allowances for *column* means.

** All confidence statements about *column* totals will be correct in each of 95% of two-way classifications to which applied: similar result holds for *row* totals, hence all statements about *all* totals will be correct in each of over 90% of all two-way classifications.

SW Sixth of Table 8

Factors from Total Range of Differences of Adjusted Values in a Two-Way Classification to Allowances for Column Totals and for Row Totals*.
(According to the Intermediate Method.)
5% ERROR RATE**

Entries to be Multipled by the Sum of Ranges between Adjusted Observations to Obtain

Allowances for {Column Totals (Use Upper Entry).
Row Totals (Use Lower Entry).

(Observations Adjusted by Subtraction of Row Means,
Ranges then Taken Within each Column.)
Number of ranges = Number of columns

Number of differences per range = Number of rows	2	3	4	5	6	7	8	9	10
15	2.40	1.59	1.22	1.00	.85	.74	.65	.59	.54
	1.68	1.08	.85	.73	.64	.58	.54	.50	.47
16	2.42	1.61	1.24	1.01	.86	.75	.66	.60	.54
	1.66	1.06	.84	.72	.64	.58	.53	.50	.47
17	2.45	1.63	1.25	1.02	.87	.76	.67	.61	.55
	1.64	1.05	.84	.71	.63	.57	.53	.49	.46
18	2.47	1.65	1.27	1.04	.88	.77	.68	.61	.56
	1.62	1.05	.83	.71	.63	.57	.53	.49	.46
19	2.50	1.67	1.28	1.05	.89	.78	.69	.62	.57
	1.60	1.04	.82	.70	.62	.57	.52	.49	.46
20	2.52	1.69	1.30	1.06	.90	.79	.70	.63	.57
	1.59	1.03	.82	.70	.62	.56	.52	.49	.46
30	2.80	1.89	1.45	1.18	1.00	.87	.78	.70	.64
	1.51	.99	.79	.68	.60	.54	.50	.48	.44
40	3.03	.2.05	1.57	1.28	1.09	.95	.85	.76	.69
	1.45	.96	.77	.66	.59	.53	.49	.46	.43
50	3.23	2.19	1.68	1.37	1.17	1.02	.91	.82	.74
	1.42	.95	.76	.65	.58	.52	.48	.45	.42
100	4.04	2.76	2.12	1.73	1.47	1.29	1.14	1.03	.94
	1.34	.90	.72	.62	.56	.50	.47	.44	.41
200	5.17	3.54	2.72	2.23	1.90	1.66	1.48	1.33	1.21
	1.28	.87	.70	.60	.54	.49	.45	.42	.40
500	7.34	5.04	3.88	3.18	2.71	2.37	2.11	1.90	1.73
	1.23	.84	.68	.58	.52	.47	.44	.41	.39
1000	9.70	6.67	5.14	4.22	3.59	3.14	2.79	2.52	2.30
	1.20	.82	.67	.57	.51	.47	.43	.40	.38

* Lower entries are also factors from mean range of differences to allowances for *row* means; upper entries are also factors from [(total range)/(number of rows)] to allowances for *column* means.

** All confidence statements about *column* totals will be correct in each of 95% of two-way classifications to which applied: similar result holds for *row* totals, hence all statements about ***all*** totals will be correct in each of over 90% of all two-way classifications.

S Cent. Sixth of Table 8

Factors from Total Range of Differences of Adjusted Values in a Two-Way Classification to Allowances for Column Totals and for Row Totals*.
(According to the Intermediate Method.)
5% ERROR RATE**

Entries to be Multipled by the Sum of Ranges between Adjusted Observations to Obtain

Allowances for { Column Totals (Use Upper Entry).
Row Totals (Use Lower Entry).

(Observations Adjusted by Subtraction of Row Means,
Ranges then Taken Within each Column.)
Number of ranges = Number of columns

Number of differences per range = Number of rows	11	12	13	14	15	16	17	18	19	20
15	.49	.46	.43	.40	.37	.354	.335	.319	.304	.290
	.45	.42	.41	.40	.38	.36	.35	.34	.33	.320
16	.50	.46	.43	.40	.38	.359	.340	.323	.308	.295
	.44	.42	.40	.39	.37	.36	.35	.34	.33	.318
17	.51	.47	.44	.41	.39	.364	.345	.328	.312	.299
	.44	.42	.40	.38	.37	.36	.35	.33	.32	.316
18	.51	.48	.44	.42	.39	.369	.350	.333	.317	.303
	.44	.42	.40	.38	.37	.35	.34	.33	.32	.314
19	.52	.48	.45	.42	.40	.374	.354	.337	.321	.307
	.43	.41	.40	.38	.37	.35	.34	.33	.32	.313
20	.53	.49	.46	.43	.40	.379	.359	.341	.325	.311
	.43	.41	.39	.38	.36	.35	.34	.33	.32	.311
30	.59	.54	.51	.47	.45	.421	.399	.380	.362	.346
	.42	.40	.38	.36	.35	.34	.33	.32	.309	.300
40	.64	.59	.55	.52	.49	.459	.435	.414	.394	.377
	.41	.39	.37	.36	.34	.33	.32	.311	.302	.294
50	.68	.63	.59	.55	.52	.492	.467	.444	.423	.404
	.40	.38	.37	.35	.34	.33	.32	.307	.298	.290
100	.87	.80	.75	.70	.66	.62	.591	.562	.535	.512
	.39	.37	.35	.34	.33	.315	.305	.296	.288	.280
200	1.12	1.03	.96	.90	.85	.80	.76	.72	.69	.66
	.38	.36	.34	.33	.32	.307	.297	.288	.280	.272
500	1.60	1.48	1.38	1.29	1.22	1.15	1.09	1.04	.99	.94
	.37	.35	.33	.32	.309	.298	.289	.280	.272	.265
1000	2.11	1.96	1.83	1.71	1.61	1.52	1.44	1.37	1.31	1.25
	.36	.34	.33	.32	.304	.293	.284	.276	.268	.261

* Lower entries are also factors from mean range of differences to allowances for *row* means; upper entries are also factors from [(total range)/(number of rows)] to allowances for *column* means.

** All confidence statements about *column* totals will be correct in each of 95% of two-way classifications to which applied: similar result holds for *row* totals, hence all statements about *all* totals will be correct in each of over 90% of all two-way classifications.

NE Sixth of Table 8

Factors from Total Range of Differences of Adjusted Values in a Two-Way Classification to Allowances for Column Totals and for Row Totals*.
(According to the Intermediate Method.)
5% ERROR RATE**

Entries to be Multiplied by the Sum of Ranges between Adjusted Observations to Obtain

Allowances for {Column Totals (Use Upper Entry). Row Totals (Use Lower Entry).

(Observations Adjusted by Subtraction of Row Means, Ranges then Taken Within each Column.)
Number of ranges = Number of columns

Number of differences per range = Number of rows	30	40	50	100	200	500	1000
2	0.250	0.189	0.153	0.080	0.042	0.0177	0.0093
	0.476	0.406	0.360	0.250	0.175	0.110	0.078
3	0.193	0.148	0.120	0.064	0.034	0.0143	0.0075
	0.374	0.320	0.284	0.199	0.139	0.088	0.062
4	0.180	0.138	0.113	0.060	0.032	0.0136	0.0071
	0.334	0.287	0.255	0.178	0.125	0.079	0.056
5	0.177	0.136	0.111	0.059	0.031	0.0134	0.0070
	0.313	0.269	0.239	0.167	0.118	0.074	0.052
6	0.177	0.136	0.111	0.059	0.031	0.0135	0.0071
	0.299	0.257	0.229	0.160	0.113	0.071	0.050
7	0.178	0.138	0.113	0.060	0.032	0.0137	0.0072
	0.290	0.249	0.222	0.155	0.109	0.069	0.049
8	0.181	0.140	0.114	0.061	0.032	0.0139	0.0073
	0.283	0.243	0.216	0.152	0.107	0.067	0.048
9	0.184	0.142	0.116	0.062	0.033	0.0141	0.0074
	0.277	0.238	0.212	0.149	0.105	0.066	0.047
10	0.187	0.144	0.118	0.063	0.033	0.0144	0.0075
	0.273	0.235	0.209	0.147	0.103	0.065	0.046
11	0.190	0.146	0.120	0.064	0.034	0.0146	0.0077
	0.269	0.231	0.206	0.145	0.102	0.064	0.045
12	0.193	0.149	0.122	0.065	0.035	0.0149	0.0078
	0.266	0.229	0.204	0.143	0.101	0.064	0.045
13	0.196	0.151	0.124	0.066	0.035	0.0151	0.0079
	0.263	0.226	0.202	0.142	0.100	0.063	0.044
14	0.199	0.154	0.126	0.067	0.036	0.0153	0.0081
	0.261	0.224	0.200	0.140	0.099	0.062	0.044

* Lower entries are also factors from mean range of differences to allowances for *row* means; upper entries are also factors from [(total range)/(number of rows)] to allowances for *column* means.

** All confidence statements about ***column*** totals will be correct in each of 95% of two-way classifications to which applied: similar result holds for *row* totals, hence all statements about ***all*** totals will be correct in each of over 90% of all two-way classifications.

SE Sixth of Table 8

Factors from Total Range of Differences of Adjusted Values in a Two-Way Classification to Allowances for Column Totals and for Row Totals*.
(According to the Intermediate Method.)
5% ERROR RATE**

Entries to be Multiplied by the Sum of Ranges between Adjusted Observations to Obtain

Allowances for { Column Totals (Use Upper Entry). / Row Totals (Use Lower Entry).

(Observations Adjusted by Subtraction of Row Means, Ranges then Taken Within each Column.)
Number of ranges = Number of columns

Number of differences per range = Number of rows

	30	40	50	100	200	500	1000
15	0.202	0.156	0.128	0.068	0.036	0.0156	0.0082
	0.259	0.223	0.198	0.139	0.098	0.062	0.044
16	0.205	0.158	0.129	0.069	0.037	0.0158	0.0083
	0.257	0.221	0.197	0.138	0.097	0.061	0.043
17	0.208	0.160	0.131	0.070	0.037	0.0160	0.0084
	0.255	0.220	0.196	0.137	0.097	0.061	0.043
18	0.210	0.163	0.133	0.071	0.038	0.0163	0.0085
	0.254	0.218	0.195	0.137	0.096	0.061	0.043
19	0.213	0.165	0.135	0.072	0.038	0.0165	0.0087
	0.252	0.217	0.194	0.136	0.096	0.060	0.043
20	0.216	0.167	0.137	0.073	0.039	0.0167	0.0088
	0.251	0.216	0.193	0.135	0.095	0.060	0.042
30	0.241	0.187	0.153	0.082	0.043	0.0187	0.0098
	0.242	0.209	0.186	0.131	0.092	0.058	0.041
40	0.263	0.203	0.167	0.089	0.047	0.0204	0.0107
	0.238	0.205	0.182	0.128	0.090	0.057	0.040
50	0.282	0.218	0.179	0.096	0.051	0.0219	0.0115
	0.234	0.202	0.180	0.126	0.089	0.056	0.040
100	0.357	0.277	0.227	0.121	0.065	0.0278	0.0146
	0.226	0.195	0.174	0.122	0.086	0.054	0.038
200	0.461	0.357	0.293	0.157	0.083	0.0359	0.0188
	0.220	0.190	0.169	0.119	0.084	0.053	0.037
500	0.66	0.511	0.419	0.224	0.119	0.0513	0.0270
	0.214	0.184	0.165	0.116	0.082	0.051	0.036
1000	0.87	0.677	0.555	0.297	0.158	0.0680	0.0357
	0.211	0.182	0.162	0.114	0.080	0.051	0.036

* Lower entries are also factors from mean range of differences to allowances for *row* means; upper entries are also factors from [(total range)/(number of rows)] to allowances for *column* means.

** All confidence statements about ***column*** totals will be correct in each of 95% of two-way classifications to which applied: similar result holds for *row* totals, hence all statements about ***all*** totals will be correct in each of over 90% of all two-way classifications.

Table 8a

Modified Version of Table 8 to Facilitate Interpolation When the Number of Ranges Exceeds 20. The Upper Entries are to be Divided by the Number of Ranges, the Lower Entries are to be Divided by $10\sqrt{\text{Number of Ranges}}$, then Used Exactly as the Corresponding Entries of Table 8. (See Table 8 for Further Instructions.)

5% ERROR RATE

Number of ranges = Number of columns

Number of differences per range = Number of rows	10	15	20	30	40	50	100	200	500	1000
2	7.5	7.4	7.4	7.5	7.6	7.7	8.0	8.3	8.9	9.3
	29.6	27.8	26.9	26.1	25.7	25.5	25.0	24.8	24.6	24.6
3	5.4	5.5	5.6	5.8	5.9	6.0	6.4	6.7	7.2	7.5
	22.5	21.5	21.0	20.5	20.3	20.1	19.9	19.7	19.6	19.6
4	5.0	5.1	5.2	5.4	5.5	5.6	6.0	6.3	6.8	7.1
	19.8	19.0	18.7	18.3	18.1	18.0	17.8	17.7	17.7	17.6
5	4.8	5.0	5.1	5.3	5.4	5.6	5.9	6.3	6.7	7.0
	18.4	17.7	17.4	17.1	17.0	16.9	16.7	16.7	16.6	16.6
6	4.8	5.0	5.1	5.3	5.5	5.6	5.9	6.3	6.8	7.1
	17.5	16.9	16.7	16.4	16.3	16.2	16.0	16.0	15.9	15.9
7	4.8	5.0	5.1	5.4	5.5	5.6	6.0	6.4	6.8	7.2
	16.9	16.4	16.1	15.9	15.8	15.7	15.5	15.5	15.4	15.4
8	4.9	5.1	5.2	5.4	5.6	5.7	6.1	6.5	6.9	7.3
	16.5	16.0	15.7	15.5	15.4	15.3	15.2	15.1	15.1	15.1
9	4.9	5.1	5.3	5.5	5.7	5.8	6.2	6.6	7.1	7.4
	16.1	15.6	15.4	15.2	15.1	15.0	14.9	14.8	14.8	14.8
10	5.0	5.2	5.4	5.6	5.8	5.9	6.3	6.7	7.2	7.5
	15.8	15.4	15.1	14.9	14.8	14.8	14.7	14.6	14.6	14.6
11	5.1	5.3	5.4	5.7	5.9	6.0	6.4	6.8	7.3	7.7
	15.6	15.1	14.9	14.7	14.6	14.6	14.5	14.4	14.4	14.4
12	5.1	5.4	5.5	5.8	6.0	6.1	6.5	6.9	7.4	7.8
	15.4	15.0	14.8	14.6	14.5	14.4	14.3	14.2	14.2	14.2
13	5.2	5.4	5.6	5.9	6.0	6.2	6.6	7.0	7.5	7.9
	15.2	14.8	14.6	14.4	14.3	14.3	14.2	14.1	14.1	14.1
14	5.3	5.5	5.7	6.0	6.1	6.3	6.7	7.1	7.7	8.1
	15.1	14.7	14.5	14.3	14.2	14.1	14.0	14.0	13.9	13.9

Table 8a (Cont'd)

Modified Version of Table 8 to Facilitate Interpolation When the Number of Ranges Exceed 20. The Upper Entries are to be Divided by the Number of Ranges, the Lower Entries are to be Divided by $10\sqrt{\text{Number of Ranges}}$, then Used Exactly as the Corresponding Entries of Table 8. (See Table 8 for Further Instructions.)

5% ERROR RATE

Number of ranges = Number of columns

Number of differences per range = Number of rows	10	15	20	30	40	50	100	200	500	1000
15	5.4	5.6	5.8	6.0	6.2	6.4	6.8	7.2	7.8	8.2
	14.9	14.5	14.3	14.2	14.1	14.0	13.9	13.9	13.8	13.8
16	5.4	5.7	5.9	6.1	6.3	6.5	6.9	7.4	7.9	8.3
	14.8	14.4	14.2	14.1	14.0	13.9	13.8	13.8	13.7	13.7
17	5.5	5.8	6.0	6.2	6.4	6.6	7.0	7.5	8.0	8.4
	14.7	14.3	14.2	14.0	13.9	13.8	13.7	13.7	13.7	13.7
18	5.6	5.8	6.0	6.3	6.5	6.7	7.1	7.6	8.1	8.5
	14.6	14.2	14.1	13.9	13.8	13.8	13.7	13.6	13.6	13.6
19	5.7	5.9	6.1	6.4	6.6	6.7	7.2	7.7	8.2	8.7
	14.5	14.2	14.0	13.8	13.7	13.7	13.6	13.5	13.5	13.5
20	5.7	6.0	6.2	6.5	6.7	6.8	7.3	7.8	8.4	8.8
	14.5	14.1	13.9	13.8	13.7	13.6	13.5	13.5	13.4	13.4
30	6.4	6.7	6.9	7.2	7.5	7.6	8.2	8.7	9.3	9.8
	13.9	13.6	13.4	13.3	13.2	13.2	13.1	13.0	13.0	13.0
40	6.9	7.3	7.5	7.9	8.1	8.3	8.9	9.5	10.2	10.7
	13.6	13.3	13.2	13.0	12.9	12.9	12.8	12.8	12.7	12.7
50	7.4	7.8	8.1	8.5	8.7	8.9	9.6	10.2	10.9	11.5
	13.4	13.1	13.0	12.8	12.8	12.7	12.6	12.6	12.6	12.6
100	9.4	9.9	10.2	10.7	11.1	11.3	12.1	12.9	13.9	14.6
	12.9	12.6	12.5	12.4	12.3	12.3	12.2	12.2	12.1	12.1
200	12.1	12.8	13.2	13.8	14.3	14.6	15.7	16.7	17.9	18.8
	12.6	12.3	12.2	12.0	12.0	12.0	11.9	11.8	11.8	11.8
500	17.3	18.2	18.9	19.8	20.4	20.9	22.4	23.9	25.7	27.0
	12.2	12.0	11.8	11.7	11.7	11.6	11.6	11.5	11.5	11.5
1000	23.0	24.2	25.0	26.2	27.1	27.7	29.7	31.6	34.0	35.7
	12.0	11.8	11.7	11.5	11.5	11.4	11.4	11.4	11.3	11.3

NW Sixth of Table 9

Factors from Total Range of Differences of Adjusted Values in a Two-Way Classification to Allowances for Column Totals and for Row Totals*.
(According to the Intermediate Method.)
1% ERROR RATE**

Entries to be Multiplied by the Sum of Ranges between Adjusted Observations to Obtain

Allowances for {Column Totals (Use Upper Entry). / Row Totals (Use Lower Entry).

(Observations Adjusted by Subtraction of Row Means, Ranges then Taken Within Each Column.)
Number of ranges = Number of columns

Number of differences per range = Number of rows	2	3	4	5	6	7	8	9	10
2	63.66	8.67	4.17	2.71	2.00	1.59	1.32	1.13	.99
	63.66	7.84	3.99	2.78	2.20	1.86	1.63	1.47	1.34
3	9.03	3.19	2.02	1.49	1.19	.99	.85	.75	.67
	10.01	3.19	2.10	1.64	1.38	1.21	1.09	.99	.92
4	5.27	2.42	1.64	1.26	1.02	.86	.75	.66	.60
	5.51	2.33	1.64	1.32	1.13	1.00	.91	.84	.78
5	4.21	2.16	1.51	1.17	.92	.82	.72	.64	.57
	4.10	1.97	1.43	1.17	1.01	.90	.82	.76	.71
6	3.76	2.05	1.46	1.14	.94	.81	.70	.63	.57
	3.41	1.77	1.31	1.08	.94	.84	.77	.71	.67
7	3.55	2.00	1.43	1.13	.94	.80	.70	.63	.57
	3.04	1.64	1.23	1.03	.89	.80	.73	.68	.64
8	3.44	1.97	1.43	1.13	.94	.81	.71	.63	.57
	2.79	1.55	1.18	.98	.86	.77	.71	.66	.61
9	3.38	1.97	1.43	1.14	.95	.81	.71	.64	.58
	2.62	1.49	1.13	.95	.83	.75	.69	.64	.60
10	3.35	1.97	1.44	1.15	.96	.82	.72	.65	.58
	2.48	1.43	1.10	.92	.81	.73	.67	.62	.58
11	3.33	1.98	1.45	1.16	.97	.83	.73	.65	.59
	2.38	1.39	1.07	.90	.79	.72	.66	.61	.57
12	3.33	2.00	1.47	1.17	.98	.84	.74	.66	.60
	2.30	1.36	1.05	.88	.78	.70	.65	.60	.56
13	3.34	2.02	1.48	1.18	.99	.85	.75	.67	.61
	2.23	1.36	1.03	.87	.77	.69	.64	.59	.55
14	3.36	2.03	1.50	1.20	1.00	.86	.76	.68	.62
	2.17	1.33	1.01	.86	.76	.68	.63	.58	.55

* Lower entries are also factors from mean range of differences to allowances for *row* means; upper entries are also factors from [(total range)/(number of rows)] to allowances for ***column*** means.

** All confidence statements about ***column*** totals will be correct in each of 99% of two-way classifications to which applied: similar result holds for ***row*** totals, hence all statements about ***all*** totals will be correct in each of over 98% of all two-way classifications.

N Cent. Sixth of Table 9

Factors from Total Range of Differences of Adjusted Values in a Two-Way Classification to Allowances for Column Totals and for Row Totals*.
(According to the Intermediate Method.)
1% ERROR RATE**

Entries to be Multiplied by the Sum of Ranges between Adjusted Observations to Obtain

Allowances for {Column Totals (Use Upper Entry). Row Totals (Use Lower Entry).

(Observations Adjusted by Subtraction of Row Means, Ranges Then Taken Within Each Column.)
Number of ranges = Number of columns

Number of differences per range = Number of rows	11	12	13	14	15	16	17	18	19	20
2	.88	.79	.72	.66	.61	.57	.53	.50	.471	.445
	1.24	1.16	1.10	1.04	.99	.95	.91	.88	.85	.82
3	.60	.55	.51	.47	.44	.41	.39	.36	.345	.328
	.87	.81	.77	.74	.71	.68	.65	.63	.61	.59
4	.54	.50	.46	.43	.40	.37	.35	.33	.317	.301
	.74	.70	.66	.63	.61	.58	.56	.54	.53	.51
5	.52	.48	.44	.41	.39	.36	.34	.32	.308	.293
	.67	.63	.60	.58	.56	.53	.52	.50	.48	.47
6	.52	.47	.44	.41	.38	.36	.34	.32	.306	.292
	.63	.60	.57	.54	.52	.50	.49	.47	.46	.44
7	.52	.48	.44	.41	.38	.36	.34	.32	.308	.294
	.60	.57	.54	.52	.50	.48	.47	.45	.44	.43
8	.52	.48	.44	.41	.39	.37	.35	.33	.311	.297
	.58	.55	.53	.50	.48	.47	.45	.44	.42	.41
9	.53	.49	.45	.42	.39	.37	.35	.33	.315	.301
	.56	.54	.51	.49	.47	.45	.44	.43	.41	.40
10	.53	.49	.46	.43	.40	.38	.35	.34	.320	.305
	.55	.52	.50	.48	.46	.44	.43	.42	.40	.39
11	.54	.50	.46	.43	.40	.38	.36	.34	.325	.310
	.54	.51	.49	.47	.45	.44	.42	.41	.40	.39
12	.55	.51	.47	.44	.41	.39	.37	.35	.330	.314
	.53	.51	.48	.46	.45	.43	.42	.40	.39	.38
13	.56	.51	.48	.44	.42	.39	.37	.35	.335	.319
	.52	.50	.48	.46	.44	.42	.41	.40	.39	.37
14	.56	.52	.48	.45	.42	.40	.38	.36	.340	.324
	.52	.49	.47	.45	.43	.42	.40	.39	.38	.37

* Lower entries are also factors from mean range of differences to allowances for *row* means; upper entries are also factors from [(total range)/(number of rows)] to allowances for *column* means.

** All confidence statements about *column* totals will be correct in each of 99% of two-way classifications to which applied: similar result holds for *row* totals, hence all statements about *all* totals will be correct in each of over 98% of all two-way classifications.

SW Sixth of Table 9

Factors from Total Range of Differences of Adjusted Values in a Two-Way Classification to Allowances for Column Totals and for Row Totals*.
(According to the Intermediate Method.)
1% ERROR RATE**

Entries to be Multiplied by the Sum of Ranges between Adjusted Observations to Obtain

Allowances for { Column Totals (Use Upper Entry).
Row Totals (Use Lower Entry).

(Observations Adjusted by Subtraction of Row Means,
Ranges then Take Within Each Column.)
Number of ranges = Number of columns

Number of differences per range = Number of rows	2	3	4	5	6	7	8	9	10
15	3.38	2.05	1.52	1.21	1.02	.88	.77	.69	.63
	2.12	1.28	1.00	.85	.75	.67	.62	.58	.54
16	3.40	2.08	1.54	1.23	1.03	.89	.78	.70	.63
	2.08	1.26	.99	.84	.74	.67	.61	.57	.54
17	3.42	2.10	1.55	1.24	.104	.90	.79	.71	.64
	2.04	1.25	.98	.83	.73	.66	.61	.57	.53
18	3.45	2.12	1.57	1.26	1.05	.91	.80	.72	.65
	2.01	1.23	.97	.82	.72	.65	.60	.56	.53
19	3.42	2.14	1.59	1.27	1.07	.92	.81	.73	.66
	1.98	1.22	.96	.81	.72	.65	.60	.56	.52
20	3.51	2.16	1.61	1.29	1.08	.93	.82	.74	.67
	1.95	1.21	.95	.80	.71	.64	.59	.55	.52
30	3.86	2.44	1.80	1.44	1.20	1.04	.91	.82	.74
	1.80	1.15	.90	.77	.68	.61	.57	.53	.50
40	4.14	2.63	1.94	1.56	1.30	1.13	.99	.89	.81
	1.71	1.10	.87	.74	.66	.60	.55	.51	.48
50	4.40	2.81	2.08	1.66	1.40	1.21	1.06	.95	.86
	1.65	1.07	.85	.73	.64	.58	.54	.50	.47
100	5.45	3.50	2.60	2.09	1.76	1.52	1.34	1.20	1.09
	1.52	1.01	.80	.69	.61	.55	.51	.48	.45
200	6.93	4.48	3.34	2.69	2.26	1.95	1.73	1.55	1.40
	1.43	.96	.77	.66	.59	.53	.49	.46	.43
500	9.79	6.36	4.75	3.83	3.22	2.79	2.46	2.21	2.01
	1.35	.91	.73	.63	.56	.51	.47	.44	.42
1000	12.90	8.41	6.28	5.06	4.26	3.69	3.26	2.93	2.66
	1.31	.89	.72	.62	.55	.50	.46	.43	.41

* Lower entries are also factors from mean range of differences to allowances for *row* means; upper entries are also factors from [(total range)/(number of rows)] to allowances for ***column*** means.

** All confidence statements about ***column*** totals will be correct in each of 99% of two-way classifications to which applied: similar result holds for ***row*** totals, hence all statements about ***all*** totals will be correct in each of over 98% of all two-classifications.

S Cent. Sixth of Table 9

Factors from Total Range of Differences of Adjusted Values in a Two-Way Classification to Allowances for Column Totals and for Row Totals*.
(According to the Intermediate Method.)
1% ERROR RATE**

Entries to be Multiplied by the Sum of Ranges between Adjusted Observations to Obtain

Allowances for { Column Totals (Use Upper Entry). / Row Totals (Use Lower Entry).

(Observations Adjusted by Subtraction of Row Means, Ranges then Take Within Each Column.)
Number of ranges = Number of columns

Number of differences per range = Number of rows	11	12	13	14	15	16	17	18	19	20
15	.57	.53	.49	.46	.43	.40	.38	.36	.345	.329
	.51	.49	.46	.45	.43	.41	.40	.39	.38	.37
16	.58	.53	.50	.46	.43	.41	.39	.37	.350	.333
	.51	.48	.46	.44	.42	.41	.40	.38	.37	.36
17	.59	.54	.50	.47	.44	.42	.39	.37	.355	.338
	.50	.48	.46	.44	.42	.41	.39	.38	.37	.36
18	.60	.55	.51	.48	.45	.42	.40	.38	.359	.343
	.50	.47	.45	.43	.42	.40	.39	.38	.37	.36
19	.60	.56	.52	.48	.45	.43	.40	.38	.364	.347
	.49	.47	.45	.43	.41	.40	.39	.37	.36	.35
20	.61	.56	.52	.49	.46	.43	.41	.39	.369	.352
	.49	.47	.45	.43	.41	.40	.38	.37	.36	.35
30	.68	.63	.58	.54	.51	.48	.45	.43	.410	.392
	.47	.45	.43	.41	.39	.38	.37	.36	.35	.34
40	.74	.68	.63	.59	.56	.52	.49	.47	.447	.426
	.46	.43	.41	.40	.38	.37	.36	.35	.34	.33
50	.79	.73	.68	.63	.60	.56	.53	.50	.479	.457
	.45	.42	.41	.39	.38	.36	.35	.34	.33	.32
100	1.00	.92	.86	.80	.75	.71	.67	.64	.61	.578
	.42	.40	.39	.37	.36	.35	.33	.32	.314	.306
200	1.29	1.19	1.10	1.03	.97	.91	.86	.82	.78	.75
	.41	.39	.37	.36	.34	.33	.32	.312	.303	.295
500	1.84	1.70	1.58	1.47	1.39	1.31	1.24	1.17	1.12	1.07
	.39	.37	.36	.34	.33	.32	.310	.300	.292	.284
1000	2.43	2.25	2.09	1.95	1.84	1.73	1.64	1.56	1.48	1.41
	.38	.37	.35	.34	.32	.313	.303	.294	.286	.279

* Lower entries are also factors from mean range of differences to allowances for *row* means; upper entries are also factors from [(total range)/(number of rows)] to allowances for *column* means.

** All confidence statements about *column* totals will be correct in each of 99% of two-way classifications to which applied: similar result holds for *row* totals, hence all statements about *all* totals will be correct in each of over 98% of all two-way classifications.

NE Sixth of Table 9

Factors from Total Range of Differences of Adjusted Values in a Two-Way Classification to Allowances for Column Totals and for Row Totals*.
(According to the Intermediate Method.)
1% ERROR RATE**
Entries to be Multiplied by the Sum of Ranges between Adjusted Observations to Obtain

Allowances for { Column Totals (Use Upper Entry). / Row Totals (Use Lower Entry).

(Observations Adjusted by Subtraction of Row Means, Ranges then Take Within Each Column.)
Number of ranges = Number of columns

Number of differences per range = Number of rows	30	40	50	100	200	500	1000
2	0.292	0.218	0.174	0.088	0.045	0.0191	0.0099
	0.648	0.548	0.483	0.332	0.232	0.145	0.102
3	0.221	0.167	0.135	0.070	0.037	0.0155	0.0081
	0.476	0.405	0.359	0.249	0.174	0.110	0.077
4	0.204	0.156	0.126	0.066	0.034	0.0146	0.0076
	0.411	0.351	0.312	0.217	0.152	0.096	0.068
5	0.200	0.153	0.124	0.065	0.034	0.0145	0.0076
	0.377	0.323	0.287	0.200	0.141	0.089	0.063
6	0.200	0.153	0.124	0.065	0.034	0.0146	0.0076
	0.356	0.305	0.271	0.190	0.133	0.084	0.059
7	0.201	0.154	0.125	0.066	0.035	0.0147	0.0077
	0.342	0.293	0.261	0.182	0.128	0.081	0.057
8	0.204	0.156	0.127	0.067	0.035	0.0150	0.0078
	0.331	0.284	0.253	0.177	0.124	0.078	0.055
9	0.206	0.158	0.129	0.068	0.036	0.0152	0.0079
	0.322	0.277	0.246	0.172	0.121	0.077	0.054
10	0.210	0.161	0.131	0.069	0.036	0.0155	0.0081
	0.316	0.271	0.241	0.169	0.119	0.075	0.053
11	0.213	0.163	0.133	0.070	0.037	0.0157	0.0082
	0.310	0.266	0.237	0.166	0.117	0.074	0.052
12	0.216	0.166	0.135	0.071	0.038	0.0160	0.0084
	0.305	0.262	0.234	0.164	0.115	0.073	0.051
13	0.220	0.169	0.137	0.072	0.038	0.0163	0.0085
	0.301	0.259	0.231	0.162	0.114	0.072	0.051
14	0.223	0.171	0.139	0.074	0.039	0.0165	0.0086
	0.298	0.256	0.228	0.160	0.113	0.071	0.050

* Lower entries are also factors from mean range of differences to allowances for *row* means; upper entries are also factors from [(total range)/(number of rows)] to allowances for *column* means.

** All confidence statements about *column* totals will be correct in each of 99% of two-way classifications to which applied: similar result holds for *row* totals, hence all statements about *all* totals will be correct in each of over 98% of all two-way classifications.

SE Sixth of Table 9

Factors from Total Range of Differences of Adjusted Values in a Two-Way Classification to Allowances for Column Totals and for Row Totals*.
(According to the Intermediate Method.)
1% ERROR RATE**

Entries to be Multiplied by the Sum of Ranges between Adjusted Observations to Obtain

Allowances for {Column Totals (Use Upper Entry).
Row Totals (Use Lower Entry).

(Observations Adjusted by Subtraction of Row Means,
Ranges then Take Within Each Column.)
Number of ranges = Number of columns

Rows: Number of differences per range = Number of rows

	30	40	50	100	200	500	1000
15	0.226	0.174	0.141	0.075	0.039	0.0168	0.0088
	0.295	0.253	0.226	0.158	0.111	0.070	0.050
16	0.230	0.176	0.144	0.076	0.040	0.0170	0.0089
	0.292	0.251	0.224	0.157	0.110	0.070	0.049
17	0.233	0.179	0.146	0.077	0.041	0.0173	0.0090
	0.290	0.249	0.222	0.156	0.110	0.069	0.049
18	0.236	0.181	0.148	0.078	0.041	0.0175	0.0092
	0.287	0.247	0.220	0.154	0.109	0.069	0.048
19	0.239	0.184	0.150	0.079	0.042	0.0178	0.0093
	0.285	0.245	0.219	0.153	0.108	0.068	0.048
20	0.242	0.186	0.152	0.080	0.042	0.0180	0.0094
	0.284	0.244	0.217	0.152	0.107	0.068	0.048
30	0.270	0.208	0.169	0.090	0.047	0.0201	0.0105
	0.271	0.233	0.208	0.146	0.103	0.065	0.046
40	0.294	0.226	0.185	0.098	0.051	0.0220	0.0115
	0.264	0.227	0.202	0.142	0.100	0.063	0.045
50	0.316	0.243	0.198	0.105	0.055	0.0236	0.0123
	0.259	0.223	0.199	0.139	0.098	0.062	0.044
100	0.400	0.308	0.251	0.133	0.070	0.0299	0.0156
	0.247	0.213	0.190	0.133	0.094	0.059	0.042
200	0.516	0.397	0.324	0.172	0.091	0.0386	0.0202
	0.238	0.205	0.183	0.128	0.091	0.057	0.040
500	0.74	0.568	0.463	0.245	0.130	0.0553	0.0289
	0.230	0.198	0.176	0.124	0.087	0.055	0.039
1000	0.98	0.752	0.614	0.325	0.172	0.0733	0.0383
	0.225	0.194	0.173	0.121	0.086	0.054	0.038

* Lower entries are also factors from mean range of differences to allowances for *row* means; upper entries are also factors from [(total range)/(number of rows)] to allowances for *column* means.

** All confidence statements about *column* totals will be correct in each of 99% of two-way classifications to which applied: similar result holds for *row* totals, hence all statements about *all* totals will be correct in each of over 98% of all two-way classifications.

Table 9a

Modified Version of Table 9 to Facilitate Interpolation when the Number of Ranges Exceeds 20. The Upper Entries are to be Divided by the Number of Ranges, the Lower Entries are to be Divided by $10\sqrt{\text{Number of Ranges}}$, then Used Exactly as the Corresponding Entries of Table 9. (See Table 9 for Further Instructions.)

1% ERROR RATE

Number of ranges = Number of columns

Number of differences per range = Number of rows	10	15	20	30	40	50	100	200	500	1000
2	9.7	9.2	8.9	8.7	8.7	8.7	8.8	9.1	9.5	9.9
	42.9	39.0	37.2	35.5	34.7	34.2	33.2	32.7	32.5	32.4
3	6.7	6.6	6.6	6.6	6.7	6.8	7.0	7.3	7.7	8.1
	29.8	27.8	26.9	26.1	25.6	25.4	24.9	24.7	24.5	24.5
4	6.0	6.0	6.0	6.1	6.2	6.3	6.6	6.9	7.3	7.6
	25.1	23.7	23.1	22.5	22.2	22.0	21.7	21.5	21.4	21.4
5	5.7	5.8	5.9	6.0	6.1	6.2	6.5	6.8	7.2	7.6
	22.7	21.6	21.1	20.7	20.4	20.3	20.0	19.9	19.8	19.8
6	5.7	5.8	5.8	6.0	6.1	6.2	6.5	6.8	7.3	7.6
	21.2	20.3	19.9	19.5	19.3	19.2	19.0	18.8	18.8	18.8
7	5.7	5.8	5.9	6.0	6.2	6.3	6.6	6.9	7.4	7.7
	20.3	19.5	19.1	18.7	18.5	18.4	18.2	18.1	18.1	18.0
8	5.7	5.8	5.9	6.1	6.2	6.3	6.7	7.0	7.5	7.8
	19.5	18.8	18.5	18.1	18.0	17.9	17.7	17.6	17.5	17.5
9	5.8	5.9	6.0	6.2	6.3	6.4	6.8	7.1	7.6	7.9
	19.0	18.3	18.0	17.7	17.5	17.4	17.2	17.2	17.1	17.1
10	5.9	6.0	6.1	6.3	6.4	6.5	6.9	7.3	7.7	8.1
	18.5	17.9	17.6	17.3	17.1	17.1	16.9	16.8	16.8	16.8
11	5.9	6.1	6.2	6.4	6.5	6.6	7.0	7.4	7.9	8.2
	18.2	17.6	17.3	17.0	16.9	16.8	16.6	16.5	16.5	16.5
12	6.0	6.2	6.3	6.5	6.6	6.8	7.1	7.5	8.0	8.4
	17.9	17.3	17.0	16.7	16.6	16.5	16.4	16.3	16.3	16.2
13	6.1	6.3	6.4	6.6	6.7	6.9	7.2	7.6	8.1	8.5
	17.6	17.0	16.8	16.5	16.4	16.3	16.2	16.1	16.0	16.0
14	6.2	6.3	6.5	6.7	6.8	7.0	7.4	7.8	8.3	8.6
	17.4	16.8	16.6	16.3	16.2	16.1	16.0	15.9	15.9	15.9

Table 9a (Cont'd)

Modified Version of Table 9 to Facilitate Interpolation When the Number of Ranges Exceeds 20. The Upper Entries are to be Divided by the Number of Ranges, the Lower Entries are to be Divided by $10\sqrt{\text{Number of Ranges}}$, then Used Exactly as the Corresponding Entries of Table 9. (See Table 9 for Further Instructions.)

1% ERROR RATE

Number of ranges = Number of columns

Number of differences per range = Number of rows	10	15	20	30	40	50	100	200	500	1000
15	6.3	6.4	6.6	6.8	6.9	7.1	7.5	7.9	8.4	8.8
	17.2	16.6	16.4	16.1	16.0	16.0	15.8	15.8	15.7	15.7
16	6.3	6.5	6.7	6.9	7.0	7.2	7.6	8.0	8.5	8.9
	17.0	16.5	16.2	16.0	15.9	15.8	15.7	15.6	15.6	15.6
17	6.4	6.6	6.8	7.0	7.1	7.3	7.7	8.1	8.6	9.0
	16.8	16.3	16.1	15.9	15.7	15.7	15.6	15.5	15.4	15.4
18	6.5	6.7	6.8	7.1	7.2	7.4	7.8	8.2	8.8	9.2
	16.7	16.2	16.0	15.7	15.6	15.6	15.4	15.4	15.3	15.3
19	6.6	6.8	6.9	7.2	7.3	7.5	7.9	8.3	8.9	9.3
	16.6	16.1	15.9	15.6	15.5	15.5	15.3	15.3	15.2	15.2
20	6.7	6.9	7.0	7.3	7.4	7.6	8.0	8.4	9.0	9.4
	16.5	16.0	15.7	15.5	15.4	15.4	15.2	15.2	15.1	15.1
30	7.4	7.6	7.8	8.1	8.3	8.5	9.0	9.4	10.1	10.5
	15.7	15.2	15.0	14.8	14.7	14.7	14.6	14.5	14.5	14.5
40	8.1	8.3	8.5	8.8	9.1	9.2	9.8	10.3	11.0	11.5
	15.2	14.8	14.6	14.4	14.4	14.3	14.2	14.2	14.1	14.1
50	8.6	8.9	9.1	9.5	9.7	9.9	10.5	11.1	11.8	12.3
	14.9	14.5	14.4	14.2	14.1	14.0	13.9	13.9	13.9	13.9
100	10.9	11.3	11.6	12.0	12.3	12.5	13.3	14.0	14.9	15.6
	14.2	13.8	13.7	13.5	13.4	13.4	13.3	13.3	13.2	13.2
200	14.0	14.5	14.9	15.5	15.9	16.2	17.2	18.1	19.3	20.2
	13.6	13.3	13.2	13.0	13.0	12.9	12.8	12.8	12.8	12.8
500	20.1	20.8	21.3	22.1	22.7	23.2	24.5	25.9	27.6	28.9
	13.1	12.8	12.7	12.6	12.5	12.5	12.4	12.4	12.3	12.3
1000	26.6	27.5	28.3	29.3	30.1	30.7	32.5	34.3	36.6	38.3
	12.8	12.6	12.4	12.3	12.2	12.2	12.1	12.1	12.1	12.1

setting $b_1 = +2$, $b_2 = -1$, $b_3 = -1$, whence the allowance for a simple comparison is expanded by the factor

$$\tfrac{1}{2}\, \Sigma \mid b_i \mid = +2.$$

Other, more complicated, statements are possible.

For those who prefer 1% to 5%, the corresponding factors for 1% are given in Table 5. For more than 20 treatments, the factors can also be obtained from Tables 4a and 5a, where the values tabled are slightly transformed to facilitate interpolation.

7. DOUBLE CLASSIFICATION, "EASY" PROCEDURE

The computation by the "easy" method of the total range for the double-classification example on lactic acid in alfalfa silage has already been illustrated in Table 2. The summary results of the first of the alternatives presented in that table are:

Treatment =	I	II	III	IV	
Totals =	214.6	219.9	158.6	172.6	
Period =	1	2	3	4	5
Totals =	63.8	144.0	196.2	195.7	166.0

Total range of differences = 76.8
(4 ranges of 5 differences each)

Reference to Table 6 shows that the factors by which the total range of differences are to be multiplied in order to convert them into allowances for simple differences in totals are $\begin{Bmatrix} .71 \\ .69 \end{Bmatrix}$. The upper value applies to totals taken in the direction of the ranges of differences, the lower to totals taken in the direction of the differences. Hence the allowances are

$$(.71)(76.8) = 55 \text{ (for treatment totals)},$$

$$(.69)(76.8) = 53 \text{ (for period totals)},$$

and, for example, we have

$$\begin{aligned}
(\text{II}) - (\text{III}) \quad &= 219.9 - 158.6 \pm 55 = 61.3 \pm 55 \\
&= \text{between 6 and 116 in total,} \\
(3) - (1) \quad &= 196.2 - 63.8 \pm 53 = 132.4 \pm 53 \\
&= \text{between 79 and 185 in total,} \\
(3) - (4) \quad &= 196.2 - 195.7 \pm 53 = 0.5 \pm 53 \\
&= \text{between } -53 \text{ and } +54 \text{ in total.}
\end{aligned}$$

Again we may make statements about contrasts, etc. Again there is a table (Table 7) for those who prefer 1%. For more than 20 ranges (columns), the factors can also be obtained from Tables 6a and 7a where the values have been transformed to facilitate interpolation.

8. DOUBLE CLASSIFICATIONS, "INTERMEDIATE" PROCEDURE

The computation of the total range of residuals has been illustrated in Table 3. The relevant summary is:

Treatment =	I	II	III	IV	
Totals =	214.6	219.9	158.6	172.6	
Period =	1	2	3	4	5
Totals =	63.8	144.0	196.2	195.7	166.0

Total range of differences = 53.4
(4 ranges of 5 differences each)

Reference to Table 8 shows that the factors required are 1.15 and 1.10. Hence the allowances for simple differences are

$$\pm(1.15)(53.4) = \pm 61 \text{ (for treatment totals),}$$

and

$$\pm(1.10)(53.4) = \pm 59 \text{ (for period totals).}$$

(We shall return to the fact that ± 61, obtained by this method for treatment totals, is not too similar to the ± 55 found above.)

Thus for example,

$$\begin{aligned}(\mathrm{II}) - (\mathrm{IV}) &= 219.9 - 172.6 \pm 61 = 47.3 \pm 61\\ &= \text{between } -14 \text{ and } +108 \text{ in total,}\\ (3) - (5) &= 196.2 - 166.0 \pm 59 = 30.2 \pm 59\\ &= \text{between } -29 \text{ and } 89 \text{ in total.}\end{aligned}$$

Again we can make statements about contrasts, or with the aid of Table 9, at a 1% error rate. For more than 20 ranges (columns), the factors can also be obtained from Tables 8a and 9a where the values have been transformed to facilitate interpolation.

MORE DETAILED STUDY OF THE EXAMPLES

9. THE NEED FOR MORE CAREFUL CONSIDERATION

No general, clearly specified, statistical procedure can be an adequate substitute for a careful consideration of the idiosyncrasies exhibited by each particular body of data. This is as true of the use of Tables 4 to 9 as it is of either simpler or more complex statistical procedures.

The two examples used for illustration were in no way specially selected. They were not selected to show unusual idiosyncrasies, nor were they selected to avoid them. In fact, however, each one shows, in its own way, a rather clear indication that our simple analysis is likely to be inadequate. It might be argued that this makes these two examples inappropriate applications of short-cut procedures. We do not feel that this is true. There is no reason to believe that they are atypical. They have been analyzed by conventional analysis of variance (Tippett, 1950, p. 107; Snedecor, 1948), without their peculiarities being noticed, although their peculiarities are certainly just as important, if not as prominent, in such a conventional analysis.

All that has happened is that short-cut methods have done a somewhat better job of encouraging muckraking — a better job of making it easy for the user's eye to notice the peculiarities which he must face (even if he does nothing about them). The peculiarities were there, and must have affected the conventional analysis to an unknown and unnoticed extent. It is almost certain to be better to know that they are present, than to overlook them entirely. It is for the good, and not for the bad, if short-cut methods assist in bringing to light such complexities. It is always well to be watchful.

10. THE WARP-BREAKAGE EXAMPLE

The following summary exhibits a striking tendency for the changes in range (from yarn to yarn) to follow the changes in mean:

Yarn	=	*AL*	*AM*	*AH*	*BL*	*BM*	*BH*
Mean	=	44.6	24.0	24.6	28.2	28.8	18.8
Range	=	45	24	33	30	26	15

This tendency suggests, quite definitely, a connection between variance and mean. A little further exploration yields:

Yarn	=	*AL*	*AM*	*AH*	*BL*	*BM*	*BH*
Max.	=	70	36	43	44	42	28
Min.	=	25	12	10	14	16	13
$\sqrt{\text{Max.}}$	=	8.4	6.0	6.6	6.6	6.5	5.3
$\sqrt{\text{Min.}}$	=	5.0	3.5	3.2	3.7	4.0	3.6
Diff.	=	3.4	2.5	3.4	2.9	2.5	1.7

Thus the square-root transformation, often appropriate to counted data, appears to provide substantial help here. The complete computation of an analysis of the square roots is given in Table 10. The results of this analysis are quite similar to those of the analysis in raw values, as may be seen when we compare summaries for the two (linearly coded) quantities

total count — 262

10 (total square root) — 480

with the following results

Yarn	*AL*	*BM*	*BL*	*AH*	*AM*	*BH*	Allowance
Coded Count	139	−3	−8	−41	−46	−93	±123
Coded √	109	−3	−8	−42	−46	−92	±116

The only noticeable effect on the yarn totals has been a moderate reduction in the highest value, that for *AL*. The constancy of the range is considerably improved by taking square roots; consequently the latter analysis seems more likely to be appropriate, and almost certain to be more precise.

Table 10

The Short-Cut Analysis Applied to the Square Roots of Tippett's Breakage Rates

	AL	AM	AH	BL	BM	BH
	5.1	4.2	6.0	5.2	6.5	4.5
	5.5	4.6	4.6	3.7	5.1	4.6
	7.3	5.4	4.9	5.4	4.4	4.9
	5.0	4.1	4.2	4.4	4.0	4.1
	8.4	3.5	3.2	5.4	6.2	3.6
	7.2	4.2	6.6	5.6	5.3	3.9
	7.1	5.9	5.3	6.4	4.6	3.9
	5.1	5.5	3.9	4.5	6.2	4.0
	8.2	6.0	5.1	6.6	5.4	5.3
Total	58.9	43.4	43.8	47.2	47.7	38.8
Range	3.4	2.5	3.4	2.9	2.5	1.7

Total range = 16.4,

Factor (Table 4) = .71,

Allowance for simple differences among totals = ±11.6.

11. THE LACTIC-ACID-IN-ALFALFA-SILAGE EXAMPLE

Here the variations in the results for the allowance to be applied to the treatment totals are striking. The use of different column orders, and the use of different approaches, leads to the following results:

Method	Order	Total Range of Diff's.	Factor	Allowances for Treatment Totals
"easy"	I-II-III-IV-I	76.8	.72	±55
	I-III-II-IV-I	101.9	.72	±73
	I-II-IV-III-I	80.5	.72	±58
	(mean)	(86.4)	—	(±62)
"intermediate"	—	53.4	1.15	±61
"intermediate"	*	64.6	1.00	±65
(use of *SSD* and s^2)	—	—	—	±66

* With rows and columns interchanged.

The most striking fact is the variation from ±55 through ±58 to ±73 for the values obtained by the "easy" method. Since only 12 degrees of freedom are available at best, variation of this magnitude between replicate experiments would in no way be surprising. But so much variation within a single experiment, due only to the order in which the columns are written down, is somewhat disturbing. In this particular example, it would clearly be unwise to use the "easy" method if there were any great conflict of opinion over the results, since the conflicting parties would be rather likely to enter upon an unprofitable discussion as to whether the order giving ±55 or the order giving ±73 is a better choice.

If these large differences were typical of double-classification experiments, the "easy" method would not be of much use. Fortunately, there are two reasons why this is not so. First, an increase in the number of columns tends to stabilize the total range. Thus, since order of columns makes no difference for 2 or 3 columns, the present example illustrates the case, that of 4 columns, most sensitive to changes of order. Second, there is almost surely more than just random variation back of the residuals in this particular example. If we look at Table 3 searchingly, we begin to detect some system to the pattern of variation. When we calculate the two-way residuals

$$\text{(observed value)} - \text{(column mean)} - \text{(row mean)} + \text{(grand mean)}$$

we find the following results:

	Treatments				
Periods	I	II	III	IV	dev.
1	−7.2	−5.7	5.0	7.8	(−22)
2	−3.1	1.0	−0.1	2.3	(−2)
3	11.6	0.2	−6.0	−5.5	(11)
4	7.3	2.5	−3.2	−6.4	(10)
5	−8.4	2.0	4.5	2.0	(3)
dev.	(5)	(6)	(−7)	(−4)	

These values show a rather consistent pattern in relation to the deviation [row mean-grand mean] and [column mean-grand mean] shown in parentheses. A simple test for nonadditivity (Tukey, 1949b) applied to the original data is significant and shows that period 1 appears to be behaving quite differently from the others. After removing period 1, the remainder of the table continues to show non-additivity. After lactic acid content is expressed logarithmically, a further reanalysis shows that the nonadditivity of the table of logarithms omitting period 1 falls far short of significance.

When the logarithms of the observations for periods 2 to 5 are analyzed, as shown in Table 11, the results are as given in Table 12. The agreement of the three orders is now quite good enough for most practical purposes. We can say this because we are not solely concerned with a nominal 5% level. If we only wished to say "significant" or "not significant," we might be concerned with the fact that the difference between period 3 and period 2 comes out

$$2.714 - 2.209 \pm .43 = \text{between } +0.07 \text{ and } +0.94$$

for one order of taking differences, and

$$2.714 - 2.209 \pm .51 = \text{between } -0.01 \text{ and } +1.02$$

for the other. If we live by a rigid, entirely arbitrary, 5% standard of significance, one of these is "significant" and the other is "not significant." From a practical and reasonable point of view, however, there is relatively little difference between the confidence statements. For almost all purposes, the "easy" method is satisfactory for this modified example. And if we have a special need for apparent precision, the extra work involved in the "intermediate" method is small.

Table 11

Logarithms of the Lactic Acid Concentration in Alfalfa Silage ($\log_{10}$ of Concentration in Tens of mg Per Gram) with Differences and Means

	Treatment									(Other Orders)	
Period	I	Diff.*	II	Diff.*	III	Diff.*	IV	Diff.*	Total	Diff.*	Diff.*
2	.574	(+56)	.630	(−163)	.467	(+71)	.538	(+38)	2.209	(−107)	(−92)
3	.814	(−57)	.740	(−179)	.561	(+38)	.599	(+215)	2.714	(−253)	(−141)
4	.784	(−27)	.757	(−165)	.592	(−4)	.588	(+196)	2.721	(−192)	(−169)
5	.576	(+116)	.692	(−96)	.596	(3)	.599	(−23)	2.463	(+20)	(−93)
Total	2.748		2.819		2.216		2.324				
Range		.173		.083		.075		.238		.273	.077
		(II-I)		(III-II)		(IV-III)		(I-IV)		(III-I)	(IV-II)

* In units of the 3rd decimal.

Table 12

Allowances for Totals Obtained by Analysis of the Data in Table 11

Method	Order	Total Range of Differences	Factor	Allowances (5%) for Totals
"easy"	I-II-III-IV-I	.569	.76	±.43
"	I-III-II-IV-I	.671	.76	±.51
"	I-II-IV-III-I	.598	.76	±.45
"	mean	(.613)	(.76)	(±.47)
"intermediate"	—	.368	1.21	±.45
"intermediate"	*	.433	1.21	±.52
(use of *SSD* and s^2)	—		—	±.50

* With rows and columns interchanged.

CHOOSING A MULTIPLE-COMPARISONS PROCEDURE

12. WHEN SHOULD MULTIPLE COMPARISONS BE MADE

The possibility of multiple comparison arises when we have both indicated or apparent values — usually mean responses for each "treatment" of a family of more than two "treatments" — and a measure of the variability of the differences of these indicated or apparent values. [In order for one measure of variability to serve, either the results for the various "treatments" must be considered equally variable, which is the case treated in this paper, or else assumed ratios of the variabilities of the various differences and contrasts must have been specified. At the other extreme multivariate methods would be required (e.g., Roy and Bose, 1953, Section 4; Scheffe, 1959, Section 8.1).] Here, as noted earlier, "treatments" may indeed be treatments, i.e. sets of conditions, may be species, varieties, races, or strains of plants or animals, may be types of raw material, fertilizer, or test tubes, may be names of operators or analysts, may be any describable or identifiable generalities whose effects it is desired to compare. But not every situation in which multiple comparison is possible is a situation in which multiple comparison is wise.

In one large class of exceptions, regression, on some specific expression or expressions of some auxiliary observable quantity, is the most appropriate beginning toward analyzing the data. Most situations involving sets of conditions which differ by the values of a single quantitative variable, or which correspond to points along a single, reasonably smooth curve in the space defined by several quantitative variables, are best treated by regression. Of course, regression on *several* expressions may give rise to new problems of multiple comparisons.

Another major exception arises when two subsets of the treatments are appropriately compared in terms of the means, for the two subsets, of the observed mean responses. When such comparisons arise, it is almost inevitable that the "treatments" of a subset are regarded as representative of, as a sample from, a larger collection of "treatments," most of which are not represented in the data. As a consequence, the correct "error term" must be based upon the differences *among* the (mean or total) *responses* to different "treatments" belonging to the same subset. In such instances, using the "error term" corresponding to routine multiple-comparisons procedures, which are based upon differences in repeated response under the same "treatment," is definitely wrong.

A third large class of exceptions arises when only a few comparisons will be made among the mean responses to the various "treatments," no matter how these mean responses may behave. In such a situation the error rate which may properly be spent on comparisons among a family of "treatments" should be spent on these few comparisons, and not wastefully spread over all possible differences among mean responses.

In most other situations where typical responses to a number of treatments are to be compared, the use of the appropriate sort of multiple-comparisons procedure is indicated.

Even in the cases of regression or subset-subset comparison, where multiple comparisons should definitely *not* be used to assess either the definiteness of sign of, or the interval of confidence for, the differences in question, there is an appropriate role for multiple-comparison thinking, and for multiple-comparison technique. This role is associated with multiple-comparison applications to residuals of mean responses, either residuals after regression or residuals from subset means. The process of taking residuals will lead to somewhat reduced variability, as compared with the original mean responses. Consequently application of conventional multiple-comparison procedures to residuals will lead to error rates lower than nominal. This effect tends to compensate for the fact that both (i) regression or comparison between subset means, and (ii) residuals, have been assigned separate error rates. Thus it is often reasonable to use conventional multiple-comparisons procedures on residuals without taking any account of the change in error rate. (Note that this use is only for the purpose of studying irregular variation. Regular variation, e.g. curvature where the regression was first thought of as straight, should first be studied by fitting a more general regression that is flexible enough to describe both expected and feared variation. And the multiple-comparisons technique is usually best applied to the residuals from the more complex fit, even if the additional terms were not significantly different from zero.)

13. WHAT SORT OF A MULTIPLE-COMPARISON PROCEDURE SHOULD BE SELECTED?

Multiple-comparison procedures are in use, or have been suggested, that have very different aims, whether we consider their "true" aims or look only at their formal structure. Some aim at a decision, a pointing-out of some action as "best" under specific assumptions about loss functions. Others aim at a conclusion, an establishment of a statement as strongly confirmed by the specific data being analyzed. Still others aim at a selection, a reduction of a list of candidates (usually as one stage of a selection process) in a way that provides safety or efficiency or some desirable compromise between the two.

Those which lead to conclusions need to be further divided into those which seek only to conclude about significance and those which seek to establish confidence statements. The former typically classify every difference between a pair of "treatment" means either as "significant" (implicitly or explicitly as "significantly >" or "significantly <") or as "significance not established, difference may be zero or of either sign." The latter typically equip each observed difference between two "treatments" and, usually, each contrast among "treatments," with a confidence interval within which it has been "established" to fall. The probability of any of these simultaneous statements being incorrect is then controlled at a known and agreed-upon level, at least when presuppositions about independence and normality of distribution hold to an adequate approximation.

There are places for methods of all these natures. By and large, however, these places are separate and do not overlap. A complete discussion of where these places are, and how they are distinguished, which went into sufficient detail to be helpful, would require much more space than it is appropriate to use here. Only an extremely sketchy discussion is thus in order. In particular, we make no attempt to review the extensive literature on multiple-decision procedures or on multiple-comparison procedures concerned solely with significance statements.

The same considerations which are appropriate in guiding an investigator to use decision, significance, confidence, or selection in comparing two simple samples, apply to the choice among the same alternative forms of answer in a multiple comparison situation. The balance between considerations, however, is noticeably altered. There are few multiple comparison methods which seek to meet a true decision aim, though many cast their formal discussion in terms of decision theory. The reason is not far to seek. True decision aims require the use of loss functions which are at least moderately reasonable. A loss function for the mis-evaluation of a single difference between two population means can be a function of a single variable. Not infrequently one can convince oneself that a particular such loss function is sensible. But a loss function in a multiple comparison situation is a function of several variables, and it is rare indeed that its reasonability is convincing.

If decision be set aside as infrequently useful, there remains the disjunction between conclusion and selection. Here the whole nature of experimentation plays a role. Is an isolated experiment the true object of concern? Or is this experiment best assessed as one phase in a continuing process of experimentation? An adequate answer here must cut extremely deeply into the foundations of experimentation.

Experiments in some fields, e.g. many orchard trials, take so long that it is felt (Pearce, 1953, p. 65) that single experiments must be assessed on their merits, almost entirely without regard to the possibility of checking upon apparent results by further experimentation. In most fields of experimentation, on the other hand, iterative development of ideas, and reliance on confirmation (or disconfirmation) in repetitions by other workers are important. Much more work on experimentation as an iterative process is certainly needed. And as this work is done, there will inevitably be an increased emphasis on selection, in a very general sense, as the approach of choice in considering what to do next.

It may be argued that selection is only a tiny part of iterative experimentation or observation. In the narrowest sense of the word this is true. Relatively few experimenters follow the plant breeders' pattern of successive stages of elimination of candidates. But selection of which hints to follow up, of which things to try out or seek for next, are key steps in iterative experimentation or iterative observation. In the broad sense, they are instances of selection, (or of internal decision, if you will) which are vital to progress.

Increased emphasis on larger sections of experimentation will not, however, eliminate or reduce the importance of the assessment of the results of a single experiment. For the results of single experiments need to be communicated to other investigators. And communication in *conclusion* terms appear to be the most effective mode of communication we know (Tukey, 1960).

In providing multiple-comparison *conclusion* procedures, then, we are helping to solve only one class of multiple-comparison problems. However, this class is important now, and will be important throughout the foreseeable future.

14. SIGNIFICANCE OR CONFIDENCE?

Having chosen to provide a conclusion procedure, shall it provide significance statements or confidence statements? The answer to this question is very important in terms of basic aims, and, as we shall see, very important to the user of the resulting technique in a way often neglected. As a matter of practical mathematical statistics, i.e. as a matter of studying distributions and preparing tables, its importance is very small. Essentially the same problems of distribution, approximation and tabulation have to be met in either case. This is so, since (i) confidence procedures include significance procedures as special cases, and (ii) the contemplation of many alternative null hypotheses of specific, non-zero, differences will convert a significance procedure into a confidence procedure. As a consequence, the same tables meet the needs of those who desire to test significance and those who desire to make confidence statements. Why, then, should one be emphasized over the other?

The basic advantage of confidence statements is their informativeness. This greater informativeness takes two forms. They answer not the one question: "Is it reasonable that the true difference is zero"? but the whole sequence of questions, one for each value of A: "Is it reasonable that the true difference is A"? And, as will be discussed below, they allow greater flexibility in choosing the standard of reasonableness.

Greater informativeness has an important indirect consequence. Since a significance statement has but two possibilities, of which one is usually considered a failure, there is severe pressure to change the nominal level and bring results up to "significance." On the other hand, changes in confidence level make relatively much less difference in the appearance of the conclusion, and the pressure to change levels is moderate. We discuss this issue further in Section 17.

The confidence procedures have a greater flexibility for major changes in level of statement. The thoughts of a user with a quantitative problem may, for example, run along these lines: "On the face of the experiment the difference is B units, statistical analysis at a 5% level shows that it lies between C units and D units, but just how sure am I that it is less than E units"? To supply an answer in the form which will help this user best, there are two choices:

(i) We can build up for him a wholly new method of analysis; or

(ii) we can show him how to bend an already available method to his needs.

If feasible, it is simpler to choose the latter.

Yes-or-no significance procedures tend to be rather unbending once calculation is completed and its details forgotten. Without redoing most of the calculation, it is hard indeed to say whether "the ith is less than the jth at 5%" can or cannot be strengthened to "the ith is less than the jth at 1%." In a confidence procedure, however, we always have a convenient basis for bending. If a 95% interval from C to D is given, we know that the interval for 80% confidence will be somewhat narrower, that the interval for 99% confidence will be somewhat wider, and that the interval for 99.9% confidence will be substantially wider. While the ratios by which this widening will take place depend, to a moderate degree, on characteristics of the situation other than the nominal levels, it is still quite easy to develop an adequate intuition about the amount of change required.

These ratios, expressed in terms of the widths of confidence intervals, are not so far from unity as we might think. (For many of us, frequent use of the F-table at different levels of significance has developed a quantitative intuition inappropriate to the present situation. This has happened because tabular F-ratios correspond to ratios of *squares* of interval-widths.) The values presented in Table 13 should be an adequate guide to our intuition.

Table 13

Relative Widths of Confidence Intervals for Significance Levels of 20%, 5%, 1% and 0.1%*

Number of Means or Totals Compared**	Significance level = Error rate = 100% — Confidence				Degrees of Freedom of Error Term
	20%	5%	1%	0.1%	
2	0.59	1.00	1.51	2.44	6
	0.62	1.00	1.40	1.98	12
	0.66	1.00	1.31	1.68	∞
3	0.62	1.00	1.45	2.29	6
	0.69	1.00	1.34	1.85	12
	0.73	1.00	1.24	1.52	∞
4	0.67	1.00	1.43	2.23	6
	0.72	1.00	1.30	1.76	12
	0.77	1.00	1.21	1.44	∞
10	0.70	1.00	1.39	2.14	6
	0.77	1.00	1.25	1.63	12
	0.85	1.00	1.13	1.29	∞
50	0.72	1.00	1.37	2.08	6
	0.81	1.00	1.22	1.56	12
	0.91	1.00	1.07	1.13	∞
∞	0.73	1.00	1.37	2.07	6
	0.82	1.00	1.21	1.54	12
	1.00	1.00	1.00	1.00	∞

* Entries are ratios of two values of $\sqrt{F}$.

** One greater than numerator degrees of freedom.

The values in this table were obtained as ratios of values of $\sqrt{F}$, but they will approximate, to an accuracy adequate for the calibration of intuitions, the appropriate ratios of widths for any multiple-comparison-by-simultaneous-confidence-statement procedure.

Some users of significance make all statements in terms of the *P*-value, the significance level at which the observed result would be "just significant." They are immediately able to judge significance of deviations from the null hypothesis at any level. They *could* develop a quantitative intuition about the significance of deviations from other contemplated values of the true difference. To do this, they would need to combine two things: the critical *P*-value *and* the apparent difference. Experience to date suggests that they are unlikely to do this.

MORE DETAILED ASPECTS OF CHOICE

15. WHAT EMPHASIS SHOULD BE GIVEN TO DIFFERENT CONTRASTS?

There are two ways to approach this question. We may list various kinds of contrasts, starting with the simple comparisons of one treatment mean with another and then discuss the relative importance of shorter or longer confidence intervals for contrasts of various types. Alternatively, one may observe that choosing emphasis on different contrasts is equivalent to ordering the different sets of outcomes as to their relative discrepancy. Here there can be pictures, and data-based intuition may come more easily into play. In the present section we state, without proof or extensive discussion, the most cogent arguments so far made for and against various patterns of emphasis and state the conclusions we have reached. In the following section we illustrate the relative discrepancy approach.

Four criteria of relative discrepancy are involved in multiple-comparisons procedures which have been seriously proposed for use in the sort of situation we are discussing. They are:

(1) sums of squares of deviations (Scheffé, 1953; see also Tukey, 1953b);

(2) ranges (Newman, 1939; Keuls, 1952; Tukey, 1953b);

(3) deviations of the extreme result from the mean of the remaining results (or equivalently from the mean of all the results). (Halperin et al. 1955; see also Tukey, 1953b);

(4) widest gaps (between adjacent ordered results) (Tukey, 1949a, 1953a, b).

Arguments of the following sorts have been, or can be, advanced for (and against) these four criteria:

The SSD:

(1) The distribution of the resulting key statistic is more robust, is less dependent on the shape of the distribution of the errors.

(2) The confidence limits placed upon those contrasts which assign substantial coefficients to *each of several or many* "treatments" are shorter than those provided by the other three criteria. (But the most important contrasts involving many non-zero coefficients are (i) estimates of regressions, and (ii) comparisons of sub-group means, neither of which is appropriately treated by such multiple-comparison procedures, the former requiring the separation of specific degrees of freedom, and the latter requiring the use of a within-subgroup error-term.)

(3) If the errors should be precisely normally distributed, its use would correspond to a likelihood-ratio criterion. (But are they precisely normally distributed? And is there any special *practical* virtue to such a likelihood-ratio?)

The range:

(4) Simple comparisons (comparisons of one "treatment" with another) are both more frequent and more important than any other contrasts. The confidence limits for simple comparisons are shortest when the range is used. (This choice is "unfair" to other contrasts. Considerations of importance should not enter. Robustness is extremely important.)

The extreme deviations:

(5) It is important, above all, to identify "treatments," which are very unusual by comparison with the remaining treatments. Ordering by extreme deviations does this. (While identification of extreme "treatments" is *sometimes* the important thing, this is not so in *most* investigations.)

The widest gap.

(6) The most effective qualitative conclusion is one which separates all "treatments" into two groups and states that those in the one group gives smaller results than those in the other. Using the gap criterion leads us to make such statements most frequently. (These are not the only useful statements. Until this use of the gap is modified to provide more general conclusions it is impractical. And in any event

a gap-emphasizing procedure is only useful in significance situations.)

Our conclusions are these:

(A) The range seems best for general use.

(B) The *SSD* should sometimes take precedence when robustness is highly essential, or when extreme non-normality is feared. (See Scheffé (1953 or 1959, Section 3.4) for techniques.)

(C) The extreme deviation should be used when the detection of single disparate "treatments" is unusually important (See Halperin et al. (1955) for techniques.)

(D) The use of procedures emphasizing gaps should be confined to situations of pure significance testing, and should await the development of procedures which appropriately combine use of gaps and emphasis on other criteria.

Short-cut procedures could be provided combining any one of these criteria of relative discrepancy with a measure of variability based upon ranges. Different tables of critical factors would be required in each case. (Since there would be a loss in robustness in using ranges for variability estimation, such a procedure is likely to be undesirable when the *SSD* is needed to assess relative discrepancy.) The short-cut procedures described in this paper are appropriate to conclusion (A) and are thus recommended for *general* use.

16. CHOICE OF EMPHASIS IN TERMS OF RELATIVE DISCREPANCY

Each confidence procedure must contain within itself a significance procedure for every conceivable null hypothesis. Often these significance procedures are equivalent to one another in some simple way. For each of the multiple-comparisons procedures we are discussing, equivalence is obtained by rigid translation ("slippage"). Thus it will suffice to consider, for our present purpose, only the nullest null hypothesis of all, the hypothesis that all treatments have the same effect.

Any significance procedure must arrange the possible results to which it applies in a rigid order. Any two sets of "treatment" means must, in a multiple-comparisons procedure, be regarded either as precisely equally discrepant, or as ordered in discrepancy. Frail human judgment, when confronted with cases where the degree of discrepancy is clearly nearly the same, often wishes to avoid making fine judgments of relative discrepancy by classing many possible outcomes as equally discrepant. But an objective, precisely specified, significance procedure cannot do this. Thus human judgment has to be applied to fine distinctions if it is to guide choice among different patterns of emphasis (of the various types of contrast).

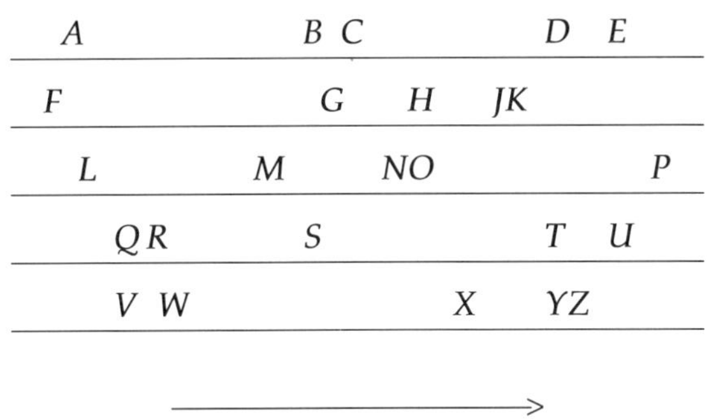

Figure 1. Alternative sets of five outcomes ("Treatment" Means) selected to illustrate different orderings of discrepancies by different criteria. (Estimated variability supposed to be the same in all cases.)

Figure 1 illustrates a number of possible outcomes when five "treatments" are intercompared. These have been carefully chosen to be of almost balanced discrepancy. In fact, the orders of relative discrepancy according to the four criteria discussed in the last section are as follows:

SSD:	*QRSTU* > *VWXYZ* > *LMNOP* > *ABCDE* > *FGHJK*
range:	*LMNOP* > *QRSTU* > *ABCDE* > *VWXYZ* > *FGHJK*
extreme deviation:	*FGHJK* > *ABCDE* > *LMNOP* > *VWXYZ* > *QRSTU*
gap:	*FGHJK* > *VWXYZ* > *ABCDE* > *QRSTU* > *LMNOP*

where ">" is to be read "is more discrepant than." The five sets of outcomes have been so chosen that the ordering between each pair of sets is in one direction for two criteria and in the other for the other two.

The question to be put to our intuitions and data-trained feelings is the following: "In the type of experimental or observational work with which I am concerned, which of these four conclusions is it most important to reach:

(1) that Q, R, S, T, U are not all alike (and perhaps that V, W, X, Y, Z are not all alike);

(2) that L is to the left of P (and perhaps that Q is to the left of U);

(3) that F is to the left of the *mean* of G, H, J and K (and perhaps that A is to the left of the *mean* of B, C, D and E)

(4) that F is to the left of *all* of G, H, J, and K.

All preference for one of these corresponds to a preference for a multiple comparisons procedure

(1) based on the use of the *SSD* (in the numerator);

(2) based on the use of the range (in the numerator);

(3) based on the use of the extreme deviation (in the numerator);

(4) based on the use of the widest gap (in the numerator).

Where lies *your* preference?

CHOICE OF ERROR RATES

17. HOW SHOULD ERROR RATES BE DESCRIBED?

Even after choosing an approach, and choosing a relative emphasis on different kinds of contrasts, on different aspects of discrepancy, there remains a choice of how to express error rates. In the present paper we have chosen to take a 5% error rate as meaning that in 19 out of 20 experiments *all* the detailed confidence statements referring to a single family of contrasts will hold simultaneously. Others would be willing to apply "5%" to a much looser statement. It might seem that such an issue was purely a matter of words, wholly without substance, that so long as an agreement which permitted unequivocal communication was reached, it did not matter what the agreement was.

Such a view disregards two extremely important considerations: the force of custom, and the need for realizing effective communication. Many investigators, and not a few editors, have allowed 5% and 1% to play an oversize role in their lives; most of these people are not going to rethink their attitudes to 5%, just because some statisticians agree upon how to define 5% for multiple comparisons. Cultural lag is real, and large; decisions about the meaning of 5% must recognize its presence.

We have adopted the more conservative standard of charging errors against the confidence procedure on the basis that if one or more confidence statements about the "treatments" of a single family fails, that family has been erroneously evaluated. Thus a 5% error rate (per family) means that all the confidence statements about 19 families out of 20 are to be correct. This is the standard of the *F*-test of significance, which was the sole standard of the earliest days of the analysis of variance. It is not the watered-down standard of 5% error rate for each of the single degrees of freedom that may be incautiously pulled out without reason (so long as these single degrees of freedom involve a suitable independence in fluctuation, however dependent they may or may not be in interpretation). It is, we feel, the appropriate standard for a conclusion procedure which is supposed to identify as many results as "established by the data" as is reasonable, that is, "in such a way as not to have to eat one's words too often." This is the natural path of growth of knowledge for the classical sciences.

We do not say that this is appropriate when 5% is really used for some other, quite different, purpose. (Though in most of these other situations the whole ideas of 20%, 5% or 1% might well be avoided.) Thus if the real purpose of statistical analysis is guidance in deciding what experiment to do next, or what observations to take next, then a 5% familywise error rate is quite often, indeed probably usually, far too stringent. The appropriate cure is not merely to change the familywise error rate. Instead, suitable methods for such guidance should be provided which are essential and explicitly different from those which are to be used to draw conclusions about what has been established.

The state of a field which has come to live almost solely by tests of significance, and where acceptance or rejection of papers can depend on whether the test of significance yields "significant at 4%" rather than "significant at 6%" is a sorry one. We can sympathize with potential authors in their desire to somehow reach the magic 5%, but we doubt that the appropriate way out is to give up the good scientific standard that: *Most of what is said to be established should turn out to behave in the long run as it is stated to behave*. And we believe in the scientific value of results which show that the difference between "treatments" *A* and *B* is *less* than some well-defined amount.

18. HOW SHOULD ERROR RATES BE CHOSEN IN COMPLEX SITUATIONS?

A number of situations have already been mentioned in which two families of comparisons are associated with a single body of data. In many double-classification situations, comparisons are made within both families of treatments separately. And we have already suggested following regression or subset-mean analyses with multiple comparisons among residuals. No great concern has been shown about the fact that there were two separate error rates, that 5% and 5% might come close to 10%, that 1% and 1% might come close to 2%. This was because only *two* families of comparisons were involved. The practical nature of the situation is very different when there are five, twenty or a hundred families.

There are those who argue that 50% instead of 5% differs from 10% instead of 5% only in amount, not in nature. True. The practical importance of the difference is substantial, however. A single factor of 2 is barely recognizable, but the combination of several factors of two is usually of vital importance.

In a complex experiment there may be many families within which comparisons are in order. How shall the overall error rate be controlled and adjusted? Statisticians have faced up to one instance of this rather recently. Factorial experiments in a 2^n pattern, and fractional replications of such patterns, have the property that the natural families for intercomparison are all of the minimum possible size, each involving a single degree of freedom. The unsatisfactoriness of spending a 5% error rate on each and every one of these single degrees of freedom had become evident to a number of statisticians ten years ago. One aspect of the introduction of the "half-normal plot" analysis for such experiments (Daniel, 1959), has been the readjustment of the use of error rates. The half-normal plot analysis, in its original form, tries to spend one 5% error rate on all the single degrees of freedom put together.

Another indication is to be drawn from the carefully thought-out approach to horticultural experimentation developed at East Malling (Pearce, 1953) where (pp. 64-66) comparisons or families of comparisons "nominated" in advance of the experiment are each assessed individually, while those comparisons, or families of comparisons, whose analysis was planned for but which were not "nominated," are pooled together and screened as a whole, individual assessments being made only where the pool is "significant."

The conclusion seems inevitable. We believe that it will slowly become widely recognized that budgeting an appropriate error rate for the various comparisons, or families of comparisons, of a complex experimental or observational pattern is a natural and appropriate part of the *planning* of the experiment, or of the observational program. Other error rates than 0% (i.e. "don't look"), 1%, and 5% will be needed in more complex situations. While no explicit provisions for such additional possibilities are provided in the present paper, the three levels of 0%, 1%, and 5%, together with intuitive feelings about the effect of changing probability levels on the width of confidence intervals, can take us a long way.

CHOICES ABOUT AND AMONG SHORT-CUTS

19. HOW SERIOUS IS THE LOSS OF STRINGENCY?

The remark that "the range is an awfully inefficient indicator of variability" is made many times a day in one form or another. Like many common remarks, it is both true and misleading. The reason why it is misleading when variability is assessed as a part of a multiple-comparisons procedure is simple, though rarely stated. *Crucial comparisons always involve ratios*

Table 14

Average Lengths of 95% Confidence Intervals Obtained with the Studentized Range.* (Average Length for Known σ Taken as 1.00 in each Case.)

Number m of "Treatments" Compared	Equivalent number of degrees of freedom available				
	$v = 6$	$v = 8$	$v = 12$	$v = 24$	$v = \infty$
2	1.20	1.14	1.09	1.04	1.00
3	1.26	1.18	1.12	1.05	1.00
5	1.32	1.23	1.14	1.07	1.00
10	1.39	1.28	1.18	1.09	1.00
20	1.45	1.33	1.21	1.10	1.00
50	1.52	1.38	1.25	1.12	1.00
100	1.55	1.41	1.27	1.14	1.00

* Entries are the product of the ratio $q\,(m, v\,; 5\%)/q\,(m\,; \infty; 5\%)$ and the average value of s on v degrees of freedom.

whose numerators have mean values which are not large compared to their standard deviations. Thus the coefficient of variation of the denominator need not be very small for *its* size to be of minor importance in determining the distribution of the ratio.

The conventional comparisons of efficiency of estimation overstate the importance of a precise denominator about as far as is reasonably possible.

Table 15

Relative Efficiency in a Double Classification when m Ranges of n Items are Used in "Easy" and "Intermediate" Methods to Compare m "Treatments"

					Behavior Relative to *SSD*			
		Equivalent Degrees of Freedom*			Efficiency of Estimation of Variability**		Efficiency of Confidence Intervals***	
m	n	"Easy"	"Int."	*SSD*	"Easy"	"Int."	"Easy"	"Int."
5	2	3.26	3.84	4	.81	.96	.81	.96
	5	12.7	15.1	16	.79	.94	.95	.988
	10	26.6	31.4	36	.74	.87	.97	.987
	20	48.4	56.4	76	.64	.74	.98	.986
	100	138.	154.	396	.35	.39	.985	.988
20	2	12.6	17.3	19	.66	.91	.88	.98
	5	49.9	69.4	76	.66	.91	.97	.989
	10	105.	144.	171	.61	.84	.98	.994
	20	192.	254.	361	.53	.70	.988	.994
	100	546.	656.	1881	.29	.35	.994	.995
100	2	61.5	88.1	99	.62	.89	.96	.992
	5	248.	359.	396	.63	.91	.990	.998
	10	523.	741.	891	.59	.83	.995	.998
	20	956.	1301.	1881	.51	.69	.997	.998
	100	2722.	3319.	9801	.28	.34	.998	.999

* The entries are 2 (average value)^2/variance.

** The entries are ratios of equivalent degrees of freedom.

*** The entries are the squared ratios (average width using *SSD*/average width using ranges)^2 which apply under normal theory.

A reasonable balance and counteractant is a table of relative average widths of confidence intervals. Such a table is easily computed for intervals obtained from the studentized range, and appears as Table 14.

For the double classification with m columns and n rows, Table 15 compares the efficiency of the total range estimates of variability used in the "easy" and "intermediate" procedures with the interaction mean square estimate used in the conventional *SSD* procedure. The stability of each estimate may be represented by an "equivalent degrees of freedom" as illustrated in Table 15 and as detailed in part 2 of this paper.

Two measures of relative efficiency are shown. One, appropriate when variability *per se* is estimated, is the ratio of the equivalent degrees of freedom (or, approximately, the reciprocal of the ratio of the variances of the two estimates). The other, appropriate when variability is estimated to allow intercomparison of, say, column means, is the square of the ratio of average widths of confidence intervals using the two estimates. The latter ratio is squared to make the two measures properly comparable.

The comparisons typified by Table 15 show that for the uses of this paper, the total range estimates are highly efficient and losses of efficiency are very small except when both the number of ranges and the number of items per range are small.

20. RECOMMENDATIONS

We have endeavored to formulate some conservative recommendations about the use of these short-cut methods. Our results run as follows:

(1) The short-cut procedure for single classifications can be safely used in almost every case. (Attention to the use of transformations to obtain reasonably constant variance will be worthwhile here, just as with other methods.)

(2) In extremely pathological cases, or where the results of the analysis are likely to come under sharp attack, the initial computation of the short-cut analysis for a single classification should later be replaced by a more conventional (multiple-comparisons) analysis based on sums of squares of deviations and the critical values of the studentized range.

(3) The "easy" short-cut procedure for double classifications is worthy of routine application as an initial analysis to data in this form.

(4) If either moderate pathology is suspected, or, the results of the analysis are likely to be moderately controversial, an "intermediate" analysis should be made to replace the "easy" one (or we should move directly to the next step).

(5) If either moderately severe pathology is expected, or if the results are likely to be solidly controversial, an analysis based on the *SSD* and the studentized range should be made to replace the short-cut analysis.

(6) In almost every case, and probably always, an analysis by even the shortest short-cut method is much better than no analysis at all.

21. NATURE AND HISTORY OF THE APPROXIMATIONS USD TO COMPUTE THE TABLES

The factors presented in Tables 4-9 are approximate percent points of statistics each of the form of a ratio of a range to a sum of ranges. The total range in the denominator appears as an indicator of variability, and may be usefully regarded as being of the form

$$\text{(constant) (estimate of the underlying variance)}^{1/2}$$

where the estimate is unbiased for normally distributed errors. It is tempting to approximate to the situation by replacing these total ranges by δs, where s_v has the distribution of a square root of a mean square on v degrees of freedom and where both v and the constant δ are suitably chosen. If this may be done, then each ratio will have, approximately, the distribution of a studentized range (times a constant), and approximate percent points of the ratio may be obtained from the available tables of percent points of the studentized range. The detailed choices and applications underlying Tables 4-9 are presented in part 2 of this paper (Kurtz et al. 1965).

The first to follow this tempting path seem to have been Patnaik (1950) for the case of the one-way classification, and Hartley (1950) for the case of the two-way classification. Both authors contented themselves with giving the constants and equivalent degrees of freedom n needed to use the studentized range. Both authors were concerned, moreover, with the problem of significance testing rather than the problem of multiple comparisons. They faced all the distribution problems, but did not produce tables designed for the greatest ease in use. (Hartley's tables require some numerical revision in view of results obtained in part 2 of this paper.)

The first appearance of complete short-cut methods was in a talk by one of the authors (Tukey) before an American Society for Quality Control conference in Baltimore in 1950. This was based on work by Lord (1947). Further developments, leading to the present papers, were presented by various of the authors, and preliminary tables appeared in Tukey (1953a).

APPROXIMATE FORMULAS

22. EXTENDED-RANGE FORMULAS

Part 2's detailed study of how to approximate the needed percent points leads to approximations of several degrees of precision. The more precise approximations, used for the calculation for the parts of Tables 4 to 9 for 20 or fewer ranges, each of 20 or fewer items, make use of the best available tables of the studentized range, and provide relatively high accuracy through moderate labor. The less precise approximations use empirical representations in terms of polynomials, square-roots, and logarithms, but use no values from special tables; these provide less precise, but entirely adequate, values with less effort. If either the number of ranges, or the number of items per range is large, approximations of the latter sort are the most precise approximations presently available.

These approximate formulas of lesser precision are likely to prove useful:

(1) for inserting the critical values of these procedures into stored-program computers; and

(2) as a means of extending the tables;

and are now given. Let m denote the number of ranges (number of columns) and n the number of items per range (number of rows). Tables 16 (for $m, n \geq 3$) and 17 (for either m or $n = 2$) specify functions $\phi_{u,\alpha}$, $\theta_{u,\alpha}$, Ψ_u so that

$$\text{Table entry} \approx \phi_{u,\alpha}(m, n)[1+\theta_{u,\alpha}(m, n)\Psi_u(m, n)].$$

Table 16

Components of Extended-Range Approximations for $m \geq 3$ and $n \geq 3$

u	$\phi_{u,\,5\%}$	$\phi_{u,\,1\%}$	$\theta_{u,\,5\%}$	$\theta_{u,\,1\%}$	Ψ_u	Approximation to Tables
1	$f_1(m)/h_1(n)$	$1.02g_1(m)/h_1(n)$	$f_2(m)$	$g_2(m)$	$h_2(n)/m$	4,5
2	$.7071f_1(m)/h_1(n)$	$.7215g_1(m)/h_1(n)$	$f_2(m)$	$g_2(m)$	$1.5h_2(n)/m$	6,7 (upper entry)
3	$.6859f_3(n)/\sqrt{m}$	$.6876g_3(n)/\sqrt{m}$	$f_2(n)$	$g_2(n)$	$1.5h_2(n)/m$	6,7 (lower entry)
4	$f_4(m)/h_1(n)$	$1.02g_4(m)/h_1(n)$	$f_2(m)$	$g_2(m)$	$h_2(n)/(m-1)$	8,9 (upper entry)
5	$.97f_3(n)/\sqrt{m-1}$	$.972g_3(n)/\sqrt{m-1}$	$f_2(n)$	$g_2(n)$	$h_2(n)/(m-1)$	8,9 (lower entry)
	$f_1(m) = (\sqrt{\log_{10} 1.5m})/m$	$g_1(m) = (\sqrt{\log_{10} 3.2})/m$				
	$f_2(k) = 1.7 + 2.6 \log_{10} k$	$g_2(k) = 3.58 + 3.2 \log_{10} k$				
	$f_3(n) = 1 + \dfrac{.501}{\log_{10}(n+0.1)}$	$g_3(n) = 1 + \dfrac{.722}{\log_{10} n}$	$h_1(n) = \sqrt{(\log_{10}(n+1.8)) - .5116} / \sqrt{n}$			
	$f_4(m) = f_1(m)\sqrt{\dfrac{m}{m-1}}$	$g_4(m) = g_1(m)\sqrt{\dfrac{m}{m-1}}$	$h_2(n) = [4.46 \log_{10}(n+1) - .78]^{-2}$			

The subscript α identifies the error rate, either 5% or 1%, and the subscript u identifies the specific situation as follows

u	5% points (entries)	1% points (entries)	procedure
1	Table 4	Table 5	one-way
2	Table 6 (upper)	Table 7 (upper)	"easy"
3	Table 6 (lower)	Table 7 (lower)	"easy"
4	Table 8 (upper)	Table 9 (upper)	"intermediate"
5	Table 8 (lower)	Table 9 (lower)	"intermediate"

Values from the first order approximation were compared to the "best" values for $m, n \le 20$, with the following range of deviations

	$\alpha = 5\%$	$\alpha = 1\%$
$m \ge 3,\ n \ge 3$	−1.7% to +1.8%	−7.8% to +3.5%
$m \ge 4,\ n \ge 4$	−.5% to +1.0%	−.3% to +2.2%
$m \ge 5,\ n \ge 3$ or $m \ge 3,\ n \ge 5$	−.4% to +1.8%	−.3% to +3.5%
$m = 2,\ n \ge 10$ or $m \ge 9,\ n = 2$	−.5% to +1.4%	−3.9% to +2.0%
$m = 2,\ n \ge 12$ or $m \ge 12,\ n = 2$	0.0% to +1.3%	−1.5% to +2.0%

The first-order approximation should suffice for direct use in practice provided, say, $(m-1)(n-1) \ge 9$. All percent points (in Tables 4 to 9) for m or $n \ge 20$ were obtained by these first-order approximations.

Table 17

Peripheral Values for Table 16

u	$\phi_{u,\,5\%}$	$\phi_{u,\,1\%}$	$\theta_{u,\,5\%}$	$\theta_{u,\,1\%}$	Ψ_u
	(Components of extended-range approximations for $m \geq 3$ and $n = 2$)				
1	$5.166f_1(m)$	$5.250g_1(m)$	$f_2(m)$	$g_2(m)$	$.57/m$
2	$3.653f_1(m)$	$3.712g_1(m)$	$f_2(m)$	$g_2(m)$	$.856/m$
3	$1.736/\sqrt{m}$	$2.283/\sqrt{m}$	2.3	4.1	$.856/m$
4	$5.166f_4(m)$	$5.250g_4(m)$	$f_2(m)$	$g_2(m)$	$.57/(m-1)$
5	$2.455/\sqrt{m-1}$	$3.228/\sqrt{m-1}$	2.3	4.1	$.57/(m-1)$
	(Components of extended-range approximations for $m = 2$ and $n \geq 3$)				
1	$.3374/h_1(n)$	$.4437/h_1(n)$	2.3	4.1	$.5h_2(n)$
2	$.2386/h_1(n)$	$.3137h_1(n)$	2.3	4.1	$h_2(n)$
3	$.485f_3(n)$	$.4862g_3(n)$	$f_2(n)$	$g_2(n)$	$h_2(n)$
4	$.4772/h_1(n)$	$.6274/h_1(n)$	2.3	4.1	$h_2(n)$
5	$.97f_3(n)$	$.9724g_3(n)$	$f_2(n)$	$g_2(n)$	$h_2(n)$

Expressions for the functions are given in Table 16.

REFERENCES

Daniel, Cuthbert, (1959). Use of half-normal plots in interpreting factorial two-level experiments, *Technometrics*, **1**, 311-42.

David, H. A. (1953). The power function of some tests based on the range, *Biometrika* **40**, 347-53.

Duncan, Acheson J. (1952). *Quality Control and Industrial Statistics.* (First edition only). Irwin, Chicago (espec. pp. 626-27).

Fisher, R. A. (1925). *Statistical Methods for Research Workers*, First edition, 1925, through Thirteenth edition, 1958. Oliver and Boyd, Edinburgh.

Federer, Walter T. (1955). *Experimental design: theory and application.* Macmillan, New York.

Gayen, A. K. (1949). The distribution of 'Student's' t in random samples of any size drawn from non-normal universes, *Biometrika* **36**, 353-69.

Halperin, Max, Greenhouse, Samuel W., Cornfield, Jerome and Zalokar, Julia (1955). Tables of percentage points for the studentized maximum absolute deviate in normal samples, *J. Am. Stat. Assoc.* **50**, 185-95.

Hartley, H. O. (1950). The use of range in analysis of variance, *Biometrika* **37**, 271-80.

Keuls, M. (1952). The use of the "studentized range" in connection with an analysis of variance, *Euphytica* **1**, 112-22.

Kurtz, T. E., Link, R. F., Tukey, J. W. and Wallace, D. L. (1965i). Short-cut multiple comparisons for balanced single and double classifications: Part 2, Derivations and approximations, *Biometrika* **52**, 485-498.

Link, R. F. and Wallace, D. L. (1952). Short cuts to allowances. Unpublished; handed out at the 19 March 1952 Blacksburg meeting of the Institute of Mathematical Statistics and the Biometric Society (ENAR).

Lord, E. (1947). The use of range in place of standard deviation in the t test, *Biometrika* **34**, 41-67 (Corrigenda 39, 442).

Lord, E. (1950). Power of the modified t test (u test) based on range, *Biometrika* **37**, 64-77.

Mosteller, Frederick and Bush, Robert, R. (1954). Selected quantitative techniques. *Handbook of Social Psychology*. (Lindzey Gardner, ed.) Chapter 8, 289-334 (espec. 304-7). Addison-Wesley, Cambridge, MA.

Natrella, Mary Gibbons (1963). *Experimental statistics*. National Bureau of Standards Handbook 91. U.S. Department of Commerce (espec. 15-6, 15-7, Table A-15).

Newman, D. (1939). The distribution of range in samples from a normal population expressed in terms of an independent estimate of standard deviation, *Biometrika* **31**, 20-30.

Patnaik, P. B. (1950). The use of mean range as an estimator of variance in statistical tests, *Biometrika* **37**, 78-87.

Pearce, S. C. (1953). Field experimentation with fruit trees and other perennial plants. *Technical Communication No. 23*, Commonwealth Bureau of Horticulture and Plantation Crops, East Malling, Kent, England.

Pearson, E. S. (1950). Some notes on the use of range, *Biometrika* **37**, 88-92.

Roy, S. N. and Bose, R. C. (1953). Simultaneous confidence interval estimation, *Ann. Math. Stat.* **24**, 513-36.

Scheffe, Henry, (1953). A method for judging all contrasts in the analysis of variance, *Biometrika* **40**, 87-104.

Scheffe, Henry (1959). *The Analysis of Variance.* John Wiley, New York.

Snedecor, G. W. (1948). Query 57, *Biometrics* **4**, 100-8.

'Student', (1908). The probable error of a mean, *Biometrika* **6**, 1-25.

Tippett, L. H. C. (1950). *Technological applications of statistics.* John Wiley, New York; Williams and Norgate, London.

Tukey, John W. (1949f). Comparing individual means in the analysis of variance, *Biometrika* **5**, 99-114. [Also in this Volume, 315-321.]

Tukey, John W. (1949h). One degree of freedom for non-additivity, *Biometrics* **5**, 232-42. [Also in *The Collected Works of John W. Tukey, Volume VII: Factorial & ANOVA 1949-1962*. Chapter 1, 1-13. Wadsworth Advanced Books and Software, Monterey, CA, 1991.]

Tukey, John W. (1953a). Some selected quick and easy methods of statistical analysis, *Trans. N.Y. Acad. Sciences (II)* **16**, 88-97.

Tukey, John W. (1953). The problem of multiple comparisons. Duplicated manuscript, 396; Princeton University. [Also in this Volume, 1-300.]

Tukey, John W. (1960e). Conclusions vs. decisions, *Technometrics* **2**, 423-34. [Also in *The Collected Works of John W. Tukey, Volume III: Philosophy and Principles of Data Analysis 1949-1964*. Chapter 6, 127-141. Wadsworth Advanced Books and Software, Monterey, CA, 1986.]

8

Short-Cut Multiple Comparisons for Balanced Single and Double Classifications: Part 2 Derivations and Approximations

T. E. Kurtz
R. F. Link
J. W. Tukey
D. L. Wallace

SUMMARY

Part 1 of this paper (Kurtz, Link, Tukey & Wallace, 1965a) presents tables of approximate percentage points of the ratio of a range to a sum of ranges and discusses their use in multiple-comparison procedures for balanced single and double classifications. The methods of approximation used in

---

The Collected Works of John W. Tukey (1993) Chapman and Hall, New York, NY.

Now at Dartmouth College [TEK].

Prepared in connection with research sponsored by the Office of Naval Research. Completion of the work has been facilitated by a grant from the National Science Foundation while some of the computations were performed under a Public Health Service facilities grant. Final revision was aided by a grant from the Army Research Office (Durham).

Reprinted from (1965) *Biometrika* **52**, 485-498 by permission of the Biometrika Trustees.

the construction of these tables are described here with emphasis on results and methods likely to be of use in other connections.

The basic approximation and the multiple comparison application are described (Sections 2-4). The distribution whose percentage points are required are set forth (Section 5). The approximations to mean range are detailed for uncorrelated ranges (Section 6) and correlated ranges (Section 7). The approximations are assembled to yield the percentage points tabled in Part 1 (Section 8). Representation of all items entering any of the approximations in terms of powers, roots and logarithms are given with bounds on error for each (Section 9), and an asymptotic approximation is derived (Section 10). Calculation of the tables of Part 1 is described (Section 11).

1. INTRODUCTION

Ranges have long been used as short-cut substitutes for sample standard deviations. Modification of Student's t by the use of a single range or a mean range in the denominator was explored by Lord (1947, 1950) who showed how very small is the loss of stringency that is exchanged for greatly eased computation.

Similar ideas were applied to the comparison of several treatments by Patnaik (1950), Hartley (1950), David (1951), and, in a parallel development that used slightly different distributional approximations and cast its answers into substantially different form for the *user*, by some of the present writers. (Tukey, 1950, 1951, 1953; Link & Wallace, 1952, from which tables have been reproduced in several places, e.g. in Natrella, 1963 and Mosteller & Bush, 1954).

The present account, of which Part 1, containing the tables of the necessary percentage points and general discussion, has appeared in another journal (Kurtz et al. 1965a) is the first definitive presentation of our tables and of the approximations and techniques used to obtain them.

2. ROOT-MEAN-SQUARE APPROXIMATIONS

Evaluating distributions of quantities involving mean ranges is difficult, and with only a few exceptions (notably the work of Lord, 1947), approximate methods are required. A natural and often used idea is to treat a mean range as if it behaved like a multiple of a standard deviation with some appropriate degrees of freedom.

Denote by $\bar{w}(m, n)$ the mean of m independent ranges, each of a sample of n independent observations from a normal population with variance σ^2. Denote by s_v^2 a mean-square estimate of σ^2 with v degrees of freedom (so that vs_v^2/σ^2 is distributed as χ^2 on v degrees of freedom). We use the approximation of treating $\bar{w}$ as if it were distributed like cs_v. We call the c the 'scale factor' and v the 'equivalent degrees of freedom' (hereafter EQ.D.F.).

The basic study by Patnaik (1950), supplemented by the comparative study by Pearson (1952), supports the general adequacy of this approximation that forms the basis for our work as well as for that of Hartley (1950) and David (1951).

Patnaik's choices of scale factor $c\,(m, n)$ and EQ.D.F. $v\,(m, n)$ are given in Table 30A of *Biometrika Tables for Statisticians* (Pearson & Hartley, 1954). We give in Table 1 a condensed representation for $v\,(m, n)$ and for $\delta(m, n) = mc\,(m, n)$, the scale factor appropriate to a *sum* of m ranges. The scale factor δ or c is chosen so that $(n\bar{w}/\delta)^2$, like s^2, is an unbiased estimate of σ^2. In Section 6, we describe our choice of v, and why it differs slightly from Patnaik's choice.

With this approximation, short-cut analysis of variance in a balanced single classification of n observations in each of m groups is now, in principle, immediate. Let $\bar{w}$ be the mean of the m within group ranges; use $[\bar{w}/c\,(m, n)]^2$ in place of the within group mean square obtained from a conventional analysis of variance and use $v\,(m, n)$ degrees of freedom in place of $m\,(n-1)$.

In Section 4, we carry the simplification one stage further, but first we describe two methods for a double classification.

3. DOUBLE CLASSIFICATIONS

The corresponding development for analysis of a randomized block experiment is more troublesome. Either the block or the treatment effects must be removed before ranges are taken, and most methods introduce correlation between the resulting ranges. Table 2 illustrates the two methods that we study in detail in application to data on lactic acid in alfalfa silage (Snedecor, 1948).

In the first, 'easy' method, form the sum $\Sigma w' = 76.8$ of ranges of differences between values in adjacent columns, including the difference between the last and first columns. With an appropriate scale factor $\delta'(m, n)$, the ratio $[(\Sigma w')/\delta'(m, n)]^2$ is an unbiased estimate of the variance of a single observation. Table 3 gives values of $\delta'(m, n)$ for $m \leq 10$; Table 1 permits easy calculation of δ' for any $m \geq 3$; Section 7 gives explicit

Table 1

Scale Factors and Equivalent Degrees of Freedom

(m ranges, each of n values. For greater accuracy, use formulae in Sections 6 and 7.)

n	$\delta(m, n)^*$	$\delta'(m, n)^\dagger$	$\delta''(m, n)^*$	$v(m, n)^\ddagger$
2	$0.32+1.128\ m$	$0.65+1.595\ m$	$-0.17+1.128\ m$	$0.124\sqrt{m}+0.876\ m$
3	$.23+1.693\ m$	$.47+2.394\ m$	$-0.58+1.693\ m$	$.26+1.185\ m$
4	$.18+2.059\ m$	$.38+2.912\ m$	$-0.83+2.059\ m$	$.30+2.738\ m$
5	$.16+2.326\ m$	$.33+3.29\ m$	$-1.01+2.326\ m$	$.34+3.62\ m$
6	$0.15+2.534\ m$	$0.30+3.58\ m$	$-1.13+2.534\ m$	$0.37+4.47\ m$
7	$.13+2.704\ m$	$.28+3.82\ m$	$-1.24+2.704\ m$	$.40+5.27\ m$
8	$.12+2.847\ m$	$.25+4.03\ m$	$-1.32+2.847\ m$	$.42+6.03\ m$
9	$.11+2.970\ m$	$.22+4.20\ m$	$-1.40+2.970\ m$	$.45+6.76\ m$
10	$.10+3.078\ m$	$.20+4.35\ m$	$-1.47+3.078\ m$	$.47+7.45\ m$
11	$0.10+3.17\ m$	$0.19+4.49\ m$	$-1.52+3.17\ m$	$0.50+8.12\ m$
12	$.10+3.26\ m$	$.18+4.61\ m$	$-1.56+3.26\ m$	$.52+8.76\ m$
13	$.09+3.34\ m$	$.17+4.72\ m$	$-1.61+3.34\ m$	$.54+9.38\ m$
14	$.08+3.41\ m$	$.17+4.82\ m$	$-1.65+3.41\ m$	$.56+9.97\ m$
15	$.09+3.47\ m$	$.16+4.91\ m$	$-1.69+3.47\ m$	$.58+10.54\ m$
16	$0.08+3.53\ m$	$0.16+4.99\ m$	$-1.72+3.53\ m$	$0.60+11.09\ m$
17	$.08+3.59\ m$	$.15+5.07\ m$	$-1.75+3.59\ m$	$.62+11.63\ m$
18	$.08+3.64\ m$	$.14+5.15\ m$	$-1.78+3.64\ m$	$.64+12.14\ m$
19	$.07+3.69\ m$	$.14+5.22\ m$	$-1.81+3.69\ m$	$.65+12.65\ m$
20	$.08+3.73\ m$	$.14+5.28\ m$	$-1.85+3.73\ m$	$.67+13.14\ m$
30	$0.05+4.09\ m$	$0.11+5.78\ m$	$-2.03+4.09\ m$	$0.82+17.39\ m$
60	$.04+4.64\ m$	$.08+6.56\ m$	$-2.33+4.64\ m$	$1.16+26.35\ m$
100	$.04+5.02\ m$	$.07+7.09\ m$	$-2.53+5.02\ m$	$1.50+34.4\ m$
|Error|	$\leq 0.5\%$ if $m \geq 3$ $\leq 0.7\%$ if $nm \geq 3$	$\leq 0.5\%$ if $m \geq 3$	$\leq 0.5\%$ if $m \geq 5$ $\leq 1.0\%$ if $n(m-2)^2 \geq 10$	$\leq 2\%$ if $m \geq 2$ Do not use for $m = 1$

* Coefficient of m is d_n. † Coefficient of m is $\sqrt{2}d_n$. ‡ Coefficient of m is $d_n^2/(2V_n)$.

formulae. For the example, $\delta'(4, 5) = 13.5$, so that the estimate of σ is $76.8/13.5 = 5.7$.

To complete the approximation, we must specify an EQ.D.F. $v'(m,n)$. As shown in Section 8, the EQ.D.F. is well approximated (if $m \geq 3$) in two stages by

$$v'(m, n) \doteq v(m', n),$$

$$m' \doteq {}^2\!/_3\, m.$$

Table 2

Lactic Acid in Alfalfa Silage (Snedecor, 1948)

	Treatment				
Block	I	II	III	IV	Mean
1	13.4	16.0	14.4	20.0	16.0
2	37.5	42.7	29.3	34.5	36.0
3	65.2	54.9	36.4	39.7	49.0
4	60.8	57.1	39.1	38.7	48.9
5	37.7	49.2	39.4	39.7	41.5

Difference between adjacent columns

	II-I	III-II	IV-III	I-IV
	2.6	−1.6	5.6	−6.6
	5.2	−13.4	5.2	3.0
	−10.3	−18.5	3.3	25.5
	−3.7	−18.0	−0.4	22.1
	11.5	−9.8	0.3	−2.0
Range:	21.8	16.9	6.0	32.1

Total range: $76.8 = \Sigma w'$

Residuals from block means

	−2.6	0.0	−1.6	4.0
	1.5	6.7	−6.7	−1.5
	16.2	5.9	−12.6	−9.3
	11.9	8.2	−9.8	−10.2
	−3.8	7.7	−2.1	−1.8
Range:	20.0	8.2	11.0	14.2

Total range $53.5 = \Sigma w''$

In the example, $m' = {}^{8}/_{3}$, so that from Table 1, $v'(4, 5) \doteq v\,({}^{8}/_{3}, 5) = 10.0$, to be compared to the 12 degrees of freedom for the interaction mean square.

Return to the example of Table 2 to apply the second 'intermediate' method, so called because it is intermediate in arithmetic difficulty between the 'easy' method and a conventional analysis of variance. Form the sum $\Sigma w'' = 53.4$ of ranges within each treatment of the residuals from block means.

Table 3

Values of $\delta'(m, n)$

$\delta'(m, n)$ is the divisor for the total of m (column-wise) ranges each of n differences between two successive columns (the mth range being of differences between the last and first columns) in an array of m columns and n rows, appropriate to obtain the square root of an estimate of σ^2 that is unbiased when the original observations come from normal populations with common variance σ^2 and population means exhibiting no interaction. For more than 10 ranges, see Table 1.

$n \backslash m$	2	3	4	5	6	7	8	9	10
2	4.00	5.41	7.01	8.61	10.21	11.81	13.41	15.01	16.60
3	5.41	7.65	10.04	12.44	14.84	17.23	19.63	22.02	24.41
4	6.33	9.12	12.03	14.94	17.85	20.77	23.68	26.59	29.50
5	7.02	10.19	13.48	16.78	20.06	23.35	26.64	29.93	33.2
6	7.56	11.04	14.63	18.21	21.80	25.38	28.96	32.5	36.1
7	8.00	11.73	15.56	19.38	23.21	27.03	30.86	34.7	38.5
8	8.38	12.32	16.35	20.37	24.40	28.43	32.5	36.5	40.5
9	8.71	12.82	17.02	21.22	25.42	29.62	33.8	38.0	42.2
10	8.99	13.26	17.62	21.97	26.32	30.67	35.0	39.4	43.7
11	9.25	13.66	18.14	22.63	27.12	31.61	36.1	40.6	45.1
12	9.48	14.01	18.62	23.23	27.83	32.4	37.1	41.7	46.3
13	9.68	14.33	19.05	23.77	28.48	33.2	37.9	42.6	47.4
14	9.87	14.62	19.44	24.26	29.08	33.9	38.7	43.5	48.3
15	10.05	14.89	19.80	24.71	29.62	34.5	39.4	44.4	49.3
16	10.21	15.14	20.14	25.13	30.13	35.1	40.1	45.1	50.1
17	10.36	15.37	20.45	25.52	30.60	35.7	40.7	45.8	50.9
18	10.51	15.59	20.74	25.89	31.03	36.2	41.3	46.5	51.6
19	10.64	15.79	21.01	26.23	31.44	36.7	41.9	47.1	52.3
20	10.76	15.98	21.27	26.55	31.8	37.1	42.4	47.7	53.0
30	11.72	17.44	23.22	29.00	34.8	40.6	46.3	52.1	57.9
60	13.24	19.76	26.32	32.9	39.4	46.0	52.6	59.1	65.7
100	14.29	21.34	28.43	35.5	42.6	49.7	56.8	63.9	71.0

m = number of ranges = number of columns; n = number of differences per range = number of rows.

With an appropriate scale factor $\delta''(m, n)$, the ratio $[(\Sigma w'')/\delta''(m, n)]^2$ is an unbiased estimate of σ^2, with an EQ.D.F. $v''(m, n)$ that is well approximated by

$$v''(m, n) \doteq v(m'', n),$$

$$m'' \doteq m - 1.$$

Table 1 permits easy calculation of δ'' and v'' except for small m and n. For the example, $\delta''(4, 5) = 8.29$, and $v''(4, 5) = 11.2$, so that the estimate of σ is $53.4/8.29 = 6.4$. Some troubles in this numerical example have been examined in Part 1 (Kurtz et al. 1965a).

The intermediate method is exactly that proposed by Hartley (1950), and values, slightly different from ours, of the EQ.D.F. and the scale factor for the mean range are given in Table 30B of *Biometrika Tables* (Pearson & Hartley, 1954).

4. SHORT-CUT CONFIDENCE PROCEDURES

The chief reason for using ranges is to shorten computations and a further step is needed to provide a complete short-cut procedure. Our procedures in Part 1 are cast entirely in terms of providing simultaneous confidence intervals for differences of totals or means, though as far as distributional computation is concerned, a significance testing formulation would be equivalent.

To set confidence limits on the difference between each pair of treatment means so that with probability 0.95 (say), the confidence statements will hold simultaneously for all pairs, we need to use as half-width of the interval:

$$\pm s_v \frac{1}{\sqrt{n}} q\,(m, v)]^{5\%},$$

where s_v is the square root of a mean-square estimate of the variance of a single observation — or one of its mean range substitutes — and $q\,(m, v)]^{5\%}$ denotes the upper 5% point of the distribution of studentized range for a range of m, studentized with v degrees of freedom. While it is straightforward to compute this half-width for the easy or intermediate estimates of variance, to require the user to do so defeats the purpose of a short-cut method.

The main tables of Part 1 provide the factors R such that, for the easy method, for example, $\pm R_{.}[\Sigma w']$ is the half-width of confidence intervals for treatment *totals*. We work with totals instead of means because the former requires slightly less arithmetic and because, in the most used portion of the table, the values of R are more nearly constant and closer to unity. The factors depend on the confidence level, on which estimate is used, and on whether we are setting limits on differences of treatment means or differences of block means.

The combination of the various approximations to obtain the factors tabled in Part 1 is detailed in Section 8.

5. THE SPECIFIC DISTRIBUTIONAL PROBLEMS

In this section, we set up without approximation the key distributional problems that arise in providing short-cut multiple comparison procedures for observations arranged in balanced single and double classifications. This will be done under the ideal assumptions of no interactions, common variance, and independently normally distributed fluctuations — i.e. when the deviations of the individual observations from contributions corresponding to the categories of the one or two classifications are sampled independently from identical normal populations.

Let $\{X_{ij}\}$ denote the observations, $\{\mu_i\}$ contributions from the first classification, $\{\eta_j\}$ contributions from the second classification (if any), and $\{Z_{ij}\}$ a set of independent normal deviates with a common variance and a common average value. Agree that replacement of a subscript by a '+' or a '.' will denote the sum or mean, respectively, over that subscript. Then

$$X_{ij} = \mu_i + \eta_j + Z_{ij} \quad (i=1,\ldots,m\,;\, j=1,\ldots,n),$$

and the three sets of ranges for $i = 1, \ldots, m$, are

$$w_i = \operatorname*{range}_{1\le j\le n} X_{ij} = \operatorname*{range}_{1\le j\le n} Z_{ij} \quad \text{(one-way)},$$

$$w_i' = \operatorname*{range}_{1\le j\le n} (X_{ij} - X_{i+1,j}) = \operatorname*{range}_{1\le j\le n} (Z_{ij} - Z_{i+1,j}) \quad \text{(easy)},$$

where

$$X_{m+1,j} = X_{1,j} \quad \text{and} \quad Z_{m+1,j} = Z_{1,j}\,,$$

$$w_i'' = \operatorname*{range}_{1\le j\le n} (X_{ij} - X_{.j}) = \operatorname*{range}_{1\le j\le n} (Z_{ij} - Z_{.j}) \quad \text{(intermediate)}.$$

We consider five confidence procedures, indexing them throughout by values of u. One ($u=1$) applies to a single classification; two ($u=2, 3$) apply to the *easy* method and two ($u=4, 5$) to the *intermediate* method for a double classification. Three ($u=1, 2, 4$) apply to comparison of classification 1 totals; two ($u=3, 5$) to classification 2 totals.

For $u = 2$, the procedure consists of the family of confidence statements:

$$\left\{ \text{for all } i,\ i' = 1, \ldots, m,\ n\,(\mu_i - \mu_{i'}) \text{ is between the limits} \right.$$

$$\left. X_{i+} - X_{i'+} \pm (R_2]^{\alpha}) \cdot \left[\sum_{i=1}^{m} w_i' \right] \right\}.$$

If and only if $(R_2]^{\alpha})$ is the upper α point of the key statistic

$$R_2 = \frac{\operatorname*{range}_{1 \le i \le m} X_{i+}}{\Sigma w_i'} = \frac{\operatorname*{range}_{1 \le i \le m} Z_{i+}}{\Sigma w_i'},$$

all statements will be correct, simultaneously, with probability (expressed in percentage) $100\% - \alpha$. We say the procedure has simultaneous confidence coefficient $100\% - \alpha$ or a family-wise error rate α.

For $u = 3$, the procedure consists of the statements:

$$\left\{ \text{for all } j,\ j' = 1, \ldots, n,\ m\,(\eta_j - \eta_{j'}) \text{ is between the limits} \right.$$

$$\left. X_{+j} - X_{+j'} \pm (R_3]^{\alpha}) \cdot \left[\sum_{i=1}^{m} w_i' \right] \right\},$$

and $(R_3]^{\alpha})$ must be the upper α point of the key statistic

$$R_3 = \frac{\operatorname*{range}_{1 \le j \le n} Z_{+j}}{\Sigma w_i'}$$

for this family of statements to have simultaneous confidence coefficient $100\% - \alpha$.

The key distributional problems are to find upper percentage points of the distribution of

$$R_u = \frac{W^{(u)}}{\sum_{i=1}^{m} w_i^{(u)}} \qquad (u = 1, \ldots, 5),$$

Table 4

Description of the Various R_u

$$R_u = \frac{W^{(u)}}{w_1^{(u)} + w_2^{(u)} + \ldots + w_m^{(u)}},$$

where

(1) $W^{(u)}$ is a range of k_u independent normal variates with mean zero and variance λ_u^2;

(2) each $w_i^{(u)}$ is the range of n independent normal variates with mean zero and variance b_u^2;

(3) $w_1^{(u)}, w_2^{(u)}, \ldots, w_m^{(u)}$ are all independent of $W^{(u)}$,

(4) the dependence between $w_i^{(u)}$ and $w_j^{(u)}$ arises only from the correlation of pairs of individually corresponding normal variates entering these ranges; the correlation between individually corresponding variates is given by the (i, j)th element of a matrix $C_u(\rho_u)$,

and the five cases are specified by

u	k_u	λ_u	b_u	ρ_u	C_u
1	m	$\sqrt{n}$	1	—	I
2	m	$\sqrt{n}$	$\sqrt{2}$	$-1/2$*	C'
3	n	$\sqrt{m}$	$\sqrt{2}$	$-1/2$*	C'
4	m	$\sqrt{n}$	$\sqrt{[1-(1/m)]}$	$-\dfrac{1}{m-1}$	C''
5	n	$\sqrt{m}$	$\sqrt{[1-(1/m)]}$	$-\dfrac{1}{m-1}$	C''

with

$$C'_{ij}(\rho) = \begin{cases} 1 & i = j, \\ \rho & |i-j| = 1 \quad \text{or} \quad m-1, \\ 0 & \text{otherwise}, \end{cases} \qquad C''_{ij}(\rho) = \begin{cases} 1 & i = j, \\ \rho & i \neq j. \end{cases}$$

* Except for $m = 2$, when $\rho_2 = \rho_3 = -1$.

where we have introduced ranges $W^{(u)}$ and $w_i^{(u)}$ by letting

$$W^{(1)} = W^{(2)} = W^{(4)} = \operatorname*{range}_{1 \le i \le m} Z_{i+} ,$$

$$W^{(3)} = W^{(5)} = \operatorname*{range}_{1 \le j \le n} Z_{+j} ,$$

$$\left.\begin{array}{l} w_i^{(1)} = w_i , \\ w_i^{(2)} = w_i^{(3)} = w_i' , \\ w_i^{(4)} = w_i^{(5)} = w_i'' . \end{array}\right\} \quad (i = 1, \ldots, m).$$

If, as we may, we take the common variance of the Z_{ij} to be unity, then the statistics R_u conform to the pattern of Table 4. (Verification of the values of k, λ, b, and C is straightforward.) The effect of correlation on the behavior of the ranges is discussed in Section 7.

6. CHOOSING THE SCALE FACTOR AND EQ.D.F.

Patnaik (1950) chooses the scale factor c and the EQ.D.F. v so that the first two moments of $\overline{w}(m, n)$ and cs_v agree. Our choices differ slightly both in principle and practice. Since we are not directly interested in the distribution of $\overline{w}$ *per se*, but only, here, in the distribution of the ratio $W/\overline{w}$, we try to determine c and v to improve the approximation to the latter.

The choice of v or c might with advantage vary with the intended use of the approximation, and did so in the computation of our original tables. At this time studentized range tables were inadequate, and we applied root-mean-square approximations to the numerator range as well as to the denominator mean range, leading to a multiple of $s'/s = \sqrt{F}$. A different choice of EQ.D.F. for the numerator and denominator applications was helpful, and that experience influenced our approach. Here, we consider only denominator mean range approximations. The ideas involved in determining approximations from analogous use has been explored more thoroughly by some of us (unpublished).

For the scale factor, we make the same choice as Patnaik, determining c so that $(\overline{w}/c)^2$, like s^2, is an unbiased estimate of σ^2. Hence

$$c\,(m, n) = \sqrt{\{\text{ave}\,[\overline{w}(m, n)]^2\}} = d_n \sqrt{\left\{1 + \frac{V_n}{md_n^2}\right\}} , \tag{6.1}$$

where d_n and V_n denote the average and variance of a range of n independent unit normal deviates (Harter, 1960).

To choose the EQ.D.F., we examined known distributional facts about ranges and mean ranges: moments, percentage points of a single range, of its reciprocal, of a ratio of two ranges. Most important are the percentage points and related facts computed by Lord (1947, 1950) for the ratio of a normal deviate and an independent mean range.

Each fact provides a basis for a choice of EQ.D.F. (some only for a single range). For example, the percentage points of Lord's statistic leads to a choice of EQ.D.F. $v = v\,(m, n)$ from the equation

$$\left.\frac{y}{\overline{w}(m, n)}\right]^{5\%} = c\,(m, n)\cdot t_v\Big]^{5\%}, \tag{6.2}$$

where y is a unit normal deviate, and for any statistic v, $v\,]^{5\%}$ denotes the upper 5% point of the distribution of v.

Approximation of the distribution of $W/\overline{w}$ by that of $W/(cs)$ is closely analogous to the approximation of Lord's statistic by t-statistic. (The analogy is exact if the numerator range W is a range of 2.) We give greatest weight to the EQ.D.F. determination based on Lord's statistic for those values of (m, n) for which Lord's most accurate computations were made.

For large m, $\overline{w}(m, n)$ is approximately normally distributed, and the limiting value of the number of degrees of freedom per range,

$$\lim_{m\to\infty} \frac{v\,(m, n)}{m} = \frac{d_n^2}{2V_n},$$

is easily established. For smaller values of m rather more equivalent degrees of freedom per range are indicated. The following empirical formula has the correct limiting behavior, fits the values best determined by (6.2) well and fits the remaining values adequately

$$v\,(m, n) \doteq \frac{md_n^2}{2V_n}\left\{1 + \frac{1}{\sqrt{m}(7 + \log_{10}(n-1))}\right\}. \tag{6.3}$$

This determination is larger by several percent than that based on matching moments and by fewer percent the choice given in Table 1 and used in our computations. The effect of the latter difference is slight.

7. MEANS OF CORRELATED RANGES

For each statistic arising from analysis of double classification data, the denominator is a sum of correlated ranges. The preceding development does not apply directly, nor can it be repeated for correlated ranges, for virtually no distributional facts are known for them.

Hartley (1950) computed enough correlations between ranges to allow approximate determination of c and v by matching moments. In Kurtz et al. (1965*b*), we obtain improved values and approximations for $\rho_w(n, \rho)$, the correlation between the two ranges in a sample of size n from a bivariate normal population with correlation ρ. We now proceed in a convenient, indirect way.

The effects of the correlations must often be quite small, since the mean absolute correlation is rarely larger than $1/(2m)$. Thus it seems best to match means of correlated ranges to means of uncorrelated ranges. Then the approximation of the preceding section, or any improved approximation that might later be obtained (say, by quadrature), can be used to complete the analysis.

We match first and second moments, lacking any other information. Consider first the 'easy' method (refer to Table 4 for the necessary specifications). Successive ranges have correlation $\rho_w(n, -1/(m-1))$, except when $m = 2$ which must be handled separately. (The 'easy' and 'intermediate' methods differ only by a factor of 2 when $m = 2$.) The mean range $\overline{w}'$ satisfies

$$\text{ave}\left[\frac{\overline{w}'(m, n)}{b_2}\right] = d_n \ ,$$

$$\text{var}\left[\frac{\overline{w}'(m, n)}{b_2}\right] = \frac{1}{m^2}\{mV_n + 2mV_n\rho_w(n, -{}^1\!/\!{}_2)\} - \frac{V_n}{m'} \ ,$$

where

$$m' = \frac{m}{1 + 2\rho_w(n, -{}^1\!/\!{}_2)} \quad (m \geq 3) \ .$$

Thus, judged by first and second moments, $\overline{w}'(m, n)/b_2$ behaves like $\overline{w}(m', n)$, a mean of m' independent ranges of size n.

Similarly, for the 'intermediate' method, $\overline{w}''(m, n)/b_4$ behaves like $\overline{w}(m'', n)$, with

$$m'' = \frac{m}{1+(m-1)\rho_w\left(n, \dfrac{-1}{m-1}\right)}.$$

When $m=2$, both $\overline{w}'(2,n)$ and $\overline{w}''(2,n)$ reduce exactly to a single range. This case, exceptional for $\overline{w}'(2,n)$, will not be explicitly considered in the rest of this section.

If we use the approximation from Kurtz et al. (1966b)

$$\rho_w(n,\rho) \doteq L_n\rho^2 + {}^2/_3\,(1-L_n)\rho^4,$$

where

$$L_n = d_n^2/\{2(n-1)V_n\},$$

then

$$m'' \doteq \left\{\begin{array}{ll} 1 \quad (m=2), & \\ (m-1)+(1-L_n)\dfrac{1-{}^2/_3\,\rho^2}{1-L_n\rho-{}^2/_3\,(1-L_n)\rho^3} & (m\geq 3), \end{array}\right\} \tag{7.1}$$

where

$$\rho = -1/(m-1).$$

Noting that the multiplier of $1-L_n$ in (7.1) lies between 0 and 1, we see that for $m \geq 3$, m'' lies above $m-1$ and below $m-L_n$, approaching the latter as $m \to \infty$.

Moreover, since $m'(m,n) = (m/3)m''(3,n)$, approximation (7.1) yields

$$m' \doteq \left\{\begin{array}{ll} 1 & (m=2) \\ \dfrac{12m}{13+5L_n} & (m\geq 3) \end{array}\right\}. \tag{7.2}$$

Table 5 provides values of L_n, of m'/m and of $m''-(m-1)$ for selected values of n and m, most of the values being based on the approximation (7.1). The approximation $m'' \doteq m-1$ is fairly good, as is, consequently, the approximation $m' \doteq {}^2/_3\, m$. The scale factors for a total of m ranges, $m\bar{w}'$ and $m\bar{w}''$, are given by

$$\delta'(m, n) = mb_2 c\,(m', n)\,, \quad \delta''(m, n) = mb_4 c\,(m'', n),$$

respectively and have been tabled in Tables 1 and 3.

Table 5

Values of L_n, of m'/m, and of $m'' - (m-1)$. (Values ending in an inferior 'a' are based on the approximation (7.1))

			$m'' - (m-1)$		
n	L_n	m'/m*	$m = 3$	$m = 6$	$m = \infty$
2	0.8760	0.6906	0.0718	0.1026	0.1240
3	.9074	.6843	.0528	.0768	.0926
4	.9126	$.683_a$	$.050_a$	$.072_a$	.0874
5	.9058	$.658_a$	$.034_a$	$.078_a$	.0942
10	0.8282	0.700_a	0.100_a	0.143_a	0.1718
15	.7528	$.716_a$	$.198_a$	$.209_a$	.2742
20	.6914	$.728_a$	$.184_a$	$.264_a$	.3086
60	0.448	0.77_a	0.31_a	0.49_a	0.552
100	.348	$.81_a$	$.44_a$	$.59_a$	.652
200	.243	$.84_a$	$.53_a$	$.70_a$	.757

*These are also values of ${}^1/_3\, m''(3, n)$.

8. APPROXIMATIONS TO THE PERCENTAGE POINTS OF THE R_u

With the root-mean-square approximations for denominator mean ranges as developed in preceding sections, the percentage points of the statistics R_u are given by

$$R_u]^\alpha \doteq \left. \frac{\lambda_u}{b_u mc(m_u, n)}\, q(k_u, v(m_u, n)) \right]^\alpha$$

in which the appropriate values of λ_u, b_u, k_u are given in Table 4 for each case ($u = 1, \ldots, 5$), $c\,(m, n)$ in equation (6.1), $v\,(m, n)$ in equation (6.3),

$$m_u = \begin{cases} m & (u = 1), \\ m' & (u = 2, 3), \\ m'' & (u = 4, 5), \end{cases}$$

and m' and m'' are given in equations (7.2) and (7.1).

This approximation for $R_u]^\alpha$ requires tables of d_n and V_n, the mean variance of a normal range, and of $q\,(k, v)]^\alpha$, the percentage points of the studentized range. Harter (1960) gives the best available values for each.

9. APPROXIMATION IN TERMS OF POWERS, ROOTS, AND LOGARITHMS

Except when both m and n are very small, various approximate formulae become useful. They permit representation of the percentage points as elementary functions not involving other tabled quantities. These approximations were essential for obtaining desired percentage points for values of m or n greater than 20. (The studentized range tables of Harter (1960) now partially obviate the necessity of such approximations.)

The approximations are mostly empirical; some constituent elements are taken directly, or with slight modifications, from Tukey (1955). Asymptotic forms have been used where convenient, but with functions of the normal range, the convergence to asymptotic forms is often so slow that purely empirical approximations are essential. The accuracy of most of the formulae for large arguments are as good as available tables of range quantities will permit. Some of the component approximations seem likely to be useful in other connections, and do some alternative approximations.

The approximation (see Tukey, 1955)

$$d_k \doteq \sqrt{\{17.06 \log_{10} [0.291(k + 2.6)]\}},$$

is accurate to within 0.1% for $k \geq 20$, but is poor for small k. We use instead

$$d_k \doteq (4.105) \sqrt{\{\log_{10} (k + 1.8) - 0.5116\}},$$

which is accurate to within 0.2% for $3 \leq k \leq 1000$ (d_k is the only range constant known accurately for large k).

Tukey (1955) gives two approximations to $d_k/\sqrt{V_k}$ excellent for $k \geq 20$ (the more complex one for $k \geq 4$). Since we need accuracy in V_k/d_k^2 for small k, we use instead the approximation

$$V_k/d_k^2 \doteq \{4.46 \log_{10}(k+1) - 0.78\}^{-2},$$

which is accurate to within 0.0005 for $3 \leq k \leq 1000$.

The following approximation to the percentage points of the studentized range appears to be acceptable for ranges of sizes 3 to 1000 or more

$$q(k, v)]^{\alpha} \doteq \sqrt{2}\,(\mid t_v \mid]^{\alpha}) A_k(\alpha) \left\{1 + \frac{B_k(\alpha)}{v}\right\},$$

with

$$A_k(\alpha) = \begin{cases} 1.488\sqrt{(\log_{10} 1.5k)} & (k \geq 3,\ \alpha = 5\%), \\ 1.15\sqrt{(\log_{10} 3.2k)} & (k \geq 3,\ \alpha = 1\%), \\ 1.0 \text{ for } k = 2, & \text{all } \alpha, \end{cases} \tag{9.1}$$

$$B_k(\alpha) = \begin{cases} \dfrac{4.0 \log_{10} k - 1.0}{1.0 + \log_{10}(100+k)} & (k \geq 3,\ \alpha = 5\%), \\ \dfrac{6.1 \log_{10} k - 1.3}{1.5 + \log_{10}(100+k)} & (k \geq 3,\ \alpha = 1\%), \\ 0.0 \text{ for } k = 2, & \text{all } \alpha. \end{cases}$$

To complete the approximation, some approximate formula for $\mid t_v \mid]^{\alpha}$ must be given. Formulae of any desired accuracy for any range of degrees of freedom can be obtained with a small effort. Several possibilities are set forth below.

The 5% and 1% points of the studentized range found from (9.1) have been compared to Harter's (1960) tables of percentage points for ranges of size up to 100. Bounds for the relative errors are as follows:

for	$4 \leq k \leq 20$	and	$v \geq 10$,	$-0.5 \leq$ % error $\leq +0.5$,
for	$3 \leq k \leq 100$	and	$v \geq 10$,	$-1.0 \leq$ % error $\leq +1.4$,
for	$3 \leq k \leq 20$	and	$v \geq 5$,	$-0.5 \leq$ % error $\leq +2.0$,
for	$3 \leq k \leq 100$	and	$v \geq 5$,	$-1.0 \leq$ % error $\leq +6.0$,

where '% error' = 100 (approximate-true)/true. In this comparison, exact values of $| t_v |]^{\alpha}$ were used, so that for any complete approximation, the relative error of the t approximation must be added.

The form chosen for the approximation becomes less satisfactory for small v as k increases, but in our applications (and in most applications), the degrees of freedom v increase along with k and the contribution of the B/v term to the final calculation is not critical.

One can choose A_k and B_k for each k to make the approximation fit as well as possible. Making the approximation fit exactly at $v = \infty$ degrees of freedom yields A_k as the ratio of the α point of the unstudentized range of size k to the α point of a range of size 2. The formulae given for A_k are good to within 0.5% of these values for $3 \leq k \leq 1000$. They are modified from those given by Tukey (1955, p. 481) to give a better fit for small k (to correct a typographical error: 1.51 in Tukey (1955) should be 1.15).

No choice of B_k is as natural as for A_k, and the value of B_k *per se* is unimportant; what matters is the accuracy of the resulting approximation to the studentized range. The formulae given fit the known values well for $3 \leq k \leq 20$ and have, for large k, both the correct limiting value and the correct asymptotic form of the second-order term (although the detailed structure has been chosen to give a satisfactory fit for smaller k, too).

The approximations

$$A_k / d_k \doteq 0.35 \left(1 + \frac{0.501}{\log_{10}(k + 0.1)}\right) \quad \text{at } 5\%,$$

$$A_k / d_k \doteq 0.267 \left(1 + \frac{0.722}{\log_{10} k}\right) \quad \text{at } 1\%$$

are accurate to within 0.5% for $3 \leq k \leq 1000$.

For percentage points of Student's t, we may use

$$| t_v |]^{5\%} \doteq | t_\infty |]^{5\%} \left(1 + \frac{1.40}{v}\right),$$

$$| t_v |]^{1\%} \doteq | t_\infty |]^{1\%} \left(1 + \frac{2.30}{v}\right).$$

The former is accurate to within 0.6% for $v \geq 7$, and 1.2% for $v = 6$; the latter to within 1.1% for $v \geq 8$. These may be improved by adding a term quadratic in $1/v$, as by

$$| t_v |]^{5\%} \doteq | t_\infty |]^5 \left(1 + \frac{1.17}{v} + \frac{1.98}{v^2}\right),$$

$$| t_v |]^{1\%} \doteq | t_\infty |]^{1\%} \left(1 + \frac{1.8}{v} + \frac{5.0}{v^2}\right),$$

where the former is accurate to within 0.2% for $v \geq 4$, and 1.0% for $v = 3$, the latter to within 0.2% for $v \geq 6$, 0.4% for $v = 5$, 1.5% for $v = 4$. (Other better approximations can easily be found, particularly by approximation to the logarithm of $| t |$.)

All necessary approximations have been given, and if we accept the form of the studentized range approximation, the desired percentage points can be written as

$$R_u]^\alpha \doteq \frac{\lambda_u \sqrt{2} \, | t_\infty |]^\alpha A_k}{m b_u d_n} \, \frac{1}{\sqrt{[1+\psi_u)]}} \left(1 + \frac{B_k}{v}\right) \frac{| t_v |]^\alpha}{| t_\infty |]^\alpha},$$

where $v = v\,(m_u, n)$ and where we have introduced $\psi_u = V_n/(m_u d_n^2)$, largely for later convenience.

For example, for the 5% point of the single-classification statistic $R_1^{(1)}$, we have

$$\lambda = \sqrt{n}\,, \quad k = m\,, \quad b = 1\,, \quad m_u = m\,, \quad | t_\infty | = 1.960,$$

and

$$R_1]^{5\%} \doteq \frac{2.77 \sqrt{n}}{m} A_m \frac{1}{d_n} \frac{1}{\sqrt{(1+\psi_1)}} \left(1 + \frac{B_n}{v}\right) \frac{| t_v |]^{5\%}}{| t_\infty |]^{5\%}}. \tag{9.2}$$

Necessary approximations for each factor can be assembled from the foregoing. Because most of the range approximations hold only for ranges of 3 or more, three formulae would be needed: one for both m and n greater than 2, one for $m = 2$, one for $n = 2$.

10. A FIRST-ORDER APPROXIMATION

Unless both m and n are small, ψ_u is small, and a simplified approximation that is linear in ψ_u is adequate for most purposes. We have

$$\frac{1}{\sqrt{(1+\psi_u)}} \doteq 1 - {}^1\!/_2\, \psi_u\,, \qquad \frac{1}{v\,(m_u, n)} \doteq 2\psi_u\,.$$

The latter approximation underestimates the degrees of freedom and amounts to using the number of degrees of freedom per range appropriate for a large number of ranges.

The simple approximations, $m' \doteq {}^2\!/_3\, m$, and $m'' \doteq m = 1$ are used here, as is the first-order approximation to t_v. Combining all the approximations into a single linear approximation in ψ_u leads to the form

$$R_u]^{\alpha} \doteq \phi_{u,\,\alpha}(m, n)\{1 + \theta_{u,\,\alpha}(m, n)\cdot\psi_u(m, n)\}. \tag{10.1}$$

For $u = 1, \ldots, 5$ and for $\alpha = 5\%$, 1%, the functions ϕ, θ, ψ are set forth in Part 1 (Kurtz et al. 1965a, Tables 16 and 17), along with a discussion of accuracy. In all practice, the approximation (10.1) should suffice for direct use provided, say $(m-1)(n-1) \geq 9$.

11. TABLES OF PART 1

The tables presented in Part 1 of this paper were computed using the approximation of Section 8, *except that*:

(1) for all entries with either $m \geq 20$ or $n \geq 20$, the first-order approximation (10.1) was used;

(2) older tables of range constants and percentage points of studentized range were used in a form discussed by Kurtz et al. (unpublished);

(3) EQ.D.F. from Table 1 rather than from (6.3) were used for $m \geq 2$;

(4) a few entries were modified to take account of more accurate values available elsewhere as special cases, as described below.

Computations were performed under a programming system that provided eight significant figures, so that the resulting tables are for all practical purposes as accurate as the approximating formula.

In a few instances, the statistics R_u coincide (up to known factors) with statistics whose percentage points have been tabled. The most important is the coincidence, for $m = 2$, of R_1 and R_2 with Lord's two- and one-sample range versions of t (Lord, 1947); this coincidence lies at the base of our choice of EQ.D.F. (Section 6). With $v(m, n)$ as given in (6.3), our computations agree with Lord's to within 0.2% except for errors of 1.8, 4.4, and 1.4% for R_1 with $m = 2$, $n = 2$ (5 and 1%), and $m = 2$, $n = 3$ (1%), respectively. We have adopted Lord's values for these three entries.

Harter's (1963) recent table of percentage points of the ratio of two ranges provides a new check, corresponding to the statistic R_3 for $m = 2$. Link's (1950) two-significant-figure tables were used in the original study and would have detected any gross differences. Using $v(m, n)$ as given in (6.3), our computations agree to within 0.5% of Harter's values at the 5% level and, for $n \leq 6$, at the 1% level. For $n \geq 7$ at the 1% level, our values tend to run about 1.5% higher than Harter's. The explanation apparently lies in the use of a single choice of denominator equivalent degrees of freedom (EQ.D.F.) for both the 5 and 1% values. In all other parts of the tables, the denominator is a mean of at least two ranges, and the effect should be smaller. We did not adjust our entries to agree with Harter's values.

REFERENCES

David, H. A. (1951). Further applications of range to the analysis of variance, *Biometrika* **38**, 393-409.

Harter, H. Leon (1960). Tables of range and studentized range, *Ann. Math. Stat.* **31**, 1122-47.

Harter, H. Leon (1963). Percentage points of the ratio of two ranges and power of the associated test, *Biometrika* **50**, 187-94.

Hartley, H. O. (1950). The use of range in analysis of variance, *Biometrika* **37**, 271-80.

Kurtz, T. E., Link, R. F., Tukey, J. W. and Wallace, D. L. (1965e). Short-cut multiple comparisons for balanced single and double classifications: Part 1, Results, *Technometrics* **7**, 95-161. [Also in this Volume, 353-430.]

Kurtz, T. E., Link, R. F., Tukey, J. W. and Wallace, D. L. (1966b). Correlation of ranges of correlated deviates, *Biometrika* **53**, 191-197.

Link, Richard F. (1950). The sampling distribution of the ratio of two ranges from independent samples, *Ann. Math. Stat.* **21**, 112-16 (amended, **23**, 289.)

Link, R. F. and Wallace, D. L. (1952). Short cuts to allowances. Unpublished; handed out at the 19 March 1952 Blacksburg meeting of the Institute of Mathematical Statistics and the Biometric Society (ENAR).

Lord, E. (1947). The use of range in place of standard deviation in the *t* test, *Biometrika* **34**, 41-67 (corrigenda, **39**, 442).

Lord, E. (1950). Power of the modified *t* test (*u* test) based on range, *Biometrika* **37**, 64-77.

Mosteller, Frederick and Bush, Robert R. (1954). Selected quantitative techniques. *Handbook of Social Psychology*. (G. Lindzey, ed.) Chapter 8, 289-334 (especially 304-7). Addison-Wesley, Cambridge, MA.

Natrella, Mary Gibbons (1963). *Experimental Statistics*. National Bureau of Standards Handbook 91, U.S. Department of Commerce (especially pp. 15-6, 15-7, Table A-15).

Patnaik, P. B. (1950). The use of mean range as an estimator of variance in statistical tests, *Biometrika* **37**, 78-87.

Pearson, E. S. (1952). Comparison of two approximations to the distribution of the range in small samples from normal populations, *Biometrika* **39**, 130-6.

Pearson, E. S. and Hartley, H. O., (eds.) (1954). *Biometrika Tables for Statisticians*. **1**. Cambridge University Press.

Snedecor, G. W. (1948). Query 57, *Biometrika* **4**, 100-8.

Tukey, John W. (1950U4). Notes on the use of the range in industrial statistics. Unpublished; handed out at the 17 November 1950 meeting of the Baltimore and Washington Sections of the American Society for Quality Control.

Tukey, John W. (1951d). Quick and dirty methods in statistics. Part II — Simple analyses for standard designs. (Note: no Part I ever appeared.) *Quality Control Conference Papers* 1951 (5th Annual Convention), Amer. Soc. Qual. Control.

Tukey, John W. (1953a). Some selected quick and easy methods of statistical analysis, *Trans. N.Y. Acad. Sciences (II)* **16**, 88-97.

Tukey, John W. (1955c). Interpolations and approximations related to the normal range, *Biometrika* **42**, 480-5.

9

Authors' Reply to Anscombe's Comments

T. E. Kurtz
R. F. Link
J. W. Tukey
D. L. Wallace

We regard our use of error rates as consistent with both the Fisher and classical Neyman-Pearson approaches. Communication of a level of uncertainty of a confidence statement is crucial; our error rates provide such information. Statements at higher or lower levels of uncertainty are implied and easily obtained. Multiple confidence statements present additional problems, but we feel our conservative choice for the interpretation of the error rate as the *maximum* level of uncertainty for any statement (however composite) is appropriate and effective, especially in those situations where no statements have been singled out in advance.

We differ with Anscombe primarily in where we would draw the line between problems to be formulated as decision problems and those to be analyzed by inferential, conclusion or communication procedures [1]. Problems in which one *action* must be chosen from among some clearly defined set of alternatives *should* be formulated as a decision problem. The difficult tasks of assessing some appropriate utility or loss functions and of carrying out an appropriate analysis ought to be faced squarely. The bulk of the problems involving the comparisons of several means *do not have clearly* delineated action purposes. We see no virtue in forcing such

The Collected Works of John W. Tukey (1993) Chapman and Hall, New York, NY.

Reprinted from (1965) *Technometrics* **7**, 169 by permission of the American Statistical Association.

problems into a decision-theoretic mold — not, at least, into any we have seen to date. We doubt that any generally suitable formulations or techniques will be available soon. We feel that procedures which *communicate* the extent of uncertainty about the different comparisons are generally applicable, and do not preclude later use of the information for more specific purposes. Our specific techniques are not the only ones that might be used, but they appear to us to be valuable for analysis of a large class of problems, including the examples in the paper.

We look forward to the development of other procedures: for decision-theoretic methods; for procedures within inferential frameworks that take account of the observed configurations of the means, or of prior knowledge or belief; for ways of balancing and expressing the uncertainties in inferences from complex experiments.

REFERENCES

[1] Tukey, J. W. (1960e). Conclusions vs. decisions, *Technometrics* **4**, 423-434. [Also in *The Collected Works of John W. Tukey, Volume III: Philosophy and Principles of Data Analysis 1949-1964*. Chapter 6, 127-141. Wadsworth Advanced Books and Software, Monterey, CA, 1986.]

10

Multiple Comparisons Through Orderly Partitions: The Maximum Subrange Procedure

Henry I. Braun
John W. Tukey

1. INTRODUCTION

Simultaneous inference deals with the problem of making useful multiple inferences in a single experiment. The inferences may be in the form of a set of confidence intervals or a set of confident inequalities (confident directions). The general aim is to control the experimentwise error rate while obtaining the greatest power. The experimentwise error rate is the probability of making one or more false statements among the ensemble of statements associated with the analysis of the experiment. There are a number of different kinds of power one can consider in this area. We will deal with two in what follows.

The problem we wish to consider is one of pairwise multiple comparisons. The prototypical situation arises when n observations are made on each of the k populations. It is desired to make inferences

The Collected Works of John W. Tukey (1993) Chapman and Hall, New York, NY.

Educational Testing Service [HIB].

Reprinted with permission, from (1983) *Principals of Modern Psychological Measurement: A Festschrift for Frederic M. Lord* (H. Wainer and S. Messick, eds.) 55-65. Lawrence Erlbaum Associates, Inc.

concerning the means of the k populations; in particular, which differences have confident signs and, by omission, which do not. To facilitate the analysis, it is natural to begin by assuming that the observations are normally distributed about the population means with some common (unknown) variance. Eliminating or weakening these assumptions is an important task not addressed in this paper.

A popular confidence interval procedure is based on the use of the studentized range statistic (Miller, 1981). If $\bar{y}_i$ denotes the sample mean for population i $(i=1,2,\ldots,k)$ and s^2 denotes the pooled estimate of the within — population variance, then the procedure constructs $\binom{k}{2}$ intervals of the form

$$\left|\bar{y}_i - \bar{y}_j\right| \pm q(k\,;\nu,\alpha)\, s/\sqrt{n}$$

for the differences $\mu_i - \mu_j$ of populations means. The population means μ_i and μ_j are declared significantly different at level α if the corresponding confidence interval does not contain zero. The constant $q\,(k\,;\nu,\alpha)$ is the upper α-percent point of the studentized-range distribution for k means and an estimate of variance based on ν degrees of freedom. Under the null hypothesis

$$H_0:\ \mu_i = \mu_2 = \ldots = \mu_k$$

the probability that at least one pair of means is falsely declared significantly different is exactly α.

A common confident-difference procedure which does not involve setting confidence intervals is the Newman-Keuls procedure (Miller, 1981). To carry it out, the sample means are ordered, yielding:

$$y_{(1)} \le y_{(2)} \le \ldots \le y_{(k)}.$$

For every pair $i \le j$, $y_{(j)} - y_{(i)}$ is compared with $q\,(j-i+1;\nu,\alpha)s/\sqrt{n}$ and if it exceeds it, the corresponding difference of population means, $\mu_j - \mu_i$, is declared confidently positive, provided that a consistency condition is not violated. The consistency condition may be stated as follows: If a stretch $(y_{(i)}, y_{(j)})$ is not declared significantly different, then no stretch $(y_{(l)}, y_{(k)})$ $(i \le l \le j-1;\ i+1 \le k \le j)$ contained in $(y_{(i)}, y_{(j)})$ may be declared significant. Under H_0: $\mu_1 = \ldots = \mu_k$ the experimentwise error rate is α. Since $q\,(k_1;\nu,\alpha) \le q\,(k_2;\nu,\alpha)$ for $k_1 \le k_2$, Newman-Keuls gains in power, in terms of confident signs, over the studentized procedure at the expense of not making confidence interval statements.

The consistency condition mentioned above may be thought of as an example of interference, in the sense that the decision about one pair of means influences the decision about other pairs. More complex forms of interference are the essence of some procedures we will discuss in Section 2.

As Welsch (1977) has noted, in the analysis of real data the hypothesis that all means are equal may not be plausible as, for example, when a few members of each of a number of treatment groups are being investigated. Null hypotheses in which the means are partitioned into blocks of equal means are more plausible. One example might be:

$$H_1:\ \mu_1 = \mu_2 << \mu_3 << \mu_4 = \mu_5 = \mu_6 << \mu_7 = \mu_8\,.$$

The two procedures we have described above may well considerably exceed their nominal error rates under such partition hypotheses. Ryan (1959, 1960) and Welsch (1977) have devised procedures which do control error rates under such hypotheses comparing differences of sample means with suitably elevated critical values. Both are confident-direction-only procedures and invoke consistency conditions as does Newman-Keuls.

2. MORE COMPLEX PROCEDURES

Peritz (1970) devised a complex modification of the Newman-Keuls procedure which successfully controlled the experimentwise error rate. Both Begun and Gabriel (1981) and Ramsey (1981) provide simplified descriptions of the Peritz procedure. Ramsey's description is most appropriate for our purposes. Define the null hypothesis corresponding to each partition of the k means to be that means in the same element of the partition are equal. If a partition contains only one element with more than one mean, then the test for the partition consists of carrying out the Newman-Keuls procedure for the means in that element. If the partition contains two or more elements with more than one mean in each, then each element is tested according to Ryan's procedure. This is similar to Newman-Keuls except that the critical value used to test a p-stretch is $q\,(p\,;\nu,\alpha_p)$ where $\alpha_p = 1-(1-\alpha)^{p/k}$. The partition hypothesis is rejected if the means in any of the components are declared significantly different by Ryan's test.

A partition hypothesis is said to be relevant to a pair of means if they are contained in the same element of the corresponding partition.

A pair of means is declared significantly different (according to Peritz) if and only if every relevant partition hypothesis has been rejected. Thus, a pair of means are Peritz significant if they are Ryan significant or if they are Newman-Keuls significant and if every subset of the remaining $(k-2)$ means is Ryan significant. Begun and Gabriel prove that Peritz's procedure does have an experimentwise error rate $\leq \alpha$.

The interesting feature of the Peritz procedure is that the configuration of all the other means is allowed to interfere with the decision about a pair of means rendered by Newman-Keuls. If the remaining means are sufficiently distant, the rejection of equality by Newman-Keuls is allowed to stand; otherwise, the hypothesis of equality is not rejected. In this manner, the excessive rejections of Newman-Keuls are brought under control.

Ramsey (1981) introduced a competitor to Peritz called the model-testing procedure (M-T). Like Peritz, the hypothesis $\mu_1 = \mu_j$ is rejected only if every relevant hypothesis is rejected. The difference is that the rejection of a partition hypothesis is based on an F-statistic criterion. Specifically, suppose a partition had p elements with m_i means $(m_i \geq 2)$ in the ith element $(i = 1, 2, \ldots, p)$. Let SS_i denote the sum of squares between sample means in the ith element and s^2 the pooled estimate of within population variance based on ν degrees of freedom. Then the calculated statistic is $T = \dfrac{\Sigma\, SS_i}{m_+ - k} / s^2$, where $m_+ = \sum_i m_i$. The corresponding partition hypothesis is rejected if T exceeds $F\,(m_+ - k\,; \nu, \alpha)$.

Ramsey carried out a small sampling experiment to investigate the power of several pairwise multiple comparison procedures. The Peritz and M-T procedures performed well, though neither completely dominated the other. Presumably the virtue of these procedures lies in their judging the significance of a difference in the light of both the error term and any other difference of means that happen to be small (although this effect is indirect). The motivation for the maximum subrange procedure (MSR) was to capture this advantage with less computation.

3. THE MAXIMUM SUBRANGE PROCEDURE

The MSR procedure compares $y_{(i)} < y_{(j)}$ in the following steps:

(i) Determine $d = y_{(j)} - y_{(i)}$; and the length of the stretch $(j - i + 1)$.

(ii) How many stretches outside $[y_i, y_j]$ of maximum span can be put down disjointly with length less than or equal to d?

(iii) Is $d\sqrt{n}/s$ greater than the corresponding critical value, Q, where the value of Q is a function of the number and span of the stretches identified in (ii)?

An example best illustrates the procedure. Consider the following ordered sequence of sample means:

$d_3 = 1.20$ (8.19 to 9.39)

2.54 3.69 7.71 8.19 8.92 9.39 14.21 16.18 16.82 17.04 21.02 22.10

$d_2 < 1.20$ (2.54 to 3.69) $d_3 < 1.20$ (16.18 to 17.04) $d_2 < 1.20$ (21.02 to 22.10)

Inference concerning the population means μ_α and μ_β corresponding to sample means 8.19 and 9.39 proceeds as follows:

(i) $d = 9.39 - 8.19 = 1.20$
$j - i + 1 = 6 - 4 + 1 = 3$

(ii) (a) Among sample means less than 8.19, the stretch of maximum span < 1.20 is the gap between 2.54 and 3.68.

(b) Among the sample means greater than 9.39, the stretch of maximum span < 1.20 is the three-stretch from 16.18 to 17.04.
In this case, among the remaining means there is another stretch less than 1.20 — namely the gap between 21.02 and 22.10

(iii) Denote the configuration of means relevant to testing $\mu_\alpha = \mu_\beta$ by (3,3,2,2), reflecting the fact that the ordered sample means corresponding to μ_α and μ_β have span 3 and that there are 3 other disjoint stretches with spans 3,2 and 2 of smaller length. The statistic is denoted as $T(3,3,2,2)$ and equals 1.20. To carry out the test, $T(3,3,2,2)\sqrt{n}/s$ is compared with $Q(3,3,2,2,\nu,\alpha)$. For example, $Q(3,3,2,2;\nu,.05) = 3.83$. If $n = 5$ and $s = 1$, the hypothesis would not be rejected.

The computation of the critical values Q is described in the Appendix. They are derived from the distribution of the studentized range. Note that 3.83 exceeds $3.31 = q(3;\infty,.05)$, the appropriate critical value for the N-K procedure. Thus like Peritz and M-T, the MSR procedure controls the excessive rejections of N-K by employing more extreme critical values. The critical value is determined for each pair by the configuration of the sample means relevant to the test, as described above.

The MSR requires far less computation than does Peritz or M-T. Instead of looking at all partitions, it considers only those conforming to the ordering of the sample means. Moreover, because it looks for disjoint stretches of maximal span, it generally does not search through all "orderly" partitions. Nonetheless, limited experimental sampling suggests its performance is comparable to Peritz or M-T.

Although the experimentwise error rate for the MSR remained below the nominal level in all the simulations, we can identify configurations of means for which the error rate will slightly exceed the nominal level. An example follows. Suppose the true state of nature is

$$\mu_1 = \mu_2 << \mu_3 = \mu_4 << \mu_5 = \mu_6$$

where the distance between the equally spaced pairs is very large in comparison to the within group variance. Thus, with near certainty, the sample means will appear in corresponding groups. Let

$$d_1 = |\bar{y}_1 - \bar{y}_2|, \; d_2 = |\bar{y}_3 - \bar{y}_4| \text{ and } d_3 = |\bar{y}_5 - \bar{y}_6|.$$

Denote the reordered absolute gaps by $f_{(1)} \le f_{(2)} \le f_{(3)}$. For a test of nominal level .05 (and assuming $\nu = \infty$), a Type I error occurs if any of the following events occur:

$$\sqrt{n}\, f_{(1)} / s > 2.77$$

or

$$\sqrt{n}\, f_{(2)} / s > 3.19$$

or

$$\sqrt{n}\, f_{(3)} / s > 3.38.$$

Now 3.38 is the upper 95 percent point of the distribution of the maximum of three studentized gaps. Consequently, $Pr\,\{\sqrt{n} f_{(3)} / s > 3.38\} = .05$ and so $Pr\,\{\text{Type I error}\} > .05$. In the present instance, it can be shown that the Type I error $\le .053$. More thorough investigations of the error rate properties of the procedure are planned.

4. SIMULATIONS

A small-scale simulation was carried out so that the power of the procedure could be compared to that of Peritz and M-T. We will report on one set of simulations involving an experimental setup with the following parameters:

Number of populations $(k) = 6$

Number of observations per population $(n) = 5$

Significance level $(\alpha) = .05$

Degrees of freedom for variance estimation $(\nu) = 24$.

The ratio of between-means variance to within-means variance was allowed to vary systematically. If the latter variance is denote by σ^2 and the former by $\tau^2 = k^{-1} \sum_{i=1}^{k} (\mu_i - \bar{\mu})^2$, then the effect size, f, is defined by $f = \tau/\sigma$.

For each fixed value of f many configurations of means are still possible. We considered four such configurations, as did Ramsey (1981):

(i) Equally spaced means: Adjacent means are equally distant.

(ii) Equally spaced null pairs: Adjacent pairs of equal means are equally distant.

(iii) Maximal configuration: Means are arranged to have the maximum range. For k even, the two extreme means are placed on either side of, and equidistant from, a central cluster of $(k-2)$ equal means.

(iv) Minimal configuration: Means are arranged to have the minimum range. For k even, two clusters of $(k/2)$ equal means each are created.

For each value of f, 1,000 simulations for each configuration were run. Each set of 1,000 simulations was begun with the same seed for the random number generator. Pseudo-random uniform deviates were obtained from a Tausworthe generator written in Assembler. Gaussian deviates were derived using the Box-Muller method. A simulation consisted of generating n independent $N(\mu_i, 1)$ observations $(i = 1, 2, \ldots, k)$ and carrying out the MSR and Peritz procedures. The M-T procedure has not been implemented by us as of this writing, and we have had to use Ramsey's experimental results (also based on 1,000 simulations) to make any comparisons with MSR. These comparisons are thus necessarily less precise than those between MSR and Peritz based on our simulations.

Ramsey has suggested that the configuration of equally spaced null pairs provides a stringent test of the Type I error characteristics of a multiple comparisons procedure. In approximately ten sets of 1,000 simulations, with values of f varying from 0.5 to 10.0, the empirical error rate of the MSR never exceeded the nominal error rate of 0.05. The same holds true for the corresponding simulations for the other configurations.

Following Ramsey, two kinds of power were considered. Any-pair power is the probability of detecting at least one true difference between pairs of means. It measures sensitivity to the largest difference in a configuration of means. All-pairs power is the probability of detecting all true differences between pairs of means. It measures sensitivity to the smallest true difference in a configuration of means. Any-pair power is of interest for small values of f while all-pairs power is of interest for large values of f $(f > 1)$.

Table 1 presents the results for a comparative study of any-pair power. Only the MSR and M-T procedures are shown because the Peritz procedure was nearly equivalent (though very slightly worse) to the MSR in the range of f values studied. Except for one instance ($f = 0.7$, maximal configuration), the MSR is uniformly more powerful than M-T. That one instance seems somewhat anomalous in view of Figure 1 which presents

Table 1

Any-Pair Power

Effect Size	Maximum Configuration		Minimum Configuration	
	MSR	M-T	MSR	M-T
.4	.31	.23	.27	.20
.5	.47	.37	.40	.33
.6	.62	.58	.55	.53
.7	.78	.87	.68	.64
.8	.90		.79	

Notes:

(i) Experimental setup: $k = 6$; $n = 5$; $\alpha = .05$; $d\nu = 24$.

(ii) Based on simulations of 1000 experiments.

(iii) M-T power taken from Ramsey (1981).

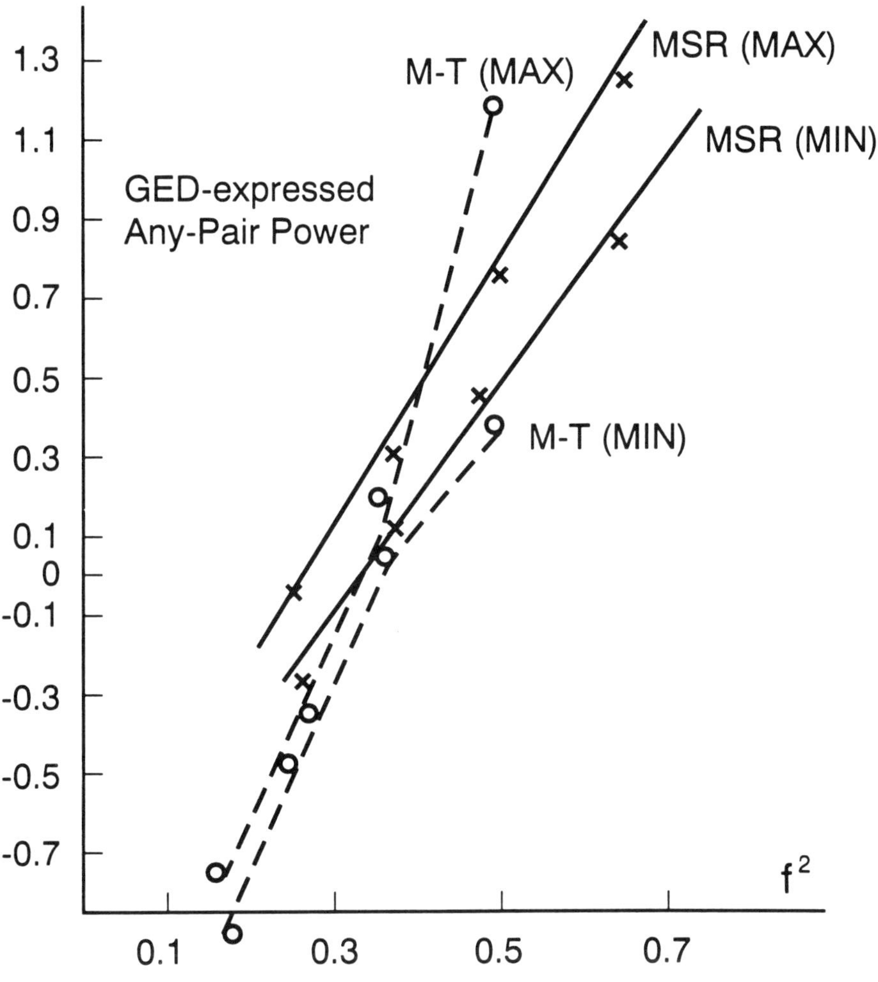

Figure 1. Power of MSR and M-T Procedures for Maximal and Minimal Configurations.

plots of the Gaussian Equivalent Deviate expressed (GED-expressed) power against f^2. In classical notation, power is denoted by $1-\beta$ where β is the Type II error. The GED-expressed power is simply $\phi^{-1}(1-\beta)$, where $\phi(\cdot)$ is the unit Gaussian cumulative, and is a useful linearizing transformation for power curves, as the figure attests.

Table 2 presents the results for a comparative study of all-pairs power. Here again the MSR is uniformly more powerful than M-T. Peritz is slightly more powerful than MSR for maximal configurations and slightly less powerful for minimal configurations.

Table 2

All-Pairs Power

Effect Size (f)	Maximum Configuration			Minimum Configuration		
	M-T	MSR	Peritz	M-T	MSR	Peritz
1.00	-	.08	.09	-	.22	.20
1.15	-	-	-	.37	-	-
1.20	.18	.22	.23	.43	.48	.44
1.25	-	-	-	.49	-	-
1.30	.27	.33	.35	.57	.62	.60
1.40	.40	.48	.50	-	.75	.74
1.50	.53	.64	.65	-	.87	.85

Notes:

(i) Experimental setup: $k = 6$; $n = 5$; $\alpha = .05$; $\nu = 24$.

(ii) Based on simulations of 1000 experiments.

(iii) M-T power taken from Ramsey (1981).

Table 3 shows the behavior of:

$$\frac{\text{GED-expressed power} + C}{f^2}$$

where C is chosen to keep the resulting numbers nearly constant over a broad range of effect sizes. The actual and relative sizes of the medians over f are shown. In relative terms, M-T is at 90.5 and 94.5, MSR is at 98 and 100, while Peritz is at 100 and 98. Clearly the differences in performance between MSR and Peritz (except for minimum configuration and $f = 1.5$) are not of practical importance.

Table 3

All-Pairs Power
(GED-Expressed and Converted)

Effect Size(f)	Maximum Configuration (GED + 2.770) / f^2			Minimum Configuration (GED + 2.120) / f^2		
	M-T	MSR	Peritz	M-T	MSR	Peritz
1.00	-	1.39	1.40	-	1.35	1.27
1.15	-	-	-	1.36	-	-
1.20	1.29	1.38	1.41	1.35	1.43	1.37
1.25	-	-	-	1.33	-	-
1.30	1.28	1.38	1.41	1.36	1.43	1.40
1.40	1.28	1.39	1.41	-	1.43	1.41
1.50	1.27	1.39	1.40	-	1.50	1.40
Median	1.28	1.39	1.41	1.36	1.43	1.40
Relative Size[4]	90.5	98	100	95	100	98

Notes:

(1) Experimental setup: $k = 6$; $n = 5$; $\alpha = .05$; $\nu = 24$.

(2) Based on simulations of 1000 experiments.

(3) M-T power taken from Ramsey (1981).

(4) Ratio of median power to maximum median power.

5. CONCLUSIONS

The maximum subrange procedure appears to be a plausible competitor to the Peritz and model-testing procedures while maintaining a substantial edge in simplicity and computational speed.

Its experimentwise error rate and power characteristics need to be evaluated more thoroughly and its properties analyzed analytically. It does seem to hold substantial promise.

Table A1

Critical Values for Maximum Subrange Procedure

Configuration	$\nu = \infty$	$\nu = 24$	Configuration	$\nu = \infty$	$\nu = 24$
10	4.47	4.92	4,4,2	4.02	4.40
9	4.39	4.81	4,4,3	3.99	4.37
8,2	4.31	4.72	4,4	3.98	4.34
8	4.29	4.68	4,3,2	3.91	4.26
7,3	4.25	4.66	4,2,2,2	3.86	4.21
7,2	4.20	4.59	4,3	3.80	4.18
7	4.17	4.54	4,2,2	3.80	4.13
6,4	4.21	4.62	4,2	3.73	4.02
6,3	4.13	4.52	4	3.63	3.90
6,2,2	4.11	4.49	3,3,3	3.87	4.22
6,2	4.07	4.43	3,3,2,2	3.83	4.17
6	4.03	4.37	3,3,2	3.76	4.08
5,5	4.19	4.60	3,2,2,2	3.70	4.02
5,4	4.10	4.48	3,3	3.67	3.97
5,3,2	4.04	4.42	3,2,2	3.60	3.90
5,3	4.00	4.36	3,2	3.48	3.74
5,2,2	3.97	4.32	3	3.31	3.53
5,2	3.91	4.25	2,2,2,2,2	3.64	3.94
5	3.86	4.17	2,2,2,2	3.52	3.81
			2,2,2	3.38	3.63
			2,2	3.17	3.37
			2	2.77	2.92

APPENDIX

The critical values for the MSR procedure are obtained by integration and interpolation. Let $F_p(\cdot)$ denote the cumulative distribution function of the range of p independent unit Gaussian random variables,

$$F_p(u) = p \int_{-\infty}^{\infty} \phi(y)\,[\Phi(y) - \Phi(y-u)]^{k-1} dy,$$

where $\phi(\cdot)$ and $\Phi(\cdot)$ are the density function and distribution function of the unit Gaussian. When the range is studentized by division by the square root of the ratio of an independent chi-square random variable on degrees-of-freedom to its degrees-of-freedom, the distribution function, denoted by H_p, may be represented as

$$H_p(u) = \int_0^\infty F_p(ux) \frac{\nu^{\nu/2} x^{\nu-1} e^{-\nu x^2/2}}{\Gamma(\nu/2) 2^{(\nu/2)-1}} \, dx \, .$$

For a fixed value of α and a configuration $(\ell_1, \ell_2, \ldots, \ell_m)$ we seek a value u_o such that $J(u_o) = 1 - \prod_{j=i}^{m} H_{\ell_j}(u_o) = \alpha$.

The desired value of u_o was obtained by evaluating $J(\cdot)$ at some thirty points and carrying out Aitken's scheme of Lagrange interpolation using the DALI subroutine of SSP. The integrals $H_p(u)$ were evaluated with a 16-point Gauss-Hermite quadrature using the DQH16 subroutine of SSP. For certain choices of p and u, the values of the integral were checked against those tabulated in Harter (1960). Table A1 displays the critical values for a test of level .05 for all configurations arising from ten means when the estimate of variance is based on 24 or ∞ degrees of freedom.

REFERENCES

Begun, J. and Gabriel, R. (1981). Closure of the Newman-Keul's multiple comparisons procedure, *J. Am. Stat. Assoc.* **76**, 241-245.

Harter, H. L. (1960). Tables of range and studentized range, *Ann. Math. Stat.* **31**, 1122-1147.

Miller, R. G. (1981). *Simultaneous Statistical Inference.* 2nd ed. Springer-Verlag, New York.

Ramsey, P. H. (1981). Power of univariate pairwise multiple comparison procedures, *Psychol. Bull.* **90**, 352-366.

Ryan, T. A. (1959). A multiple comparisons in psychological research, *Psychol. Bull.* **56**, 26-47.

Ryan, T. A. (1960). Significance tests for multiple comparison of proportions, variance, and other statistics, *Psychol. Bull.* **57**, 318-328.

Welsch, R. E. (1977). Stepwise multiple comparison procedures, *J. Am. Stat. Assoc.* **76**, **72**, 566-575.

11

Reminder Sheets for "Discussion of Paper on Multiple Comparisons by Henry Scheffé"

John W. Tukey

There are four ways in which my opinions on this matter seem to differ with my friend, the speaker, namely:

(1) estimation vs. decision;

(2) w-procedures vs. F-procedures;

(3) rational vs. irrational group means;

(4) designing for prescribed accuracy.

I shall try to commit just enough of my views on each to paper to serve as a reminder.

First, there is the question of choice between a confidence region and a collection of significance statements. Even when two are compared, it is often far more important to know that A and B agree within 3 units or that A may be 50 units greater than B, than to know that they are or are not significantly different. All four combinations are possible, as the following confidence intervals illustrate:

$$(-3, 3) \qquad (1, 3)$$
$$(-3, 50) \qquad (1, 50).$$

The Collected Works of John W. Tukey (1993) Chapman and Hall, New York, NY.

Prepared in connection with research sponsored by the Office of Naval Research.

Previously unpublished manuscript (circa 1951).

For many practical purposes, knowledge of the row is far more important than knowledge of the column.

The results of most experiments can be identified as intended (i) to guide judgment in a particular situation or (ii) to establish a bench mark for future use in unspecified ways. Clearly case (i) calls for a decision procedure. Conversely case (ii) calls for estimation, not only of a "best" value, but also of reasonable limits. The appropriate decision procedure for case (i) will almost never correspond to significance at 5% or 1%! It seems to me, then, that the role of 5% and 1% is an estimation role, and not a decision role.

As a clincher on point one, note that a confidence procedure always includes a significance procedure, and that a confidence statement always contains a significance statement, but not *vice-versa*.

Perhaps here the difference between us is only a matter of emphasis.

Before discussing the second point, we need to set up the two procedures in a common form. If $y = \Sigma c_i x_i$ is a linear expression in the determinations x_i, with mean value $\eta = \Sigma c_i \eta_i$ then we wish to make statements of the form

$$| y - \eta | \leq (\text{norm}\{y\})(HSF)(t\sqrt{2})s$$

where the value of the ***hindsight factor***, *HSF*, will depend on:

1) the norm used;

2) whether y is allowed to vary over contrasts with $\Sigma c_i = 0$, or over all linear expressions.

We seek to have ***all*** such statements correct in 95% of the repetitions of this analysis on m x_i's, provided that s^2 is an unbiased estimate of the variance of each with *DDF* degrees of freedom. The two norms with which we shall be concerned are:

$$\text{norm}_w\{y\} = \max\left\{\sum_{c_i>0} c_i\,,\ -\sum_{c_i<0} c_i\right\}$$

and

$$\text{norm}_F\{y\} = \left(\tfrac{1}{2}\,\Sigma c_i^2\right)^{1/2}.$$

The values of the product (*HSF*) ($t\sqrt{2}$) are then upper 5% points of the following quantities:

contrasts, norm_w:	studentized range of m with *DDF* degrees of freedom,
all, norm_w:	studentized augmented range of m with *DDF* degrees of freedom,
contrasts, norm_F:	square root of $(m-1)$ times F with $m-1$ and *DDF* degrees of freedom,
all, norm_F:	square root of m times F with m and *DDF* degrees of freedom.

Here the augmented range of m is the range of the $m+1$ quantities: zero and m independent unit normal deviates.

Comparative values are given in approximate form in Table 1.

The values of *HSF* give a correct impression of the strengths of the confidence statements for simple comparisons, since

$$\text{norm}_w\{x_j - x_k\} = 1 = \text{norm}_F\{x_j - x_k\},$$

and favor the w-method unduly otherwise, since if y is not a simple comparison

$$\text{norm}_w\{y\} > \text{norm}_F\{y\}.$$

If we follow Scheffé in taking the square of the length of the confidence interval as the measure of relative efficiency, and take $DDF = \infty$ for convenience (for finite *DDF* the comparison is more favorable to the w-procedure than for $DDF = \infty$), we obtain Table 2. Hence the one-way and two-way discrepancies are

$$y_1 = x_1 - x_{(i)}$$

and

$$y_2 = x_{ij} - x_{(i)j} - x_{i(j)} + x_{(i)(j)},$$

Table 1

Comparison of Variance *HSF*'s

Norm Scope Notation	Norm_w Contrasts	*HSF*	Norm_w All	Norm_F Contrasts *FHSF* (m)	Norm_F All *FHSF* $(m+1)$
$m=2$	$1.00+0.0/DDF$		$1.02+0.2/DDF$	$1.00+0.0/DDF$	$1.25+0.3/DDF$
3		$1.20+0.4/DDF$		$1.25+0.3/DDF$	$1.42+0.8/DDF$
4		$1.32+0.6/DDF$		$1.42+0.8/DDF$	$1.57+1.0/DDF$
5		$1.40+0.8/DDF$		$1.57+1.0/DDF$	$1.70+1.3/DDF$
6		$1.46+1.0/DDF$		$1.70+1.3/DDF$	$1.81+1.6/DDF$
7		$1.51+1.2/DDF$		$1.81+1.6/DDF$	$1.91+1.9/DDF$
8		$1.55+1.4/DDF$		$1.91+1.9/DDF$	$2.01+2.1/DDF$
9		$1.59+1.5/DDF$		$2.01+2.1/DDF$	$2.10+2.4/DDF$
10		$1.62+1.6/DDF$		$2.10+2.4/DDF$	$2.18+2.7/DDF$
11		$1.65+1.7/DDF$		$2.18+2.7/DDF$	$2.26+3.0/DDF$
12		$1.67+1.8/DDF$		$2.26+3.0/DDF$	$2.34+3.2/DDF$
13		$1.70+1.9/DDF$		$2.34+3.2/DDF$	$2.41+3.4/DDF$
14		$1.72+2.0/DDF$		$2.41+3.4/DDF$	$2.48+3.6/DDF$
15		$1.74+2.1/DDF$		$2.48+3.6/DDF$	$2.55+3.8/DDF$
16		$1.75+2.2/DDF$		$2.55+3.8/DDF$	$2.62+4.0/DDF$
17		$1.77+2.3/DDF$		$2.62+4.0/DDF$	$2.68+4.2/DDF$
18		$1.78+2.4/DDF$		$2.68+4.2/DDF$	$2.74+4.4/DDF$
19		$1.80+2.4/DDF$		$2.74+4.4/DDF$	$2.80+4.6/DDF$
20		$1.81+2.5/DDF$		$2.80+4.6/DDF$	$2.86+4.8/DDF$
30		—		$3.33+6.4/DDF$	$3.38+6.6/DDF$
40		—		$3.76+8.2/DDF$	$3.80+8.4/DDF$
50		—		$4.16+9.4/DDF$	$4.20+9.6/DDF$
60		—		$4.52+10.7/DDF$	$4.55+10.8/DDF$
80		—		$5.11+13.3/DDF$	$5.14+13.4/DDF$
100		—		$5.67+15.1/DDF$	$5.70+15.2/DDF$

where (i) indicates the arithmetic mean over all values of the subscript *except* i, and the next two columns refer to such contrasts as

$$y_3 = \frac{x_1+x_2}{2} - \frac{x_3+x_4+\ldots+x_m}{m-2}$$

Table 2

Efficiency of w-procedure Relative to F-procedure
(in % for contrasts only)

Number of Determinations	Simple Comparisons	Discrepancies One-way	Discrepancies Two-way	Two vs. Rest	Extreme Case
2	100	100	100	—	100
3	109	82	—	82	82
4	118	79	59	59	59
5	128	79	—	53	53
10	169	96	53	54	34
15	206	110	49	63	28
20	240	127	51(→)	67	24

(→) = use value to right for 4×5

and (for $m = 2k$)

$$y_4 = \frac{x_1 + x_2 + \ldots + x_k}{k} - \frac{x_k + 1x_k + 2 + \ldots + x_{2k}}{k}.$$

Now we have some basis for choice.

For $m = 20$, to take an extreme example, we must ask ourselves if we think that 240% efficiency on simple comparisons outweighs 24% efficiency in the extreme case. Those who accept Scheffé's position that the most important contrasts are comparisons of means of small groups will say "no." The writer feels rather strongly that the correct answer is "yes"! The reasons for this difference lie in the third point of difference in opinion between Scheffé and the writer.

In many a conventional analysis-of-variance situation, the observations are arranged according to various classifications, a procedure which singles out various group means. Thus there arise various "lines," in the analysis, the means entering into a line "above" another being means of the means or values in that other. It is the custom to appropriate (in the legislative sense) separate error rates for each line. Thus in a replicated three-way classification, where 7 lines are usually tested, say at the 5% level, there is a thirty-odd percent chance of finding at least one line significant when all classifications are meaningless. To some, such expenditure of error rates may seem very reckless and spendthrift. In 2^3, 2^4, etc. designs it usually seems spendthrift to the writer.) But the cure for reckless spending is not the elimination of separate appropriations! Some of the reasons for this run as follows:

(1) In many cases the appropriate probability model is such that different "error variances" should be used for different lines.

(2) The group means (rational group means) selected by the classifications are undoubtedly more important than the means of groups selected in other ways (irrational group means) and we should spend more of our precious error rate on comparisons between and among the first kind of group means.

This is the writer's view.

I have been unable to decide from reading Scheffé's paper what he intends to do with such a replicated 3-way design. If he intends to appropriate one error rate and calculate one

$$(HSF)\,(t\,\sqrt{2})s$$

to be used in all lines of the analysis, I feel that this is a *very unwise* procedure! If he proposes to treat each line in the analysis of variance separately, as I would hope, then I feel that the important group means have been given suitably narrow allowances, and I claim that comparisons between the means of *other* groups are unimportant.

Thus I conclude that I wish to use the w-procedure as the procedure of choice unless:

(1) the accuracies of the various comparisons are enough different to require explicit treatment, or,

(2) the fluctuations involved are very seriously nonnormal (and on the day in question I prefer the known insensitivity to nonnormality of F, to the hypothetical insensitivity of the studentized range).

In these exceptional cases I should prefer the F-procedure.

Finally, we come to designing for prescribed accuracy. We shall call the factor

$$(HSF)\,(t\sqrt{2})s$$

by which each norm is multiplied in order to find that ±, the *allowance* (or more precisely, perhaps, the allowance for a simple comparison). I prefer to design experiments in terms of a *confident bound for the allowance*. Thus, for example, we might require 90% confidence that comparisons in yield of grain will be known to ± 3 bushels per acre or better, or 50% confidence that comparison in yield of a valuable product by chemical processes will be known to $\pm 3\%$ or better — in the sense that the allowance will be 3 bushels per acre or less in 90% of such field experiments, or 3% or less in 50% of such pilot plant programs.

This is easy to do, since we have only to find an upper confidence limit for s by whatever standard procedure is suitable.

Not only is it simple, but it corresponds, in the writer's judgment, more closely to what the user wants.

Index to Volume VIII of the Collected Works